Advanced Structured Materials

Volume 255

Common engineering materials are reaching their limits in many applications, and new developments are required to meet the increasing demands on engineering materials. The performance of materials can be improved by combining different materials to achieve better properties than with a single constituent, or by shaping the material or constituents into a specific structure. The interaction between material and structure can occur at different length scales, such as the micro, meso, or macro scale, and offers potential applications in very different fields.

This book series addresses the fundamental relationships between materials and their structure on overall properties (e.g., mechanical, thermal, chemical, electrical, or magnetic properties, etc.). Experimental data and procedures are presented, as well as methods for modeling structures and materials using numerical and analytical approaches. In addition, the series shows how these materials engineering and design processes are implemented and how new technologies can be used to optimize materials and processes.

Advanced Structured Materials is indexed in Google Scholar and Scopus.

Holm Altenbach · Nicholas Fantuzzi ·
António J. M. Ferreira
Editors

Composite Structures and Technology

In Memoriam to Prof. Cristóvão Manuel Mota Soares

Springer

Editors
Holm Altenbach
Faculty of Mechanical Engineering
Otto von Guericke University Magdeburg
Magdeburg, Sachsen-Anhalt, Germany

Nicholas Fantuzzi
DICAM Department
University of Bologna
Bologna, Italy

António J. M. Ferreira
Faculty of Engineering
University of Porto
Porto, Portugal

ISSN 1869-8433 ISSN 1869-8441 (electronic)
Advanced Structured Materials
ISBN 978-3-032-24655-4 ISBN 978-3-032-24656-1 (eBook)
https://doi.org/10.1007/978-3-032-24656-1

This Springer imprint is published by the registered company Springer Nature Switzerland AG
The registered company address is: Gewerbestrasse 11, 6330 Cham, Switzerland

Preface

This Book is dedicated to Professor Cristóvão Manuel Mota Soares, late Associate Professor at Instituto Superior Técnico, University of Lisbon, deceased on July 3, 2025.

Professor Cristóvão Mota Soares was born on the Island of São Miguel, in Azores, Portugal, on May 4, 1941. His life was forever marked by the years of his youth spent in the Azores Islands of São Miguel, Terceira and Graciosa, where he gained experience in the art of whaling, following his family tradition.

Professor Cristóvão Mota Soares obtained a B.Sc. in Electrical and Mechanical Engineering at the Instituto Industrial do Porto, Portugal, in 1963, and shortly after began his professional experience, working for Austin Motor Company, Ford Lusitania, Auto-Sueco (Volvo) and PROFABRIL, among others. Later on, at the University of Aston, Birmingham, United Kingdom, he obtained a M.Sc. in Solid Mechanics in 1972, and a Ph.D. in Mechanical Engineering in 1977.

Upon returning to Portugal, Professor Cristóvão Mota Soares began his academic career in 1980 at Instituto Superior Técnico, University of Lisbon, as an Assistant Professor, and later, in 1988 as an Associate Professor. Throughout his academic career, as a researcher, Professor Cristóvão Mota Soares became renowned for his expertise in mechanics of composite materials and structures, smart technologies in structural engineering and multidisciplinary design optimization, including finite element modelling and analysis of plates and shells, viscoelastic sandwich structures and functionally graded structures.

Overall, Professor Cristóvão Mota Soares authored about 190 papers in International Journals, developing high-level research with significant impact in his field of expertise, over the years, reaching an h-index of 46.

Professor Cristóvão Mota Soares was also distinguished with Awards, including the Award for Lifetime Achievement in Composites at the IV ECCOMAS Thematic Conference on Mechanical Response of Composites, in Azores, Portugal, in September 2013, and the Excellence in Research Award at the 16th International Conference on Composite Structures, in June 2011.

Over the years, Professor Cristóvão Mota Soares made a difference to the lives of many researchers in his field of expertise by his scientific supervision of over 20

M.Sc. Theses and 13 Ph.D. Theses in Mechanical and Aerospace Engineering, as well as Naval Architecture and Ocean Engineering. Ultimately, he left a legacy by contributing to the start of the academic career of many researchers by his outstanding example.

Fellow researchers of Professor Cristóvão Mota Soares can, undoubtedly, remember his enthusiastic attendance and participation in International Conferences in his field of expertise, where he dearly enjoyed being mindful of innovative research developments, in the company of long-time friends and colleagues, who delighted him immensely. He was forever eager for the next spring and summer time filled with a series of International Conferences.

Magdeburg, Germany *Holm Altenbach*
Bologna, Italy *Nicholas Fantuzzi*
Porto, Portugal *António Joaquim Mendes Ferreira*
February 2026

Contents

List of Contributors

Michele Bacciocchi
DESID Department, Università di San Marino, Via Consiglio dei Sessanta 99, Dogana, 47891, Republic of San Marino,
e-mail: michele.bacciocchi@unirsm.sm

Inês J.C. Barbosa
IDMEC, Instituto Superior de Engenharia de Lisboa, Instituto Politécnico de Lisboa & CIMOSM, ISEL, IPL – Centro de Investigação em Modelação e Optimização de Sistemas Multifuncionais, Av. Conselheiro Emídio Navarro 1, 1959-007 Lisboa, Portugal,
e-mail: ines.barbosa@isel.pt

Joaquim I. Barbosa
IDMEC, Instituto Superior de Engenharia de Lisboa, Instituto Politécnico de Lisboa & CIMOSM, ISEL, IPL – Centro de Investigação em Modelação e Optimização de Sistemas Multifuncionais, Av. Conselheiro Emídio Navarro 1, 1959-007 Lisboa, Portugal,
e-mail: joaquim.barbosa@isel.pt

Dongfeng Cao
State Key Laboratory of Materials Synthesis and Processing, Wuhan University of Technology, Wuhan 430070, P. R. China,
e-mail: cao_dongf@whut.edu.cn

Erasmo Carrera
Mul2 Lab, Politecnico di Torino, Corso Duca degli Abruzzi 24, 10129 Torino, Italy,
e-mail: erasmo.carrera@polito.it

André R.D. Carvalho
IDMEC, Instituto Superior de Engenharia de Lisboa, Instituto Politécnico de Lisboa & CIMOSM, ISEL, IPL – Centro de Investigação em Modelação e Optimização de Sistemas Multifuncionais, Av. Conselheiro Emídio Navarro 1, 1959-007 Lisboa, Portugal,
e-mail: andre.carvalho@isel.pt

Amaury Alvarez Cruz
Universidade Federal do Rio de Janeiro, RJ, CP 21941-617, Brasil,
e-mail: amaury@ic.ufrj.br

Oscar Luis Cruz-González
Aix Marseille Univ, CNRS, Centrale Marseille, IRPHE UMR 7342, Marseille, France,
e-mail: oscar-luis.cruz-gonzalez@univ-amu.fr

Jesús Cuartero
Department of Mechanical Engineering, University of Zaragoza, 50009 Zaragoza, Spain,
e-mail: jcs@unizar.es

Rossana Dimitri
University of Salento, Edificio La Stecca - S.P. 6, Lecce - Monteroni, Lecce, Italy,
e-mail: rossana.dimitri@unisalento.it

Qingfeng Duan
State Key Laboratory of Materials Synthesis and Processing, Wuhan University of Technology, Wuhan 430070, P. R. China,
e-mail: absinthe1223@163.com

Yoanh Espinosa-Almeyda
Instituto de Ingeniería y Tecnología, Universidad Autónoma de Ciudad Juárez, Cd. Juárez, Chihuahua, CP 32310, México,
e-mail: yoanh.espinosa@uacj.mx

Nicholas Fantuzzi
DICAM Department, Università di Bologna, Viale del Risorgimento 2, Bologna, 40136, Italy,
e-mail: nicholas.fantuzzi@unibo.it

Gaizka Farrera-Rivera
Instituto de Ingeniería y Tecnología, Universidad Autónoma de Ciudad Juárez, Cd. Juárez, Chihuahua, CP 32310, México,
e-mail: gaizka.farrera@outlook.com

Susana C. F. Fernandes
Department of Mechanical Engineering, University of Zaragoza, 50009 Zaragoza, Spain,
e-mail: sfernandes@umaia.pt

António J.M. Ferreira
FEUP Department, Universidade do Porto, Rua Dr. Roberto Frias, Porto, 4200-465, Portugal, e-mail: ferreira@gcloud.fe.up.pt

Tanmaye Y. Heblekar
Texas A&M University, MS 3123, College Station, TX 77843, USA,
e-mail: tanmayeyashodan.h@tamu.edu

Haixiao Hu
Hubei Key Laboratory of Theory and Application of Advanced Materials Mechanics, School of Science, Wuhan University of Technology, Wuhan 430070, P. R. China,
e-mail: hhx1987@whut.edu.cn

Shuxin Li
State Key Laboratory of Materials Synthesis and Processing, Wuhan University of Technology, Wuhan 430070, P. R. China,
e-mail: lishuxin@whut.edu.cn

M. Amélia R. Loja
IDMEC, Instituto Superior de Engenharia de Lisboa, Instituto Politécnico de Lisboa & CIMOSM, ISEL, IPL – Centro de Investigação em Modelação e Optimização de Sistemas Multifuncionais, Av. Conselheiro Emídio Navarro 1, 1959-007 Lisboa, Portugal,
e-mail: amelia.loja@isel.pt

Boris Mederos-Madrazo
Instituto de Ingeniería y Tecnología, Universidad Autónoma de Ciudad Juárez, Cd. Juárez, Chihuahua, CP 32310, México,
e-mail: boris.mederos@uacj.mx

Jose Merodio
Universidad Politécnica de Madrid, E.T.S.I. de Sistemas Informáticos, Departamento de Matemática Aplicada a las TIC, Madrid, CP 28031, Spain,
e-mail: merodioj@gmail.com

Ana M.A. Neves
Faculty of Engineering, University of Porto, Rua Dr. Roberto Frias, s/n 4200-465 Porto, Portugal,
e-mail: ananeves@fe.up.pt

Carlos Quesada-González
Universidad Politécnica de Madrid, E.T.S.I. de Sistemas Informáticos, Departamento de Matemática Aplicada a las TIC, Madrid, CP 28031, Spain,
e-mail: carlos.quesada@upm.es

Junuthula N. Reddy
Texas A&M University, MS 3123, College Station, TX 77843, USA,
e-mail: jnreddy@tamu.edu

Reinaldo Rodríguez-Ramos
Departamento de Matemáticas, Facultad de Matemática y Computación, Universidad de La Habana, La Habana, CP 10400, Cuba & PPG-MCCT, Universidade Federal Fluminense, RJ, CP 27255-125, Brasil,
e-mail: reinaldo@matcom.uh.cu, reinaldorr@id.uff.br

Daniele Scano
Mul2 Lab, Politecnico di Torino, Corso Duca degli Abruzzi 24, 10129 Torino, Italy,
e-mail: daniele.scano@polito.it

Arun R. Srinivasa
Texas A&M University, MS 3123, College Station, TX 77843, USA,
e-mail: asrinivasa@tamu.edu

Francesco Tornabene
University of Salento, Edificio La Stecca - S.P. 6, Lecce - Monteroni, Lecce, Italy,
e-mail: francesco.tornabene@unisalento.it

Riccardo Vescovini
DAER Department, Politecnico di Milano, Via La Masa 34, Milan, 20156, Italy,
e-mail: riccardo.vescovini@polimi.it

Matteo Viscoti
University of Salento, Edificio La Stecca - S.P. 6, Lecce - Monteroni, Lecce, Italy,
e-mail: matteo.viscoti@unisalento.it

Cheng A. Yan
DAER Department, Politecnico di Milano, Via La Masa 34, Milan, 20156, Italy,
e-mail: chengangelo.yan@polimi.it

Chapter 1
Nonlinear Analysis of Composites with Two-Dimensional Finite Elements and Arbitrary Displacement Fields

Erasmo Carrera and Daniele Scano

Abstract This paper presents an approach for developing two-dimensional plate theories within the Unified Formulation framework in quasi-static nonlinear analysis. In nonlinear structural analysis, achieving accurate stress predictions with computationally efficient models is particularly valuable, as these simulations are inherently computationally demanding. The proposed methodology enables each displacement component to be represented by an independent expansion function, permitting the integration of different kinematic theories within a single model. Furthermore, it allows for the combined use of Equivalent Single Layer and Layer-Wise approaches. The structural model is implemented in a finite element context using an enhanced version of the Unified Formulation. The governing equations are linearized using the Newton–Raphson method, and the Crisfield arc-length method is employed. The accuracy of the proposed models is validated against literature data for isotropic and composite plates, focusing on large-deflection and post-buckling responses.

Key words: Nonlinear plate analysis, Unified formulation, Higher-order models, Composites, Equivalent-single layer, Layer-wise

1.1 Introduction

The geometrically nonlinear static analysis of elastic plate and shell structures remains a topic of considerable importance in structural mechanics. Highly flexible plate and shell structures continue to be employed in a diverse array of structural components [20, 21], including space membranes and solar sails [18], wing structures and rotor blades in aeronautical engineering. Furthermore, complex soft tissues and organs, as well as sophisticated biological instruments, are often modelled or

Erasmo Carrera · Daniele Scano (corresponding author)
Mul2 Lab, Politecnico di Torino, Corso Duca degli Abruzzi 24, 10129 Torino, Italy, e-mail: erasmo.carrera@polito.it,daniele.scano@polito.it

H. Altenbach et al. (eds.), *Composite Structures and Technology*,
Advanced Structured Materials 255,
https://doi.org/10.1007/978-3-032-24656-1_1

composed of highly flexible structural elements [2]. Moreover, flexible mechanical phononic crystals and metamaterials have been proposed the tune of acoustic and elastic waves [4, 13]. Highly flexible structures are prone to instability, such as buckling and post-buckling, due to large displacements and rotations. However, some critical aerospace components are specifically designed to maintain significant load-carrying capacity beyond the buckling point [14, 36].

In particular, the study focuses on composite laminated structures. A multilayered plate and its reference coordinate system are illustrated in Fig. 1.1. The reference plane Ω_0 lies in the x-y plane. Along the thickness direction (the z-axis), the plate cross-section may be subdivided into several distinct domains, denoted by the superscript k. The displacement field for a generic mathematical domain through the thickness is expressed by the following vector:

$$\mathbf{u}^k(x, y, z) = \left\{ u_x^k(x, y, z),\, u_y^k(x, y, z),\, u_z^k(x, y, z) \right\}^\top \tag{1.1}$$

The anisotropy of composite structures gives rise to complex mechanical phenomena, including shear continuity between layers and a zig-zag displacement distribution through the thickness, described by the C_z^0-requirements, see Carrera [6] for more details. Additionally, significant interactions between in-plane and out-of-plane strain components must be accounted for in accurate structural modelling.

From the preceding considerations, it is clear that cost for the nonlinear analysis is high, an effect that is further exacerbated when evaluating multilayered composites. This work, therefore, aims to achieve two goals: first, to maintain accurate predictions of both displacements and out-of-plane stresses; and second, to implement kinematic and modelling strategies that effectively reduce the system's total degrees of freedom.

Here, a brief review of the nonlinear method for the plate is given. The first-order shear deformation theory (FSDT), originally developed by Reissner and Mindlin [23, 31] for linear elastic plates, has been extended to geometrically nonlinear bending problems using von Kármán strain–displacement relations [34]. However, FSDT assumes uniform shear distribution across the thickness, requiring shear correction factors to adjust transverse shear stiffness. Consequently, various higher-order shear deformation theories (HSDT) and refined plate theories incorporating von Kármán nonlinearity have been proposed to improve solution accuracy in large-deflection

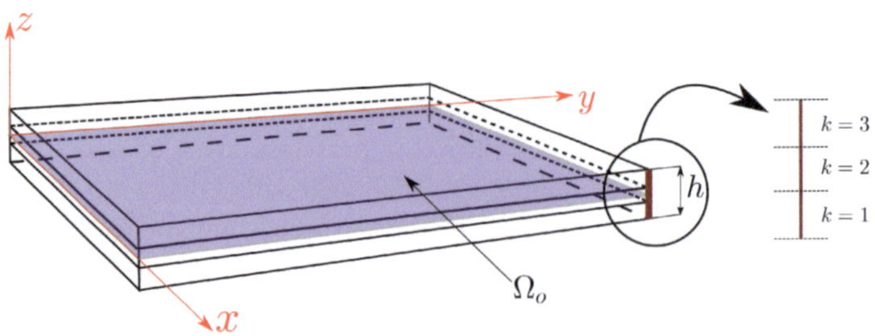

Fig. 1.1: Cartesian reference system for a multilayered plane structure. k indicates the mathematical sub-domain.

and nonlinear bending analysis of plate structures [16]. It should be emphasized that most studies employ the von Kármán nonlinear plate theory, which inherently assumes that only the nonlinear terms in the strain–displacement relation associated with the in-plane partial derivatives of the transverse displacement must be retained. While this assumption can yield physically acceptable results for thin plates and shells undergoing moderate rotations, its applicability is constrained to those specific regimes [10]. To accurately predict geometrically nonlinear response and stress distributions in laminated structures, several studies based on the HSDT have focused on performing full three-dimensional geometrically nonlinear analyses. These models retain all nonlinear terms of the Green–Lagrange strain tensor [17]. For instance, Alijani and Amabili[1] developed and applied a high-order shear deformation and thickness-stretching plate theory within a total Lagrangian description to study the effects of rotation and thickness-stretching nonlinearities on static bending and forced vibrations. More recently, Rivera *et al.* [19] proposed a 12-parameter shell finite element based on third-order thickness-stretching kinematics.

The previous models can be subdivided into two main modelling approaches. The first is the Equivalent Single-Layer (ESL) approach, widely adopted in the literature for its modeling simplicity, which employs kinematic variables that remain independent of the number of material layers. The second is the Layer-Wise (LW) approach, which is used for applications requiring higher-fidelity stress analysis. This method assigns distinct kinematic variables to each layer, at the expense of increased computational cost. A comprehensive review and comparison of both methodologies can be found in the work of Maji and Mahato [22].

Carrera and coworkers proposed an innovative unified modeling framework known as the Carrera Unified Formulation (CUF), and the details may be found in [5, 7, 9]. In CUF, governing equations are invariant, i.e., they maintain the same mathematical structure regardless of the specific structural theory used (e.g., classical, LW, or ESL). As a result, defining a structural model in CUF requires only the type and order of the kinematic expansion, along with the chosen modeling approach. Furthermore, Wu *et al.* [35] proposed a refined plate quasi-static formulation, where the total Lagrangian description is used with Green–Lagrange strain tensor.

The previous ad-hoc models were able to calculate the structural behaviour. But conventional approaches require costly reformulation for each new theory. Unified frameworks such as the CUF overcome this by allowing diverse models to be integrated within a single theoretical structure. However, current methods still have limitations: they often cannot assess the individual contributions of kinematic expansion terms, and they fail to include both ESL and LW approaches into a single unified model.

To overcome previous limitations, Carrera and coworkers [32, 33] developed a novel, consistent formulation based on CUF principles. This approach allows users to adaptively select the precision of the kinematic theory and to balance computational cost with predictive accuracy. Thus far, expansions have been based primarily on either Taylor or Lagrange expansions. Taylor expansions (ESL models), using polynomial terms of the form z^p, are efficient for constructing advanced structural theories in 2D analysis [8, 28]. They allow the flexible addition of higher-order terms to capture complex mechanical phenomena [24]. Lagrange expansions approximate the thickness using Lagrange points [27], interpolating displacements from these discrete points across layers. Mixed ESL/LW can independently study displacement components using Taylor or Lagrange expansions. This method can efficiently approximate the full three-dimensional displacement field while maintaining computational efficiency,see also [33] for more information.

To the authors' knowledge, a unified method for automatically generating reduced plate models in the nonlinear regime has not been developed. Thus, the present work significantly enhances the Unified Formulation's capabilities for nonlinear quasi-static analysis. A related approach has recently been developed by Carrera and Scano [12] for beam formulations.

The content of the work may be summarized as follows:

a) In Sect. 1.2, a concise review of the Unified Formulation for plate cases is presented within the framework of the finite element method.
b) Section 1.3 provides a detailed derivation of the governing equations and illustrates the procedure for obtaining the explicit expressions of the Fundamental Nuclei (FN), which represent the individual components of the system matrices.
c) The assembly procedure for the finite element matrices and arrays, applicable to both linear and nonlinear cases, is described in Sect. 1.4. This discussion focuses on the formulation of the tangent stiffness matrix, presented without loss of generality.
d) Numerical analyses of various plate structures under different boundary conditions are presented and discussed in Sect. 1.5.
e) Section 1.6 traces the most important conclusions of the work, and proposes some future developments for the efficient nonlinear analyses of plates.

1.2 Unified Formulation for Plates

While various models from the literature were reviewed previously, a generalized displacement field is introduced within the finite element framework. A compact notation will be presented to facilitate the derivation of the governing equations. The Carrera Unified Formulation (CUF) offers a powerful, consistent mathematical framework that can represent all such models in a unified manner. Within CUF, the displacement field is expanded using thickness-defined base functions $F_\tau^k(z)$, allowing each displacement component to be expressed as a general series of arbitrary functions over the domain. By integrating the finite element method (FEM) into the CUF framework, numerical solutions can be efficiently obtained:

$$
\begin{aligned}
u_x^k &= N_i(x,y)F_{u_x\tau}^k(z)q_{x_{\tau i}}^k \quad \text{with } \tau = 1,\dots,M_{u_x} \text{ and } i = 1,\dots,N_n \\
u_y^k &= N_i(x,y)F_{u_y\tau}^k(z)q_{y_{\tau i}}^k \quad \text{with } \tau = 1,\dots,M_{u_y} \text{ and } i = 1,\dots,N_n \qquad (1.2) \\
u_z^k &= N_i(x,y)F_{u_z\tau}^k(z)q_{z_{\tau i}}^k \quad \text{with } \tau = 1,\dots,M_{u_z} \text{ and } i = 1,\dots,N_n
\end{aligned}
$$

$F^k_{u_x\tau}$, $F^k_{u_y\tau}$, and $F^k_{u_z\tau}$ represent the expansion functions, where the subscript τ follows Einstein's summation convention. The corresponding numbers of expansion terms are M_{u_x}, M_{u_y}, and M_{u_z}. The shape functions are denoted by N_i, with the index i also summed over, and N_n represents their total number. The proposed formulation is general and allows for the use of arbitrary shape functions for the plate approximation. Herein, Lagrange-type finite elements with nine nodes (Q9) are adopted. A comprehensive treatment of the elements is provided by Bathe [3].

To simplify the notation, the following compact form is used:

$$u_l^k = N_i F^k_{u_l\tau} q^k_{l_{\tau i}} = \sum_{i=1}^{N_n} \sum_{\tau=1}^{M_{u_l}} N_i F^k_{u_l\tau} q^k_{l_{\tau i}} \tag{1.3}$$

$$\delta u_m^k = N_j F^k_{u_m s} \delta q^k_{m_{sj}} = \sum_{j=1}^{N_n} \sum_{s=1}^{M_{u_m}} N_j F^k_{u_m s} \delta q^k_{m_{sj}} \tag{1.4}$$

where $l, m \in \{x, y, z\}$. Importantly, the repeated index l (equivalently, m) in this specific term is not summed over; it serves as a free index.

1.3 Nonlinear Governing Equations

This section derives the nonlinear governing equations under the assumption of purely geometrical nonlinearities. While the full CUF derivation is available in earlier work [35], the equations for the modified formulation have been retrieved here for the sake of completeness.

Note that the superscript k is omitted in the following equations for brevity. To align with the notation used throughout this paper, matrices (and the relative components) are indicated by a superscript (e.g., $\mathbf{K}^{\top}$), rather than a subscript (e.g., $\mathbf{K}_T$) as commonly seen in CUF literature. In this convention, the transpose operator is indicated as $(\cdot)^{\top}$.

1.3.1 Preliminaries

The derivation starts from the definitions of the second Piola-Kirchhoff stresses, **S**, and the Green-Lagrange strain, **E**, which are written in the vectorial form as given below:

$$\mathbf{S} = \left\{ \sigma_{xx} \; \sigma_{yy} \; \sigma_{zz} \; \sigma_{yz} \; \sigma_{xz} \; \sigma_{xy} \right\}^{\top}, \quad \mathbf{E} = \left\{ \epsilon_{xx} \; \epsilon_{yy} \; \epsilon_{zz} \; \epsilon_{xz} \; \epsilon_{yz} \; \epsilon_{xy} \right\}^{\top} \tag{1.5}$$

Considering only geometrically nonlinear effects, a linear constitutive relation between the nonlinear strains and the corresponding stresses is employed:

$$\mathbf{S} = \mathbf{C}\mathbf{E} \tag{1.6}$$

The explicit expression of the material matrix $\mathbf{C}$ is not given here for the sake of brevity, but it may be found in the influential work [30]. The kinematic (or geometric) relations define the nonlinear strain-displacement equations as follows:

$$\mathbf{E} = \mathbf{E}_{\mathrm{l}} + \mathbf{E}_{\mathrm{nl}} = (\mathbf{b}_{\mathrm{l}} + \mathbf{b}_{\mathrm{nl}})\mathbf{u} \tag{1.7}$$

where $\mathbf{b}_{\mathrm{l}}$ and $\mathbf{b}_{\mathrm{nl}}$ denote the linear and nonlinear differential operators, respectively; their full expressions are provided in [25, 35].

1.3.2 Internal and External Force Vectors

Having defined the constitutive and kinematic relations, the nonlinear governing equations can be derived. This work employs the Principle of Virtual Displacements, which states that:

$$\delta L_{\mathrm{int}} - \delta L_{\mathrm{ext}} = 0 \tag{1.8}$$

where δL_{int} and δL_{ext} denote the virtual variation of internal energy and the virtual work of external loads, respectively. The corresponding internal force vector $\mathbf{F}^{\mathrm{int}}$ and external load vector $\mathbf{p}$ can be then used in the present formulation:

$$\mathbf{F}^{\mathrm{int}} - \mathbf{p} = \mathbf{0} \tag{1.9}$$

First, the internal force vector $\mathbf{F}^{\mathrm{int}}$ is derived. Specifically, it is obtained from the virtual strain energy δL_{int}, given by

$$\delta L_{\mathrm{int}} = \int_V \delta\mathbf{E}^{\top}\mathbf{S}\mathrm{d}V =< \delta\mathbf{E}^{\top}\mathbf{S} > \tag{1.10}$$

where $< (\cdot) >= \int_V (\cdot)\mathrm{d}V$.

Expressing the strain vector $\mathbf{E}$ (Eq. (1.7)) in terms of the generalized nodal unknowns $\mathbf{q}_{\tau i}$ gives:

$$\mathbf{E} = (\mathbf{B}^{\mathrm{l}}_{\tau i} + \mathbf{B}^{\mathrm{nl}}_{\tau i})\mathbf{q}_{\tau i} \tag{1.11}$$

The linear and nonlinear strain-displacement matrices, $\mathbf{B}^{\mathrm{l}}_{\tau i}$ and $\mathbf{B}^{\mathrm{nl}}_{\tau i}$, are given by:

$$\mathbf{B}^{\mathrm{l}}_{\tau i} = \begin{bmatrix} F_{u_x\tau}N_{i,x} & 0 & 0 \\ 0 & F_{u_y\tau}N_{i,y} & 0 \\ 0 & 0 & F_{u_z\tau,z}N_i \\ F_{u_x\tau,z}N_i & 0 & F_{u_z\tau}N_{i,x} \\ 0 & F_{u_y\tau,z}N_i & F_{u_z\tau}N_{i,y} \\ F_{u_x\tau}N_{i,y} & F_{u_y\tau}N_{i,x} & 0 \end{bmatrix} \tag{1.12}$$

and

$$\mathbf{B}^{\mathrm{nl}}_{\tau i} = \frac{1}{2}\begin{bmatrix} u_{x,x} F_{u_x\tau} N_{i,x} & u_{y,x} F_{u_y\tau} N_{i,x} & u_{z,x} F_{u_z\tau} N_{i,x} \\ u_{x,y} F_{u_x\tau} N_{i,y} & u_{y,y} F_{u_y\tau} N_{i,y} & u_{z,y} F_{u_z\tau} N_{i,y} \\ u_{x,z} F_{u_x\tau,z} N_i & u_{y,z} F_{u_y\tau,z} N_i & u_{z,z} F_{u_z\tau,z} N_i \\ u_{x,x} F_{u_x\tau,z} N_i + u_{x,z} F_{u_x\tau} N_{i,x} & u_{y,x} F_{u_y\tau,z} N_i + u_{y,z} F_{u_y\tau} N_{i,x} & u_{z,x} F_{u_z\tau,z} N_i + u_{z,z} F_{u_z\tau} N_{i,x} \\ u_{x,y} F_{u_x\tau,z} N_i + u_{x,z} F_{u_x\tau} N_{i,y} & u_{y,y} F_{u_y\tau,z} N_i + u_{y,z} F_{u_y\tau} N_{i,y} & u_{z,y} F_{u_z\tau,z} N_i + u_{z,z} F_{u_z\tau} N_{i,y} \\ u_{x,x} F_{u_x\tau} N_{i,y} + u_{x,y} F_{u_x\tau} N_{i,x} & u_{y,x} F_{u_y\tau} N_{i,y} + u_{y,y} F_{u_y\tau} N_{i,x} & u_{z,x} F_{u_z\tau} N_{i,y} + u_{z,y} F_{u_z\tau} N_{i,x} \end{bmatrix} \tag{1.13}$$

Similarly, the virtual variation $\delta\mathbf{E}$ is expressed in terms of the virtual variations of the nodal unknowns:

$$\delta\mathbf{E} = \delta\left(\left(\mathbf{B}^{\mathrm{l}}_{sj} + \mathbf{B}^{\mathrm{nl}}_{sj}\right)\mathbf{q}_{sj}\right) = \left(\mathbf{B}^{\mathrm{l}}_{sj} + 2\mathbf{B}^{\mathrm{nl}}_{sj}\right)\delta\mathbf{q}_{sj} \tag{1.14}$$

Applying the transpose operation, the following is obtained:

$$\delta\mathbf{E}^{\top} = \delta\mathbf{q}^{\top}_{sj}\left(\mathbf{B}^{\mathrm{l}}_{sj} + 2\mathbf{B}^{\mathrm{nl}}_{sj}\right)^{\top} \tag{1.15}$$

From this, the internal strain energy is derived:

$$\begin{aligned} \delta L_{\mathrm{int}} &= <\delta\mathbf{E}^{\top}\mathbf{S}> = <\delta\mathbf{q}^{\top}_{sj}\left(\mathbf{B}^{\mathrm{l}}_{sj} + 2\mathbf{B}^{\mathrm{nl}}_{sj}\right)^{\top}\mathbf{S}> \\ &= \delta q_{u_x sj} F^{\mathrm{int}}_{u_x sj} + \delta q_{u_y sj} F^{\mathrm{int}}_{u_y sj} + \delta q_{u_z sj} F^{\mathrm{int}}_{u_z sj} \end{aligned} \tag{1.16}$$

Thus, the internal force vector $\mathbf{F}^{\mathrm{int}}$ is formulated via the three fundamental nuclei (FN):

$$\begin{aligned} F^{\mathrm{int}}_{u_x sj} =& < F_{u_x s,x} N_j \left(\sigma_{xx} + \sigma_{xx} u_{x,x} + \sigma_{xy} u_{x,y} + \sigma_{xz} u_{x,z}\right) > \\ &+ < F_{u_x s,z} N_j \left(\sigma_{xz} + \sigma_{xz} u_{x,x} + \sigma_{yz} u_{x,y} + \sigma_{zz} u_{x,z}\right) > \\ &+ < F_{u_x s} N_{j,y} \left(\sigma_{xy} + \sigma_{xy} u_{x,x} + \sigma_{yy} u_{x,y} + \sigma_{yz} u_{x,z}\right) > \end{aligned} \tag{1.17}$$

$$\begin{aligned} F^{\mathrm{int}}_{u_y sj} =& < F_{u_y s,x} N_j \left(\sigma_{xy} + \sigma_{xx} u_{y,x} + \sigma_{xy} u_{y,y} + \sigma_{xz} u_{y,z}\right) > \\ &+ < F_{u_y s,z} N_j \left(\sigma_{yz} + \sigma_{xz} u_{y,x} + \sigma_{yz} u_{y,y} + \sigma_{zz} u_{y,z}\right) > \\ &+ < F_{u_y s} N_{j,y} \left(\sigma_{yy} + \sigma_{xy} u_{y,x} + \sigma_{yy} u_{y,y} + \sigma_{yz} u_{y,z}\right) > \end{aligned} \tag{1.18}$$

$$\begin{aligned} F^{\mathrm{int}}_{u_z sj} =& < F_{u_z s,x} N_j \left(\sigma_{xz} + \sigma_{xx} u_{z,x} + \sigma_{xy} u_{z,y} + \sigma_{xz} u_{z,z}\right) > \\ &+ < F_{u_z s,z} N_j \left(\sigma_{zz} + \sigma_{xz} u_{z,x} + \sigma_{yz} u_{z,y} + \sigma_{zz} u_{z,z}\right) > \\ &+ < F_{u_z s} N_{j,y} \left(\sigma_{yz} + \sigma_{xy} u_{z,x} + \sigma_{yy} u_{z,y} + \sigma_{yz} u_{z,z}\right) > \end{aligned} \tag{1.19}$$

Second, the external force vector $\mathbf{p}$ must be evaluated. This is derived from the virtual work of the applied point loads, expressed as

$$\delta L_{\mathrm{ext}} = \delta\mathbf{u}^{\top}\mathbf{p} = \delta u_x p_{u_x} + \delta u_y p_{u_y} + \delta u_z p_{u_z} \tag{1.20}$$

By applying the virtual variations within the CUF and FEM framework (see Eq. (1.4)), the following expression is obtained:

$$\delta L_{\text{ext}} = \delta q_{x_{sj}} N_j F_{u_x s} p_{u_x} + \delta q_{y_{sj}} N_j F_{u_y s} p_{u_y} + \delta q_{z_{sj}} N_j F_{u_z s} p_{u_z} \tag{1.21}$$

where:

$$\begin{aligned} p_{u_x sj} &= N_j F_{u_x s} p_{u_x} \\ p_{u_y sj} &= N_j F_{u_y s} p_{u_y} \\ p_{u_z sj} &= N_j F_{u_z s} p_{u_z} \end{aligned} \tag{1.22}$$

1.3.3 Newton-Raphson Linearization of Governing Equations

The geometrical nonlinear FE analysis is based on Eq. (1.9), which is solved using the incremental Newton-Raphson method. The nonlinear problem is equivalently expressed as the minimization of a residual function, following the approach of Reddy [29]:

$$\boldsymbol{\varphi}_{\text{res}}(\mathbf{q} + \delta\mathbf{q}, \mathbf{p} + \delta\mathbf{p}) = \boldsymbol{\varphi}_{\text{res}}(\mathbf{p}, \mathbf{q}) + \frac{\partial \boldsymbol{\varphi}_{\text{res}}}{\partial \mathbf{q}} \delta\mathbf{q} + \frac{\partial \boldsymbol{\varphi}_{\text{res}}}{\partial \mathbf{p}} \delta\lambda \mathbf{p}_{\text{ref}} = 0 \tag{1.23}$$

where $\boldsymbol{\varphi}_{\text{res}}$ is the residual nodal force vector, $\mathbf{q}$ is the vector of unknown displacements, and $\mathbf{p}$ is the external force vector. Here, $\frac{\partial \boldsymbol{\varphi}_{\text{res}}}{\partial \mathbf{q}} = \mathbf{K}^{\top}$ is the tangent stiffness matrix, and $-\frac{\partial \boldsymbol{\varphi}_{\text{res}}}{\partial \mathbf{p}} = \mathbf{I}$, with $\mathbf{I}$ being the identity matrix. The load is assumed to vary proportionally to a reference vector $\mathbf{p}_{\text{ref}}$ such that $\mathbf{p} = \lambda \mathbf{p}_{\text{ref}}$.

Equation (1.23) can thus be rewritten as:

$$\mathbf{K}^{\top} \delta\mathbf{q} = \delta\lambda \mathbf{p}_{\text{ref}} - \boldsymbol{\varphi}_{\text{res}} \tag{1.24}$$

Since the load factor λ is treated as an additional unknown, an auxiliary constraint equation $c(\delta\mathbf{q}, \delta\lambda)$ is introduced, leading to the coupled system:

$$\begin{cases} \mathbf{K}^{\top} \delta\mathbf{q} = \delta\lambda \mathbf{p}_{\text{ref}} - \boldsymbol{\varphi}_{\text{res}} \\ c(\delta\mathbf{q}, \delta\lambda) = 0 \end{cases} \tag{1.25}$$

In this work, the Crisfield arc-length method, a path-following technique, is employed [26].

Derivation of $< \delta\mathbf{E}^{\top}\delta\mathbf{S} >$ First, the linearization process begins with the constitutive relation $\mathbf{S} = \mathbf{C}\mathbf{E}$:

$$\delta\mathbf{S} = \delta\left(\mathbf{C}\mathbf{E}\right) = \mathbf{C}\delta\mathbf{E} = \mathbf{C}\left(\mathbf{B}^{\text{l}}_{\tau i} + 2\mathbf{B}^{\text{nl}}_{\tau i}\right)\delta\mathbf{q}_{\tau i} \tag{1.26}$$

The contributions $K^{0}_{u_m u_l s\tau ji}$ and $K^{T_1}_{u_m u_l s\tau ji}$ are derived from the linearized governing equations, given as follows:

$$
\begin{aligned}
< \delta \mathbf{E}^{\top} \delta \mathbf{S} > &= \delta \mathbf{q}_{sj}^{\top} < \left(\mathbf{B}_{\mathrm{l}}^{sj} + 2\mathbf{B}_{\mathrm{nl}}^{sj}\right)^{\top} \mathbf{C} \left(\mathbf{B}_{l}^{\tau i} + 2\mathbf{B}_{\mathrm{nl}}^{\tau i}\right) > \delta \mathbf{q}_{\tau i} \\
&= [\delta q_{u_x sj} \left(K^{0}_{u_x u_x s\tau ji}\delta q_{u_x \tau i} + K^{0}_{u_x u_y s\tau ji}\delta q_{u_y \tau i} + K^{0}_{u_x u_z s\tau ji}\delta q_{u_z \tau i}\right) \\
&+ \delta q_{u_y sj} \left(K^{0}_{u_y u_x s\tau ji}\delta q_{u_x \tau i} + K^{0}_{u_y u_y s\tau ji}\delta q_{u_y \tau i} + K^{0}_{u_y u_z s\tau ji}\delta q_{u_z \tau i}\right) \\
&+ \delta q_{u_z sj} \left(K^{0}_{u_z u_x s\tau ji}\delta q_{u_x \tau i} + K^{0}_{u_z u_y s\tau ji}\delta q_{u_y \tau i} + K^{0}_{u_z u_z s\tau ji}\delta q_{u_z \tau i}\right)] \\
&+ 2[\delta q_{u_x sj} \left(K^{l\mathrm{nl}}_{u_x u_x s\tau ji}\delta q_{u_x \tau i} + K^{l\mathrm{nl}}_{u_x u_y s\tau ji}\delta q_{u_y \tau i} + K^{l\mathrm{nl}}_{u_x u_z s\tau ji}\delta q_{u_z \tau i}\right) \\
&+ \delta q_{u_y sj} \left(K^{l\mathrm{nl}}_{u_y u_x s\tau ji}\delta q_{u_x \tau i} + K^{l\mathrm{nl}}_{u_y u_y s\tau ji}\delta q_{u_y \tau i} + K^{l\mathrm{nl}}_{u_y u_z s\tau ji}\delta q_{u_z \tau i}\right) \\
&+ \delta q_{u_z sj} \left(K^{l\mathrm{nl}}_{u_z u_x s\tau ji}\delta q_{u_x \tau i} + K^{l\mathrm{nl}}_{u_z u_y s\tau ji}\delta q_{u_y \tau i} + K^{l\mathrm{nl}}_{u_z u_z s\tau ji}\delta q_{u_z \tau i}\right)] \\
&+ [\delta q_{u_x sj} \left(K^{\mathrm{nl}l}_{u_x u_x s\tau ji}\delta q_{u_x \tau i} + K^{\mathrm{nl}l}_{u_x u_y s\tau ji}\delta q_{u_y \tau i} + K^{\mathrm{nl}l}_{u_x u_z s\tau ji}\delta q_{u_z \tau i}\right) \\
&+ \delta q_{u_y sj} \left(K^{\mathrm{nl}l}_{u_y u_x s\tau ji}\delta q_{u_x \tau i} + K^{\mathrm{nl}l}_{u_y u_y s\tau ji}\delta q_{u_y \tau i} + K^{\mathrm{nl}l}_{u_y u_z s\tau ji}\delta q_{u_z \tau i}\right) \\
&+ \delta q_{u_z sj} \left(K^{\mathrm{nl}l}_{u_z u_x s\tau ji}\delta q_{u_x \tau i} + K^{\mathrm{nl}l}_{u_z u_y s\tau ji}\delta q_{u_y \tau i} + K^{\mathrm{nl}l}_{u_z u_z s\tau ji}\delta q_{u_z \tau i}\right)] \\
&+ 2[\delta q_{u_x sj} \left(K^{\mathrm{nlnl}}_{u_x u_x s\tau ji}\delta q_{u_x \tau i} + K^{\mathrm{nlnl}}_{u_x u_y s\tau ji}\delta q_{u_y \tau i} + K^{\mathrm{nlnl}}_{u_x u_z s\tau ji}\delta q_{u_z \tau i}\right) \\
&+ \delta q_{u_y sj} \left(K^{\mathrm{nlnl}}_{u_y u_x s\tau ji}\delta q_{u_x \tau i} + K^{\mathrm{nlnl}}_{u_y u_y s\tau ji}\delta q_{u_y \tau i} + K^{\mathrm{nlnl}}_{u_y u_z s\tau ji}\delta q_{u_z \tau i}\right) \\
&+ \delta q_{u_z sj} \left(K^{\mathrm{nlnl}}_{u_z u_x s\tau ji}\delta q_{u_x \tau i} + K^{\mathrm{nlnl}}_{u_z u_y s\tau ji}\delta q_{u_y \tau i} + K^{\mathrm{nlnl}}_{u_z u_z s\tau ji}\delta q_{u_z \tau i}\right)] \\
&- [\delta q_{u_x sj} \left(K^{0}_{u_x u_x s\tau ji}\delta q_{u_x \tau i} + K^{0}_{u_x u_y s\tau ji}\delta q_{u_y \tau i} + K^{0}_{u_x u_z s\tau ji}\delta q_{u_z \tau i}\right) \\
&+ \delta q_{u_y sj} \left(K^{0}_{u_y u_x s\tau ji}\delta q_{u_x \tau i} + K^{0}_{u_y u_y s\tau ji}\delta q_{u_y \tau i} + K^{0}_{u_y u_z s\tau ji}\delta q_{u_z \tau i}\right) \\
&+ \delta q_{u_z sj} \left(K^{0}_{u_z u_x s\tau ji}\delta q_{u_x \tau i} + K^{0}_{u_z u_y s\tau ji}\delta q_{u_y \tau i} + K^{0}_{u_z u_z s\tau ji}\delta q_{u_z \tau i}\right)] \\
&+ [\delta q_{u_x sj} \left(K^{T_1}_{u_x u_x s\tau ji}\delta q_{u_x \tau i} + K^{T_1}_{u_x u_y s\tau ji}\delta q_{u_y \tau i} + K^{T_1}_{u_x u_z s\tau ji}\delta q_{u_z \tau i}\right) \\
&+ \delta q_{u_y sj} \left(K^{T_1}_{u_y u_x s\tau ji}\delta q_{u_x \tau i} + K^{T_1}_{u_y u_y s\tau ji}\delta q_{u_y \tau i} + K^{T_1}_{u_y u_z s\tau ji}\delta q_{u_z \tau i}\right) \\
&+ \delta q_{u_z sj} \left(K^{T_1}_{u_z u_x s\tau ji}\delta q_{u_x \tau i} + K^{T_1}_{u_z u_y s\tau ji}\delta q_{u_y \tau i} + K^{T_1}_{u_z u_z s\tau ji}\delta q_{u_z \tau i}\right)]
\end{aligned}
\tag{1.27}
$$

Specifically, the nine terms $K^{0}_{u_m u_l s\tau ji}$ constitute the linear part of the stiffness, whose explicit forms are detailed in Appendix. The nonlinear contribution $K^{T_1}_{u_m u_l s\tau ji}$ consists of three distinct components: $2 \times K^{l\mathrm{nl}}_{u_m u_l s\tau ji}$, $K^{\mathrm{nl}l}_{u_m u_l s\tau ji}$, and $2 \times K^{\mathrm{nlnl}}_{u_m u_l s\tau ji}$.

First, the nine FNs $K^{\mathrm{nl}l}_{u_m u_l s\tau ji}$ represent the first-order nonlinear stiffness terms in the formulation, expressed below in compact notation.

For $l = x$:

$$\begin{aligned} K^{\mathrm{nl}l}_{u_m u_x s\tau ji} &= \left\langle \mathbf{u}_{,x}[m]C_{11}N_jN_iF_{u_m s,z}F_{u_x\tau,z}\right\rangle + \left\langle \mathbf{u}_{,x}[m]C_{16}N_{j,x}N_{i,y}F_{u_m s}F_{u_x\tau}\right\rangle \\ &+ \left\langle \mathbf{u}_{,x}[m]C_{16}N_{j,y}N_{i,x}F_{u_m s}F_{u_x\tau}\right\rangle + \left\langle \mathbf{u}_{,x}[m]C_{66}N_{j,y}N_{i,y}F_{u_m s}F_{u_x\tau}\right\rangle \\ &+ \left\langle \mathbf{u}_{,x}[m]C_{44}N_jN_iF_{u_m s,z}F_{u_x\tau,z}\right\rangle + \left\langle \mathbf{u}_{,y}[m]C_{16}N_{j,x}N_{i,x}F_{u_m s}F_{u_x\tau}\right\rangle \\ &+ \left\langle \mathbf{u}_{,y}[m]C_{66}N_{j,x}N_{i,y}F_{u_m s}F_{u_x\tau}\right\rangle + \left\langle \mathbf{u}_{,y}[m]C_{16}N_{j,x}N_{i,x}F_{u_m s}F_{u_x\tau}\right\rangle \\ &+ \left\langle \mathbf{u}_{,y}[m]C_{12}N_{j,y}N_{i,x}F_{u_m s}F_{u_x\tau}\right\rangle + \left\langle \mathbf{u}_{,y}[m]C_{26}N_{j,y}N_{i,y}F_{u_m s}F_{u_x\tau}\right\rangle \\ &+ \left\langle \mathbf{u}_{,y}[m]C_{45}N_jN_iF_{u_m s,z}F_{u_x\tau,z}\right\rangle + \left\langle \mathbf{u}_{,z}[m]C_{44}N_{j,x}N_iF_{u_m s}F_{u_x\tau,z}\right\rangle \\ &+ \left\langle \mathbf{u}_{,z}[m]C_{45}N_{j,y}N_iF_{u_m s}F_{u_x\tau,z}\right\rangle + \left\langle \mathbf{u}_{,z}[m]C_{13}N_jN_{i,x}F_{u_m s,z}F_{u_x\tau}\right\rangle \\ &+ \left\langle \mathbf{u}_{,z}[m]C_{36}N_jN_{i,y}F_{u_m s,z}F_{u_x\tau}\right\rangle \end{aligned} \tag{1.28}$$

For $l = y$:

$$\begin{aligned} K^{\mathrm{nl}l}_{u_m u_y s\tau ji} &= \left\langle \mathbf{u}_{,x}[m]C_{16}N_{j,x}N_{i,x}F_{u_m s}F_{u_y\tau}\right\rangle + \left\langle \mathbf{u}_{,x}[m]C_{12}N_{j,y}N_{i,x}F_{u_m s}F_{u_y\tau}\right\rangle \\ &+ \left\langle \mathbf{u}_{,x}[m]C_{66}N_{j,y}N_{i,x}F_{u_m s}F_{u_y\tau}\right\rangle + \left\langle \mathbf{u}_{,x}[m]C_{26}N_{j,y}N_{i,y}F_{u_m s}F_{u_y\tau}\right\rangle \\ &+ \left\langle \mathbf{u}_{,x}[m]C_{45}N_jN_iF_{u_m s,z}F_{u_y\tau,z}\right\rangle + \left\langle \mathbf{u}_{,y}[m]C_{66}N_{j,x}N_{i,x}F_{u_m s}F_{u_y\tau}\right\rangle \\ &+ \left\langle \mathbf{u}_{,y}[m]C_{26}N_{j,x}N_{i,y}F_{u_m s}F_{u_y\tau}\right\rangle + \left\langle \mathbf{u}_{,y}[m]C_{26}N_{j,y}N_{i,x}F_{u_m s}F_{u_y\tau}\right\rangle \\ &+ \left\langle \mathbf{u}_{,y}[m]C_{22}N_{j,y}N_{i,y}F_{u_m s}F_{u_y\tau}\right\rangle + \left\langle \mathbf{u}_{,y}[m]C_{55}N_jN_iF_{u_m s,z}F_{u_y\tau,z}\right\rangle \\ &+ \left\langle \mathbf{u}_{,z}[m]C_{45}N_{j,x}N_iF_{u_m s}F_{u_y\tau,z}\right\rangle + \left\langle \mathbf{u}_{,z}[m]C_{55}N_{j,y}N_iF_{u_m s}F_{u_y\tau,z}\right\rangle \\ &+ \left\langle \mathbf{u}_{,z}[m]C_{36}N_jN_{i,x}F_{u_m s,z}F_{u_y\tau}\right\rangle + \left\langle \mathbf{u}_{,z}[m]C_{23}N_jN_{i,y}F_{u_m s,z}F_{u_y\tau}\right\rangle \end{aligned} \tag{1.29}$$

For $l = z$:

$$\begin{aligned} K^{\mathrm{nl}l}_{u_m u_z s\tau ji} &= \left\langle \mathbf{u}_{,x}[m]C_{13}N_{j,x}N_iF_{u_m s}F_{u_z\tau,z}\right\rangle + \left\langle \mathbf{u}_{,x}[m]C_{36}N_{j,y}N_iF_{u_m s}F_{u_z\tau,z}\right\rangle \\ &+ \left\langle \mathbf{u}_{,x}[m]C_{44}N_jN_{i,x}F_{u_m s,z}F_{u_z\tau}\right\rangle + \left\langle \mathbf{u}_{,x}[m]C_{44}N_jN_{i,y}F_{u_m s,z}F_{u_z\tau}\right\rangle \\ &+ \left\langle \mathbf{u}_{,y}[m]C_{36}N_{j,x}N_iF_{u_m s}F_{u_z\tau,z}\right\rangle + \left\langle \mathbf{u}_{,y}[m]C_{23}N_{j,y}N_iF_{u_m s}F_{u_z\tau,z}\right\rangle \\ &+ \left\langle \mathbf{u}_{,y}[m]C_{45}N_jN_{i,x}F_{u_m s,z}F_{u_z\tau}\right\rangle + \left\langle \mathbf{u}_{,y}[m]C_{55}N_jN_{i,y}F_{u_m s,z}F_{u_z\tau}\right\rangle \\ &+ \left\langle \mathbf{u}_{,z}[m]C_{44}N_{j,x}N_iF_{u_m s,x}F_{u_z\tau}\right\rangle + \left\langle \mathbf{u}_{,z}[m]C_{45}N_{j,x}N_{i,y}F_{u_m s}F_{u_z\tau}\right\rangle \\ &+ \left\langle \mathbf{u}_{,z}[m]C_{45}N_{j,y}N_{i,x}F_{u_m s}F_{u_z\tau}\right\rangle + \left\langle \mathbf{u}_{,z}[m]C_{55}N_{j,y}N_{i,y}F_{u_m s}F_{u_z\tau}\right\rangle \\ &+ \left\langle \mathbf{u}_{,z}[m]C_{33}N_jN_iF_{u_m s,z}F_{u_z\tau,z}\right\rangle \end{aligned} \tag{1.30}$$

Second, the components $K^{l\mathrm{nl}}_{u_m u_l s\tau ji}$ can be directly obtained from $K^{\mathrm{nl}l}_{u_m u_l s\tau ji}$ through the symmetry relation $K^{\mathrm{nl}l}_{u_m u_l s\tau ji} = \frac{1}{2}K^{l\mathrm{nl}}_{u_l u_m \tau sij}$.

Third, the generic scalar Fundamental Nucleus $K^{\mathrm{nlnl}}_{u_m u_l s\tau ji}$ of the second-order nonlinear stiffness matrix is given below:

$$
\begin{aligned}
&2 \times K^{\mathrm{nlnl}}_{u_m u_l s \tau j i} = \\
&\left\langle \mathbf{u}_{,x}[m]\mathbf{u}_{,x}[l] C_{11} N_{j,x} N_{i,x} F_{u_m s} F_{u_l \tau} \right\rangle + \left\langle \mathbf{u}_{,x}[m]\mathbf{u}_{,x}[l] C_{44} N_{j} N_{i} F_{u_m s,z} F_{u_l \tau,z} \right\rangle \\
&\left\langle \mathbf{u}_{,x}[m]\mathbf{u}_{,x}[l] C_{16} N_{j,y} N_{i,x} F_{u_m s} F_{u_l \tau} \right\rangle + \left\langle \mathbf{u}_{,x}[m]\mathbf{u}_{,x}[l] C_{66} N_{j,x} N_{i} F_{u_m s} F_{u_l \tau,z} \right\rangle \\
&\left\langle \mathbf{u}_{,x}[m]\mathbf{u}_{,x}[l] C_{44} N_{j} N_{i} F_{u_m s,z} F_{u_l \tau,z} \right\rangle + \left\langle \mathbf{u}_{,x}[m]\mathbf{u}_{,y}[l] C_{16} N_{j,x} N_{i,x} F_{u_m s} F_{u_l \tau} \right\rangle \\
&\left\langle \mathbf{u}_{,x}[m]\mathbf{u}_{,y}[l] C_{12} N_{j,x} N_{i,y} F_{u_m s} F_{u_l \tau} \right\rangle + \left\langle \mathbf{u}_{,x}[m]\mathbf{u}_{,y}[l] C_{66} N_{j,y} N_{i,x} F_{u_m s} F_{u_l \tau} \right\rangle \\
&\left\langle \mathbf{u}_{,x}[m]\mathbf{u}_{,y}[l] C_{26} N_{j,y} N_{i,y} F_{u_m s} F_{u_l \tau} \right\rangle + \left\langle \mathbf{u}_{,x}[m]\mathbf{u}_{,y}[l] C_{45} N_{j} N_{i} F_{u_m s,z} F_{u_l \tau,z} \right\rangle \\
&\left\langle \mathbf{u}_{,x}[m]\mathbf{u}_{,z}[l] C_{13} N_{j,x} N_{i} F_{u_m s} F_{u_l \tau,z} \right\rangle + \left\langle \mathbf{u}_{,x}[m]\mathbf{u}_{,z}[l] C_{36} N_{j,y} N_{i} F_{u_m s} F_{u_l \tau,z} \right\rangle \\
&\left\langle \mathbf{u}_{,x}[m]\mathbf{u}_{,z}[l] C_{44} N_{j} N_{i,x} F_{u_m s,z} F_{u_l \tau} \right\rangle + \left\langle \mathbf{u}_{,x}[m]\mathbf{u}_{,z}[l] C_{45} N_{j} N_{i,y} F_{u_m s,z} F_{u_l \tau} \right\rangle \\
&\left\langle \mathbf{u}_{,y}[m]\mathbf{u}_{,x}[l] C_{16} N_{j,x} N_{i,x} F_{u_m s} F_{u_l \tau} \right\rangle + \left\langle \mathbf{u}_{,y}[m]\mathbf{u}_{,x}[l] C_{66} N_{j,x} N_{i,y} F_{u_m s} F_{u_l \tau} \right\rangle \\
&\left\langle \mathbf{u}_{,y}[m]\mathbf{u}_{,x}[l] C_{12} N_{j,y} N_{i,x} F_{u_m s} F_{u_l \tau} \right\rangle + \left\langle \mathbf{u}_{,y}[m]\mathbf{u}_{,x}[l] C_{26} N_{j,y} N_{i,y} F_{u_m s} F_{u_l \tau} \right\rangle \\
&\left\langle \mathbf{u}_{,y}[m]\mathbf{u}_{,x}[l] C_{45} N_{j} N_{i} F_{u_m s,z} F_{u_l \tau,z} \right\rangle + \left\langle \mathbf{u}_{,y}[m]\mathbf{u}_{,y}[l] C_{66} N_{j,x} N_{i,x} F_{u_m s} F_{u_l \tau} \right\rangle \\
&\left\langle \mathbf{u}_{,y}[m]\mathbf{u}_{,y}[l] C_{26} N_{j,x} N_{i,y} F_{u_m s} F_{u_l \tau} \right\rangle + \left\langle \mathbf{u}_{,y}[m]\mathbf{u}_{,y}[l] C_{66} N_{j,y} N_{i,x} F_{u_m s} F_{u_l \tau} \right\rangle \\
&\left\langle \mathbf{u}_{,y}[m]\mathbf{u}_{,y}[l] C_{22} N_{j,y} N_{i,y} F_{u_m s} F_{u_l \tau} \right\rangle + \left\langle \mathbf{u}_{,y}[m]\mathbf{u}_{,y}[l] C_{55} N_{j} N_{i} F_{u_m s,z} F_{u_l \tau,z} \right\rangle \\
&\left\langle \mathbf{u}_{,y}[m]\mathbf{u}_{,z}[l] C_{36} N_{j,x} N_{i} F_{u_m s} F_{u_l \tau,z} \right\rangle + \left\langle \mathbf{u}_{,y}[m]\mathbf{u}_{,z}[l] C_{23} N_{j,y} N_{i} F_{u_m s} F_{u_l \tau,z} \right\rangle \\
&\left\langle \mathbf{u}_{,y}[m]\mathbf{u}_{,z}[l] C_{45} N_{j} N_{i,x} F_{u_m s,z} F_{u_l \tau} \right\rangle + \left\langle \mathbf{u}_{,y}[m]\mathbf{u}_{,z}[l] C_{55} N_{j} N_{i,y} F_{u_m s,z} F_{u_l \tau} \right\rangle \\
&\left\langle \mathbf{u}_{,z}[m]\mathbf{u}_{,x}[l] C_{44} N_{j,x} N_{i} F_{u_m s} F_{u_l \tau,z} \right\rangle + \left\langle \mathbf{u}_{,z}[m]\mathbf{u}_{,x}[l] C_{45} N_{j,y} N_{i} F_{u_m s} F_{u_l \tau,z} \right\rangle \\
&\left\langle \mathbf{u}_{,z}[m]\mathbf{u}_{,x}[l] C_{13} N_{j} N_{i,x} F_{u_m s,z} F_{u_l \tau} \right\rangle + \left\langle \mathbf{u}_{,z}[m]\mathbf{u}_{,x}[l] C_{36} N_{j} N_{i,y} F_{u_m s,z} F_{u_l \tau} \right\rangle \\
&\left\langle \mathbf{u}_{,z}[m]\mathbf{u}_{,y}[l] C_{45} N_{j,x} N_{i} F_{u_m s} F_{u_l \tau,z} \right\rangle + \left\langle \mathbf{u}_{,z}[m]\mathbf{u}_{,y}[l] C_{55} N_{j,y} N_{i} F_{u_m s} F_{u_l \tau,z} \right\rangle \\
&\left\langle \mathbf{u}_{,z}[m]\mathbf{u}_{,y}[l] C_{36} N_{j} N_{i,x} F_{u_m s,z} F_{u_l \tau} \right\rangle + \left\langle \mathbf{u}_{,z}[m]\mathbf{u}_{,y}[l] C_{23} N_{j} N_{i,y} F_{u_m s,z} F_{u_l \tau} \right\rangle \\
&\left\langle \mathbf{u}_{,z}[m]\mathbf{u}_{,z}[l] C_{44} N_{j,x} N_{i,x} F_{u_m s} F_{u_l \tau} \right\rangle + \left\langle \mathbf{u}_{,z}[m]\mathbf{u}_{,z}[l] C_{45} N_{j,x} N_{i,y} F_{u_m s} F_{u_l \tau} \right\rangle \\
&\left\langle \mathbf{u}_{,z}[m]\mathbf{u}_{,z}[l] C_{45} N_{j,y} N_{i,x} F_{u_m s} F_{u_l \tau} \right\rangle + \left\langle \mathbf{u}_{,z}[m]\mathbf{u}_{,z}[l] C_{55} N_{j,y} N_{i,y} F_{u_m s} F_{u_l \tau} \right\rangle \\
&\left\langle \mathbf{u}_{,z}[m]\mathbf{u}_{,z}[l] C_{33} N_{j} N_{i} F_{u_m s,z} F_{u_l \tau,z} \right\rangle
\end{aligned} \tag{1.31}
$$

In the expressions above, $\mathbf{u}_{,x}[m]$ represents the m-th component of the vector $\frac{\partial \mathbf{u}}{\partial x}$; e.g., $u_{,x}[y] = u_{y_{,x}}$. Analogously, $\mathbf{u}_{,x}[l]$ is the l-th component of the vector $\frac{\partial \mathbf{u}}{\partial y}$, and so forth.

Derivation of $< \delta(\delta \mathbf{E})^{\mathsf{T}} \mathbf{S} >$ The final contribution to the tangent stiffness matrix arises from the linearization of the nonlinear geometric relations. According to Crisfield [15]:

$$\delta(\delta\mathbf{E}) = \left\{\begin{matrix} (\delta u_{x,x})_{\text{v}}\,\delta u_{x,x} + (\delta u_{y,x})_{\text{v}}\,\delta u_{y,x} + (\delta u_{z,x})_{\text{v}}\,\delta u_{z,x} \\ \left(\delta u_{x,y}\right)_{\text{v}}\delta u_{x,y} + \left(\delta u_{y,y}\right)_{\text{v}}\delta u_{y,y} + \left(\delta u_{z,y}\right)_{\text{v}}\delta u_{z,y} \\ (\delta u_{x,z})_{\text{v}}\,\delta u_{x,z} + (\delta u_{y,z})_{\text{v}}\,\delta u_{y,z} + (\delta u_{z,z})_{\text{v}}\,\delta u_{z,z} \\ \left[(\delta u_{x,x})_{\text{v}}\,\delta u_{x,z} + \delta u_{x,x}\,(\delta u_{x,z})_{\text{v}}\right] + \left[(\delta u_{y,x})_{\text{v}}\,\delta u_{y,z} + \delta u_{y,x}\,(\delta u_{y,z})_{\text{v}}\right] \\ + \left[(\delta u_{z,x})_{\text{v}}\,\delta u_{z,z} + \delta u_{z,x}\,(\delta u_{z,z})_{\text{v}}\right] \\ \left[\left(\delta u_{x,y}\right)_{\text{v}}\delta u_{x,z} + \delta u_{x,y}\,(\delta u_{x,z})_{\text{v}}\right] + \left[\left(\delta u_{y,y}\right)_{\text{v}}\delta u_{y,z} + \delta u_{y,y}\,(\delta u_{y,z})_{\text{v}}\right] \\ + \left[\left(\delta u_{z,y}\right)_{\text{v}}\delta u_{z,z} + \delta u_{z,y}\,(\delta u_{z,z})_{\text{v}}\right] \\ \left[(\delta u_{x,x})_{\text{v}}\,\delta u_{x,y} + \delta u_{x,x}\left(\delta u_{x,y}\right)_{\text{v}}\right] + \left[(\delta u_{y,x})_{\text{v}}\,\delta u_{y,y} + \delta u_{y,x}\left(\delta u_{y,y}\right)_{\text{v}}\right] \\ + \left[(\delta u_{z,x})_{\text{v}}\,\delta u_{z,y} + \delta u_{z,x}\left(\delta u_{z,y}\right)_{\text{v}}\right] \end{matrix}\right\} \tag{1.32}$$

Here, the variation is denoted by the subscript "v". The linearized displacement variables and the corresponding virtual displacements are expressed as $\delta u_l = N_i F_{u_l\tau}\,\delta q_{u_l\tau i}$ and $(\delta u_m)_{\text{v}} = N_j F_{u_m s}\,\delta q_{u_m s j}$, respectively. Thus, the following expressions for the real and virtual displacement are given:

$$\delta(\delta\mathbf{E}) = \mathbf{B}^{\text{nl}*}\begin{Bmatrix} \delta q_{x_{sj}}\delta q_{x_{\tau i}} \\ \delta q_{y_{sj}}\delta q_{y_{\tau i}} \\ \delta q_{z_{sj}}\delta q_{z_{\tau i}} \end{Bmatrix} \tag{1.33}$$

$$\delta(\delta\mathbf{E})^{\top} = \begin{Bmatrix} \delta q_{x_{sj}}\delta q_{x_{\tau i}} \\ \delta q_{y_{sj}}\delta q_{y_{\tau i}} \\ \delta q_{z_{sj}}\delta q_{z_{\tau i}} \end{Bmatrix}^{\top}\left(\mathbf{B}^{\text{nl}*}\right)^{\top} \tag{1.34}$$

where:

$$\mathbf{B}^{\text{nl}*}(:,1) = \begin{bmatrix} F_{u_x\tau}F_{u_x s}N_{i,x}N_{j,x} \\ F_{u_x\tau}F_{u_x s}N_{i,y}N_{j,y} \\ F_{u_x\tau,z}F_{u_x s,z}N_iN_j \\ F_{u_x\tau}F_{u_x s,z}N_{i,x}N_j + F_{u_x\tau,z}F_{u_x s}N_iN_{j,x} \\ F_{u_x\tau,z}F_{u_x s}N_iN_{j,y} + F_{u_x\tau}F_{u_x s,z}N_{i,y}N_j \\ F_{u_x\tau}F_{u_x s}N_{i,x}N_{j,y} + F_{u_x\tau}F_{u_x s}N_{i,y}N_{j,x} \end{bmatrix} \tag{1.35}$$

$$\mathbf{B}^{\text{nl}*}(:,2) = \begin{bmatrix} F_{u_y\tau}F_{u_y s}N_{i,x}N_{j,x} \\ F_{u_y\tau}F_{u_y s}N_{i,y}N_{j,y} \\ F_{u_y\tau,z}F_{u_y s,z}N_iN_j \\ F_{u_y\tau}F_{u_y s,z}N_{i,x}N_j + F_{u_y\tau,z}F_{u_y s}N_iN_{j,x} \\ F_{u_y\tau,z}F_{u_y s}N_iN_{j,y} + F_{u_y\tau}F_{u_y s,z}N_{i,y}N_j \\ F_{u_y\tau}F_{u_y s}N_{i,x}N_{j,y} + F_{u_y\tau}F_{u_y s}N_{i,y}N_{j,x} \end{bmatrix} \tag{1.36}$$

$$\mathbf{B}^{\mathrm{nl}*}(:,3) = \begin{bmatrix} F_{u_z\tau}F_{u_zs}N_{i,x}N_{j,x} \\ F_{u_z\tau}F_{u_zs}N_{i,y}N_{j,y} \\ F_{u_z\tau,z}F_{u_zs,z}N_iN_j \\ F_{u_z\tau}F_{u_zs,z}N_{i,x}N_j + F_{u_z\tau,z}F_{u_zs}N_iN_{j,x} \\ F_{u_z\tau,z}F_{u_zs}N_iN_{j,y} + F_{u_z\tau}F_{u_zs,z}N_{i,y}N_j \\ F_{u_z\tau}F_{u_zs}N_{i,x}N_{j,y} + F_{u_z\tau}F_{u_zs}N_{i,y}N_{j,x} \end{bmatrix} \tag{1.37}$$

From these relations, the contribution of the nonlinear geometric terms can be derived as follows:

$$< \delta(\delta\mathbf{E})^\top\mathbf{S} >=< \begin{Bmatrix} \delta q_{x_{sj}}\delta q_{x_{\tau i}} \\ \delta q_{y_{sj}}\delta q_{y_{\tau i}} \\ \delta q_{z_{sj}}\delta q_{z_{\tau i}} \end{Bmatrix}^\top \left(\mathbf{B}^{\mathrm{nl}*}\right)^\top \mathbf{S} > \tag{1.38}$$

thus,

$$\begin{aligned} < \delta(\delta\mathbf{E})^\top\mathbf{S} > &= \delta\mathbf{q}_{sj}^\top < \mathrm{diag}\left(\left(\mathbf{B}^{\mathrm{nl}*}\right)^\top \mathbf{S}\right) > \delta\mathbf{q}_{\tau i} \\ &= \delta\mathbf{q}_{sj}^\top < \mathrm{diag}\left(\left(\mathbf{B}^{\mathrm{nl}*}\right)^\top (\mathbf{S}_{\mathrm{l}} + \mathbf{S}_{\mathrm{nl}})\right) > \delta\mathbf{q}_{\tau i} \\ &= \delta q_{u_x sj}\left(K^{\sigma_{\mathrm{l}}}_{u_xu_xs\tau ji} + K^{\sigma_{\mathrm{nl}}}_{u_xu_xs\tau ji}\right)\delta q_{u_x\tau i} \\ &+ \delta q_{u_y sj}\left(K^{\sigma_{\mathrm{l}}}_{u_yu_ys\tau ji} + K^{\sigma_{\mathrm{nl}}}_{u_yu_ys\tau ji}\right)\delta q_{u_y\tau i} \\ &+ \delta q_{u_z sj}\left(K^{\sigma_{\mathrm{l}}}_{u_zu_zs\tau ji} + K^{\sigma_{\mathrm{nl}}}_{u_zu_zs\tau ji}\right)\delta q_{u_z\tau i} \\ = \delta q_{u_x sj}K^{\sigma}_{u_xu_xs\tau ji}\delta q_{u_x\tau i} &+ \delta q_{u_y sj}K^{\sigma}_{u_yu_ys\tau ji}\delta q_{u_y\tau i} + \delta q_{u_z sj}K^{\sigma}_{u_zu_zs\tau ji}\delta q_{u_z\tau i} \end{aligned} \tag{1.39}$$

The equation above defines the nonlinear *geometrical* contribution to the tangent stiffness matrix. It comprises a linear part, $\mathbf{S}_{\mathrm{l}} = \mathbf{C}\mathbf{E}_{\mathrm{l}}$, and a nonlinear part, $\mathbf{S}_{\mathrm{nl}} = \mathbf{C}\mathbf{E}_{\mathrm{nl}}$. Consequently, the complete FN can be written as: $K^{\sigma}_{u_mu_ls\tau ji} = K^{\sigma_{\mathrm{l}}}_{u_mu_ls\tau ji} + K^{\sigma_{\mathrm{nl}}}_{u_mu_ls\tau ji}$.

Only three scalar FNs $K^{\sigma}_{u_mu_ls\tau ji}$ are nonzero; their explicit forms are given below:

$$\begin{aligned} K^{\sigma}_{u_xu_xs\tau ji} &= < \sigma_{xx}F_{u_x\tau}F_{u_xs}N_{i,x}N_{j,x} > + < \sigma_{yy}F_{u_x\tau}F_{u_xs}N_{i,y}N_{j,y} > \\ &+ < \sigma_{zz}F_{u_x\tau,z}F_{u_xs,z}N_iN_j > + < \sigma_{xy}F_{u_x\tau}F_{u_xs}N_{i,x}N_{j,y} > \\ &+ < \sigma_{xy}F_{u_x\tau}F_{u_xs}N_{i,y}N_{j,x} > + < \sigma_{xz}F_{u_x\tau}F_{u_xs,z}N_{i,x}N_j > \\ &+ < \sigma_{xz}F_{u_x\tau,z}F_{u_xs}N_iN_{j,x} > + < \sigma_{yz}F_{u_x\tau,z}F_{u_xs}N_iN_{j,y} > \\ &+ < \sigma_{yz}F_{u_x\tau}F_{u_xs,z}N_{i,y}N_j > \end{aligned} \tag{1.40}$$

$$
\begin{aligned}
K^{\sigma}_{u_y u_y s\tau ji} = &< \sigma_{xx} F_{u_y\tau} F_{u_y s} N_{i,x} N_{j,x} > + < \sigma_{yy} F_{u_y\tau} F_{u_y s} N_{i,y} N_{j,y} > \\
&+ < \sigma_{zz} F_{u_y\tau,z} F_{u_y s,z} N_i N_j > + < \sigma_{xy} F_{u_y\tau} F_{u_y s} N_{i,x} N_{j,y} > \\
&+ < \sigma_{xy} F_{u_y\tau} F_{u_y s} N_{i,y} N_{j,x} > + < \sigma_{xz} F_{u_y\tau} F_{u_y s,z} N_{i,x} N_j > \\
&+ < \sigma_{xz} F_{u_y\tau,z} F_{u_y s} N_i N_{j,x} > + < \sigma_{yz} F_{u_y\tau,z} F_{u_y s} N_i N_{j,y} > \\
&+ < \sigma_{yz} F_{u_y\tau} F_{u_y s,z} N_{i,y} N_j >
\end{aligned}
\tag{1.41}
$$

$$
\begin{aligned}
K^{\sigma}_{u_z u_z s\tau ji} = &< \sigma_{xx} F_{u_z\tau} F_{u_z s} N_{i,x} N_{j,x} > + < \sigma_{yy} F_{u_z\tau} F_{u_z s} N_{i,y} N_{j,y} > \\
&+ < \sigma_{zz} F_{u_z\tau,z} F_{u_z s,z} N_i N_j > + < \sigma_{xy} F_{u_z\tau} F_{u_z s} N_{i,x} N_{j,y} > \\
&+ < \sigma_{xy} F_{u_z\tau} F_{u_z s} N_{i,y} N_{j,x} > + < \sigma_{xz} F_{u_z\tau} F_{u_z s,z} N_{i,x} N_j > \\
&+ < \sigma_{xz} F_{u_z\tau,z} F_{u_z s} N_i N_{j,x} > + < \sigma_{yz} F_{u_z\tau,z} F_{u_z s} N_i N_{j,y} > \\
&+ < \sigma_{yz} F_{u_z\tau} F_{u_z s,z} N_{i,y} N_j >
\end{aligned}
\tag{1.42}
$$

In summary, the FNs of the tangent stiffness matrix are expressed as the sum of all contributions derived above:

$$
K^{\top}_{u_m u_l s\tau ji} = K^{0}_{u_m u_l s\tau ji} + K^{T_1}_{u_m u_l s\tau ji} + K^{\sigma}_{u_m u_l s\tau ji}
\tag{1.43}
$$

1.4 Assembly

This research introduces a novel assembly strategy for the fundamental nuclei (FNs) of FE matrices and arrays, which represents a significant difference from the classical CUF methodology. The most important innovation is the scalar-level representation of the FN components. The stiffness term $K^{\top}_{u_m u_l s\tau ji}$, the internal force component $F^{\text{int}}_{u_m sj}$, and the load term $p_{u_m sj}$ are treated as independent scalar quantities rather than being organized within the a 3×3 sub-matrix structure. As detailed in the previous section, this approach reduces the complete stiffness (or a 3×1 structure) characterization to only nine independent scalar parameters, while the internal force and load vectors each require three scalar terms per node. This reformulation allows a computational efficiency saving compared to the original CUF framework [5, 9]. In the latter, a single expansion theory was applied to all displacement components, thereby limiting kinematic flexibility.

Figure 1.2 illustrates the complete assembly process of the tangent stiffness matrix, from the scalar FN to the full structural matrix. The assembly procedure for the internal and external load vectors follows a similar pattern, albeit with a single index dimension. The process begins with the computation of the invariant scalar kernel $K^{\top}_{u_m u_l s\tau ji}$. Subsequent expansions are then performed in the following sequence:

1. The expansion over the indices s and τ yields the submatrix $\mathbf{K}^{\top}_{u_m u_l ji}$.
2. The expansion over the displacement components u_m and u_l produces the matrix $\mathbf{K}^{\top}_{ji}$.

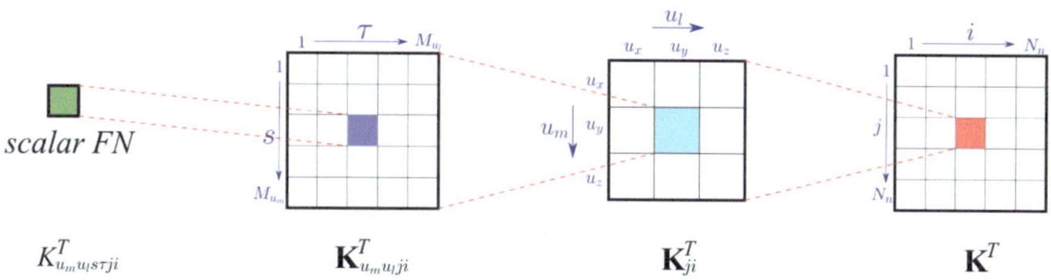

Fig. 1.2: General procedure of the assembly for the stiffness matrix.

3. Finally, the expansion over the nodal indices j and i assembles the complete tangent stiffness matrix $\mathbf{K}^\top$.

Several detailed examples of the general assembly procedure for the novel CUF are available in the literature. The beam and plate formulations for linear analysis were presented in [11, 32], while the assembly for the nonlinear beam case was explored in [12]. In fact, the procedure of assembly is the same for beams and plates.

1.5 Numerical Results

The present section studies the equilibrium curves and the stress trends. The first benchmark studies the postbuckling behaviour of an isotropic plate, while the second case investigates the bending loading of a two-layered composite cantilevered structure.

A consistent acronym system is employed to explain the kinematics over the thickness. For Taylor-based (ESL) expansions, the notation TEn_{u_x}-TEn_{u_y}-TEn_{u_z} is used, where n_{u_l} indicates the polynomial order for each displacement component u_l. Regarding Lagrange-based (LW) expansions, the format LEp-LEp-LEp is adopted, where p specifies the number of Lagrange points per CUF element in the thickness direction. The formulation also allows for mixed ESL/LW theories enabling refined kinematics such as LE4-TE1-TE2. Furthermore, other types of models are described during the presentation of the benchmarks.

1.5.1 Preliminary Postbuckling Analysis of an Isotropic Plate

This preliminary benchmark, taken from Wu *et al.* [35], examines the postbuckling behaviour of an isotropic plate. As illustrated in Fig. 1.3, the slender plate is characterised by a length of $a = 30$ [cm], a width of $b = 6$ [cm], and a thickness of $h = 0.6$ [cm]. The structure is composed of an isotropic material with a Young's modulus of $E = 75$ [GPa] and a Poisson ratio of $\nu = 0.316$. The plate is subjected to an in-plane compressive load P applied at the point $[a, b/2, 0]$, as illustrated in Fig. 1.3. Two

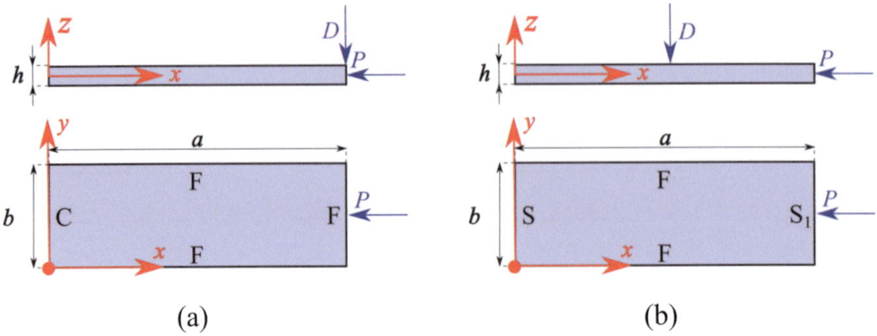

Fig. 1.3: Diagram of two boundary conditions for the isotropic plate loaded by a compressive point load P : (a) one clamped edge and three free edges (CFFF); (b) (immovable simply-supported)-free-(movable simply-supported)-free (SFS_1F).

distinct boundary conditions are considered: clamped and simply-supported. For the clamped condition 'C', all displacement components are restrained at the edge $x = 0$. For the simply-supported cases, there are two possibilities. The immovable simply-supported condition 'S' enforces $u_x = u_y = u_z = 0$ at $z = 0$ and $x = 0$, whereas the movable simply-supported condition 'S_1' imposes $u_y = u_z = 0$ at $z = 0$ and $x = a$. Also, the free edges 'F' conditions are used. Small transverse point loads D are applied to impose a stable branch. For the CFFF conditions, the defect is applied at $[a, b/2, h/2]$, while for the SFS_1F conditions, the load is posed at the centre $[a/2, b/2, h/2]$.

Table 1.1 summarizes the kinematic theories in this study. For each displacement component, it lists the specific expansion and corresponding number of terms, M_{u_l}, as well as the Degrees of Freedom (DOF). All models use a 20×5 Q9 Lagrange element mesh. The first entry gives the reference solution with an LE3 expansion for each displacement. Subsequent rows show characteristics of various TE models.

Table 1.1: Postbuckling analysis of isotropic plates with different boundary conditions. Details of the structural models: types of expansions, the number of terms, and DOF.

u_x	u_y	u_z	$M_{u_x}/M_{u_y}/M_{u_z}$	DOF
		Reference solutions		
LE3	LE3	LE3	3/3/3	4059
		TE Models		
TE2	TE2	TE2	3/3/3	4059
TE2	TE1	TE2	3/2/2	3608
TE1	TE1	TE2	2/2/3	3157
TE2	0	TE2	3/0/3	2706
TE2	z	TE2	3/1/3	3157

Here, the out-of-plane displacement u_z is always modelled by a parabolic expansion to avoid the Poisson locking effect. In TE2-0-LE9, u_y is disregarded, while TE2-z-LE9 considers only the linear term.

Figure 1.4 shows the equilibrium curves for both boundary conditions, comparing the reference model and the complete TE2-TE2-TE2 model. The adimensionalised transverse displacement, u_z/h, is evaluated at $[a, b/2, 0]$ for the CFFF condition and at $[a/2, b/2, 0]$ for the SFS_1F condition. The applied load P is normalized differently: for CFFF, $P(4a^2)/(\pi^2 EI_b)$; for SFS_1F, $P(a^2)/(\pi^2 EI_b)$, where $I_b = bh^3/12$ is the second moment of area.

The numerical results yield several key conclusions:

- Considering the computation of equilibrium curves, almost all reduced models follow the trend of the reference solution with only minor discrepancies. These models accurately capture the postbuckling behavior. The models exhibit consistent performance under both considered boundary conditions.
- In particular, TE2-0-TE2 is the least accurate model, as it differs from the reference models and poorly captures the adimensional critical load. The trends of TE1-TE1-TE2 closely match the reference curves at the beginning, but they diverge as the load increases. Finally, the most convenient model is TE2-z-TE2.
- Thus, the analyses demonstrate that the u_y displacement component can be adequately represented by a single linear term. In contrast, u_x and u_z components have a significant influence, and bi-parabolic expansions must be adopted.

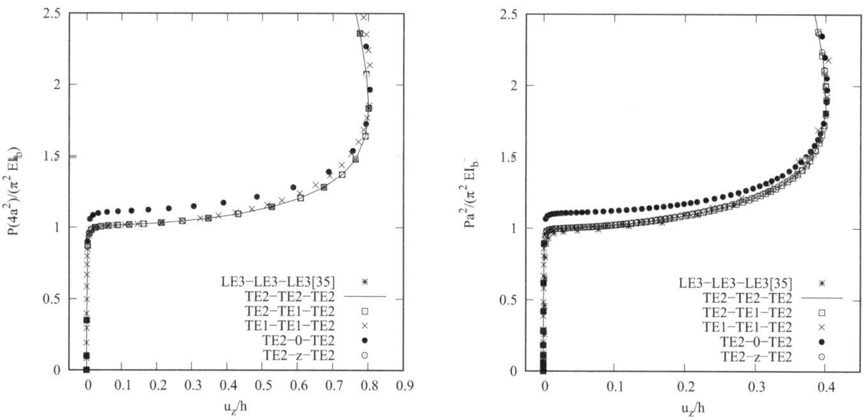

(a) CFFF conditions

(b) SFS_1F conditions

Fig. 1.4: Postbuckling equilibrium curves for various geometrically nonlinear models.

1.5.2 Bending Behaviour of a Two-Layered Cantilevered Plate

In the second benchmark, the large-displacement response of a two-layered asymmetric plate is analysed. The geometric configuration is shown in Fig. 1.5. The plate dimensions are: length $L = 9$ [m], width $b = 1$ [m], and total thickness $h = 0.6$ [m]. Both layers consist of the orthotropic graphite/epoxy composite AS4/3501-06, with the following material properties: $E_1 = 144.8$ [GPa], $E_2 = E_3 = 9.65$ [GPa], $\nu_{12} = \nu_{13} = 0.3$, $\nu_{23} = 0.3$, $G_{12} = G_{13} = 4.14$ [GPa], and $G_{23} = 3.45$ [GPa], while the stacking sequence is $[0°/90°]$. Regarding the boundary conditions, the plate is clamped along the edge at $y = 0$ and subjected to a uniform pressure p_0 applied in the positive z-direction. For the comparison, two reference models are considered. The first one is a CUF-based beam model, where LE16 expansions are used for all three displacement components [12]. Then, a refined three-dimensional solid model developed in ABAQUS is adopted [25].

The kinematic theories employed in the current analyses are summarized in Table 1.2. For all simulations, a uniform mesh of 5×15 Q9 elements is used in the x-y reference plane. The adopted expansions for each displacement component and the corresponding total number of terms M_{u_l}, and DOF are presented. The first rows correspond to the reference solutions, i.e., the CUF beam model with LE16-LE16-

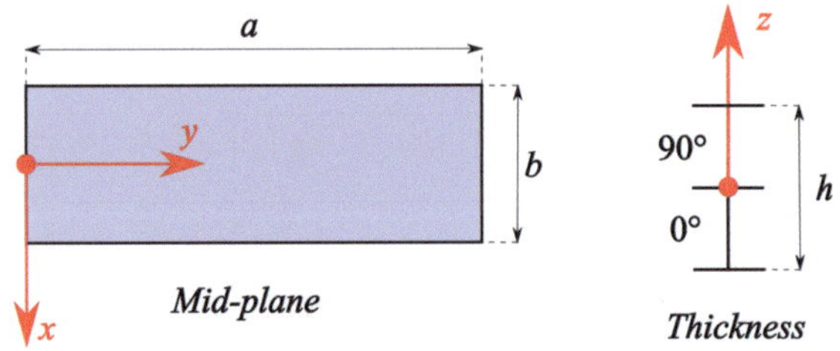

Fig. 1.5 Geometrical properties of an asymmetric laminated plate subjected to large bending deflections.

Table 1.2: Two-layered plate under a constant bending pressure. Details of the structural models.

u_x	u_y	u_z	$M_{u_x}/M_{u_y}/M_{u_z}$	DOF
		Reference solutions		
LE16	LE16	LE16	28/28/28	5124
	3D-ABAQUS		—	573675
		Uniform model		
LE4	LE4	LE4	7/7/7	3225
		ESL/LW Models		
TE2	LE4	TE2	3/7/3	2440
TE2	LE4	TE3	3/7/4	2170
0	LE4	LE4	0/7/7	2170
0	LE4	TE3	0/7/4	1705

LE16 kinematics and the three-dimensional ABAQUS solid model. Note that the M_{u_l} values are high for the LE16-LE16-LE16 case because it is a beam formulation; they are not applicable for the solid reference. The subsequent row presents the uniform plate model LE4-LE4-LE4. The other rows detail various mixed ESL/LW models. In all cases, the axial displacement u_y is modelled using the LW expansion LE4. The in-plane displacement, u_x, is either approximated by a second-order model or not considered. Different expansion orders are employed for the out-of-plane displacement, u_z.

The equilibrium paths are shown in Fig. 1.6, where the normalized transverse displacements u_z/h are plotted against the dimensionless load, which is calculated as: $p_0 a^4/(E_2 h^4)$. The displacements are computed at the location $[0, L, 0]$. Also, the linear response of the uniform LE4 model is provided for completeness. Then, the distributions of the in-plane stress σ_{yy} and the shear stress σ_{yz} are computed along the thickness at the coordinate $(0, L/2, z)$. The adimensionalised stress quantities are defined by:

$$\overline{\sigma}_{yy} = \frac{\sigma_{yy}}{p_0} \qquad \overline{\sigma}_{yz} = \frac{\sigma_{yz}}{p_0} \tag{1.44}$$

These stress distributions are evaluated at three distinct points along the equilibrium path. Figure 1.7 illustrates the trends in the linear regime. Figure 1.8 shows

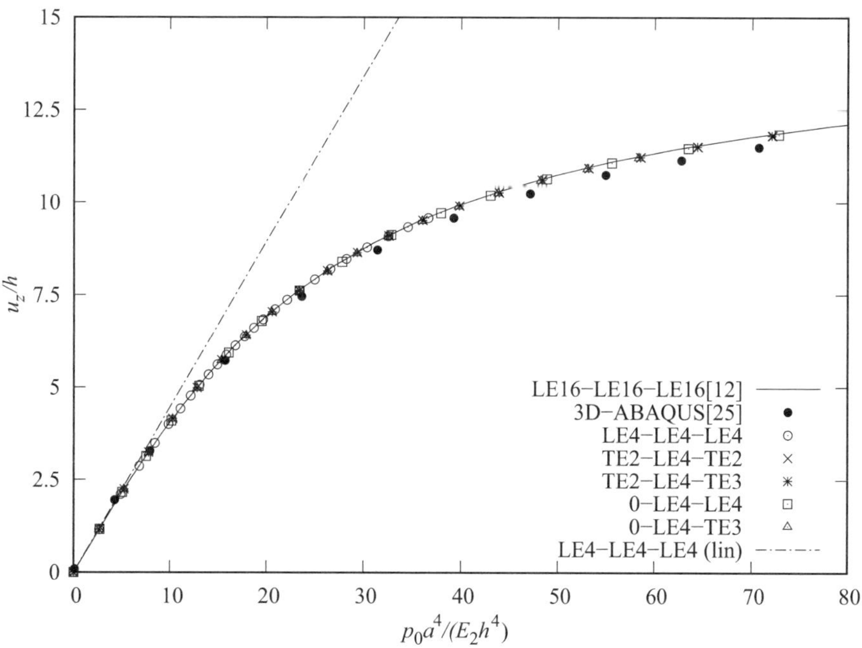

Fig. 1.6: Two-layer cross-ply plate subjected to large deformations. Equilibrium curves.

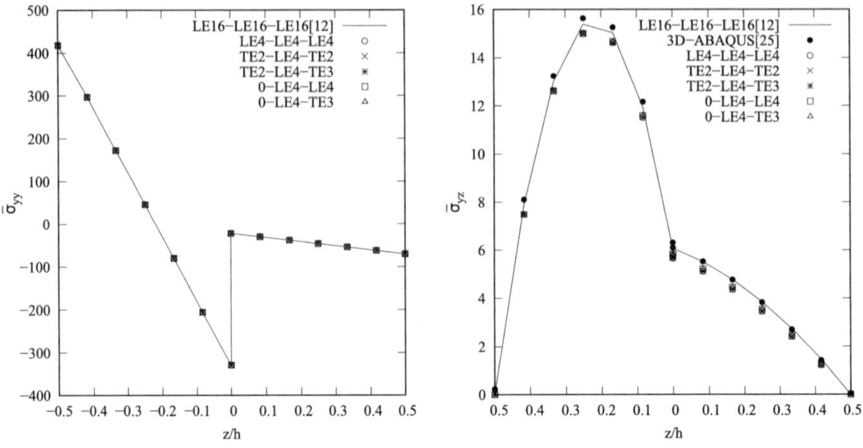

(a) (b)

Fig. 1.7: Two-layer cross-ply plate subjected to large deformations. In-plane (a) and shear stresses (b) calculated in $[0, L/2, z]$. Linear regime.

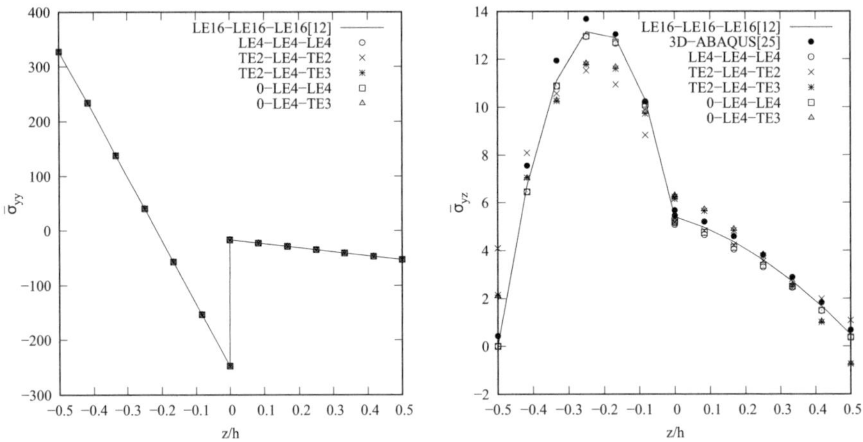

(a) (b)

Fig. 1.8: Two-layer cross-ply plate subjected to large deformations. In-plane (a) and shear stresses (b) calculated in $[0, L/2, z]$. Nonlinear regime, $p_0 = 3 \times 10^6$ [Pa].

the stresses at the load $p_0 = 3 \times 10^6$ [Pa], while the stresses under a load of $p_0 = 9 \times 10^6$ [Pa] are plotted in Fig. 1.9.

From the analyses conducted, the principal findings are summarized as follows:

- In predicting the equilibrium curve for the transverse displacement, the ESL/LW models show excellent agreement with the reference CUF solutions, i.e., the LE16-LE16-LE16 (beam), the LE4-LE4-LE4 (plate), and solid models. Among

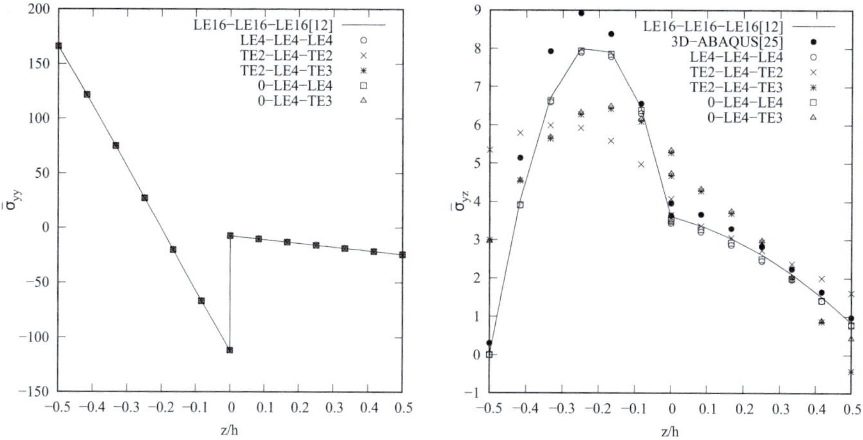

(a) (b)

Fig. 1.9: Two-layer cross-ply plate subjected to large deformations. In-plane (a) and shear stresses (b) calculated in $[0, L/2, z]$. Nonlinear regime, $p_0 = 9 \times 10^6$ [Pa].

these, the 0-LE4-TE3 model is the most computationally efficient, requiring only 1705 DOF. Furthermore, the linear strain assumption remains valid only within a restricted load range.

- In the linear regime (see Fig. 1.7 b), all mixed ESL/LW models show trends similar to those of the LE16-LE16-LE16, 3D-ABAQUS, and LE4-LE4-LE4 models.
- In the nonlinear regime, the in plane stresses are always perfectly captured by all the models adopted, whereas the shear stress may differ across models. The transverse displacement variable, u_z, is the main source of the discrepancies. This effect becomes more evident for higher loads. In contrast, it seems that the influence of the displacement component, u_x, is negligible in this particular case. Furthermore, in the nonlinear range, slight differences become noticeable between the highest-fidelity CUF models and the ABAQUS reference solution.

1.6 Conclusions

This paper introduces a nonlinear plate framework based on the CUF, aiming to reduce the computational cost of nonlinear analyses for composite plate structures. Highly refined models are often needed in the literature to accurately capture complex equilibrium paths and stress distributions, especially shear stresses, but the derivation of efficient ad-hoc models usually requires re-formulating FE matrices for each new case, which may become particularly heavy in nonlinear quasi-static formulations. To overcome these challenges, a unified approach that enables flexible selection of plate models within a single, consistent framework is proposed.

For these reasons, the work introduces a generalized framework that enhances the CUF by enabling each displacement component to adopt independent kinematic expansions and modelling approaches. The formulation incorporates the complete Green–Lagrange strain tensor to accurately model geometric nonlinearity and capture out-of-plane stresses.

For the verification of the models, two case studies from the literature have been considered, which focussed on the post-buckling and large deflections. Based on the presented results and analyses, the principal conclusions of this work are as follows:

- The choice of models depends on the desired outcome, whether displacements or stresses. Indeed, the accurate evaluation of out-of-plane stresses in multilayered structures may require higher-order models based on the LW approach.
- Integrating Taylor and Lagrange expansions within the same model serves as a powerful computational accelerator, while maintaining a good accuracy at the same time.
- Furthermore, at each equilibrium point, the meaning of different terms varies, requiring adjustments to the most accurate models to adapt to specific conditions. For instance, a displacement component may have no influence in the linear regime but become significant at higher load levels.

This paper has demonstrated the successful development of a unified formulation for nonlinear analysis, which enables the independent assignment of kinematic theories to individual displacement components and supports the use of combined ESL/LW models.

Future works should extend the present formulation to incorporate thermal and electric fields within the nonlinear framework, and further investigate the performance of ad-hoc models and mixed ESL/LW approaches in multiphysics cases.

Appendix

The explicit expressions of the linear Fundamental Nuclei are written as follows:

$$
\begin{aligned}
K^0_{u_x u_x s\tau ji} = &\left\langle C_{11} N_{j,x} N_{i,x} F_{u_x s} F_{u_x \tau} \right\rangle + \left\langle C_{44} N_j N_i F_{u_x s,z} F_{u_x \tau,z} \right\rangle \\
&+ \left\langle C_{16} N_{j,x} N_{i,y} F_{u_x s} F_{u_x \tau} \right\rangle + \left\langle C_{16} N_{j,y} N_{i,x} F_{u_x s} F_{u_x \tau} \right\rangle \\
&+ \left\langle C_{66} N_{j,y} N_{i,y} F_{u_x s} F_{u_x \tau} \right\rangle
\end{aligned} \tag{1.45}
$$

$$
\begin{aligned}
K^0_{u_x u_y s\tau ji} = &\left\langle C_{12} N_{j,x} N_{i,y} F_{u_x s} F_{u_y \tau} \right\rangle + \left\langle C_{45} N_j N_i F_{u_x s,z} F_{u_y \tau,z} \right\rangle \\
&+ \left\langle C_{16} N_{j,x} N_{i,x} F_{u_x s} F_{u_y \tau} \right\rangle + \left\langle C_{26} N_{j,y} N_{i,y} F_{u_x s} F_{u_y \tau} \right\rangle \\
&+ \left\langle C_{66} N_{j,y} N_{i,x} F_{u_x s} F_{u_y \tau} \right\rangle
\end{aligned} \tag{1.46}
$$

$$K^0_{u_xu_zs\tau ji} = \left\langle C_{13}N_{j,x}N_iF_{u_xs}F_{u_z\tau,z}\right\rangle + \left\langle C_{44}N_jN_{i,x}F_{u_xs,z}F_{u_z\tau}\right\rangle + \left\langle C_{45}N_jN_{i,y}F_{u_xs,z}F_{u_z\tau}\right\rangle + \left\langle C_{36}N_{j,y}N_iF_{u_xs}F_{u_z\tau,z}\right\rangle \tag{1.47}$$

$$K^0_{u_yu_xs\tau ji} = \left\langle C_{12}N_{j,y}N_{i,x}F_{u_ys}F_{u_x\tau}\right\rangle + \left\langle C_{45}N_{j,x}N_iF_{u_ys}F_{u_x\tau,z}\right\rangle + \left\langle C_{16}N_{j,x}N_{i,x}F_{u_ys}F_{u_x\tau}\right\rangle + \left\langle C_{26}N_{j,y}N_{i,y}F_{u_ys}F_{u_x\tau}\right\rangle + \left\langle C_{66}N_{j,x}N_{i,y}F_{u_ys}F_{u_x\tau}\right\rangle \tag{1.48}$$

$$K^0_{u_yu_ys\tau ji} = \left\langle C_{22}N_{j,y}N_{i,y}F_{u_ys}F_{u_y\tau}\right\rangle + \left\langle C_{55}N_jN_iF_{u_ys,z}F_{u_y\tau,z}\right\rangle + \left\langle C_{26}N_{j,x}N_{i,y}F_{u_ys}F_{u_y\tau}\right\rangle + \left\langle C_{26}N_{j,y}N_{i,x}F_{u_ys}F_{u_y\tau}\right\rangle + \left\langle C_{66}N_{j,x}N_{i,x}F_{u_ys}F_{u_y\tau}\right\rangle \tag{1.49}$$

$$K^0_{u_yu_zs\tau ji} = \left\langle C_{23}N_{j,y}N_iF_{u_ys}F_{u_z\tau,z}\right\rangle + \left\langle C_{55}N_jN_{i,y}F_{u_ys,z}F_{u_z\tau}\right\rangle = \left\langle C_{45}N_jN_{i,x}F_{u_ys,z}F_{u_z\tau}\right\rangle + \left\langle C_{45}N_{j,x}N_iF_{u_ys}F_{u_z\tau,z}\right\rangle \tag{1.50}$$

$$K^0_{u_zu_xs\tau ji} = \left\langle C_{13}N_jN_{i,x}F_{u_zs,z}F_{u_x\tau}\right\rangle + \left\langle C_{44}N_{j,x}N_iF_{u_zs}F_{u_x\tau,z}\right\rangle + \left\langle C_{45}N_{j,y}N_iF_{u_zs}F_{u_x\tau,z}\right\rangle + \left\langle C_{36}N_jN_{i,y}F_{u_zs,z}F_{u_x\tau}\right\rangle \tag{1.51}$$

$$K^0_{u_zu_ys\tau ji} = \left\langle C_{23}N_jN_{i,y}F_{u_zs,z}F_{u_y\tau}\right\rangle + \left\langle C_{55}N_{j,y}N_iF_{u_zs}F_{u_y\tau,z}\right\rangle + \left\langle C_{45}N_{j,x}N_iF_{u_zs}F_{u_y\tau,z}\right\rangle + \left\langle C_{36}N_jN_{i,x}F_{u_zs,z}F_{u_y\tau}\right\rangle \tag{1.52}$$

$$K^0_{u_zu_zs\tau ji} = \left\langle C_{33}N_jN_iF_{u_zs,z}F_{u_z\tau,z}\right\rangle + \left\langle C_{55}N_{j,y}N_{i,y}F_{u_zs}F_{u_z\tau}\right\rangle + \left\langle C_{45}N_{j,x}N_{i,y}F_{u_zs}F_{u_z\tau}\right\rangle + \left\langle C_{45}N_{j,y}N_{i,x}F_{u_zs}F_{u_z\tau}\right\rangle + \left\langle C_{44}N_{j,x}N_iF_{u_zs,x}F_{u_z\tau}\right\rangle \tag{1.53}$$

with $\langle(\cdot)\rangle = \int_V (\cdot)\mathrm{d}V = \int_V (\cdot)\mathrm{d}x\mathrm{d}y\mathrm{d}z$.

References

[1] Alijani F, Amabili M (2014) Non-linear static bending and forced vibrations of rectangular plates retaining non-linearities in rotations and thickness deformation. International Journal of Non-Linear Mechanics 67:394-404, DOI 10.1016/j.ijnonlinmec.2014.10.003

[2] Amabili M (2018) Nonlinear Mechanics of Shells and Plates in Composite, Soft and Biological Materials. Cambridge University Press, DOI 10.1017/9781316422892

[3] Bathe K-J (1996) Finite Element Procedure. Prentice Hall, Upper Saddle River, New Jersey, USA

[4] Bertoldi K, Vitelli V, Christensen J, van Hecke M (2017) Flexible mechanical metamaterials. Nature Reviews Materials 2(11):17066 DOI 10.1038/natrevmats.2017.66

[5] Carrera E (2001) Developments, ideas, and evaluations based upon Reissner's Mixed Variational Theorem in the modeling of multilayered plates and shells. Applied Mechanics Review 54(4):301-329, DOI 10.1115/1.1385512

[6] Carrera E (2003) Historical review of zig-zag theories for multilayered plates and shells. Applied Mechanics Review 56(3):287-308, DOI 10.1115/1.1557614

[7] Carrera E (2000) Single- vs multilayer plate modelings on the basis of Reissner's mixed theorem. AIAA Journal 38(2):342-352, DOI 10.2514/2.962

[8] Carrera E (2003) Theories and finite elements for multilayered plates and shells: a unified compact formulation with numerical assessment and benchmarking. Archives of Computational Methods in Engineering 10(3):215-296, DOI 10.1007/BF02736224

[9] Carrera E, Cinefra M, Zappino E, Petrolo M(2014) Finite Element Analysis of Structures Through Unified Formulation. John Wiley & Sons, Chichester, UK, DOI 10.1002/9781118536643

[10] Carrera E, Parisch H (1997) An evaluation of geometrical nonlinear effects of thin and moderately thick multilayered composite shells. Composite Structures 40(1):11-24, DOI 10.1016/S0263-8223(97)00145-1

[11] Carrera E, Scano D (2024) Beam finite element models with node dependent kinematics and arbitrary displacement fields over the cross-section. Mechanics of Advanced Materials and Structures 1-16, DOI 10.1080/15376494.2024.2434198

[12] Carrera E, Scano D (2026) Nonlinear composite beam finite elements with arbitrary cross-sectional displacement fields. Composite Structures 377:119883, DOI 10.1016/j.compstruct.2025.119883

[13] Chen Y, Wu B, Su Y, Chen W (2019) Tunable two-way unidirectional acoustic diodes: design and simulation. Journal of Applied Mechanics 86(3):031010, DOI 10.1115/1.4042321

[14] Chia CK (1980) Nonlinear Analysis of Plates. McGraw-Hill, New York, USA

[15] Crisfield MA(1996) Non-Linear Finite Element Analysis of Solids and Structures. Wiley, Chichester, West Sussex, UK

[16] Civalek O (2006) Harmonic differential quadrature-finite differences coupled approaches for geometrically nonlinear static and dynamic analysis of rectangular plates on elastic foundation. Journal of Sound and Vibration 294(4-5):966-980, DOI 10.1016/j.jsv.2005.12.041

[17] Dash P, Singh BN. (2010) Geometrically nonlinear bending analysis of laminated composite plate. Communications in Nonlinear Science and Numerical Simulation 15(10):3170-3181, DOI 10.1016/j.cnsns.2009.11.017
[18] Gdoutos E, Truong A, Pedivellano A, Royer F, Pellegrino S (2020) Ultralight deployable space structure prototype. AIAA Scitech 2020 Forum DOI 10.2514/6.2020-0692
[19] Gutierrez Rivera M, Reddy, JN, Amabili M (2016) A new twelve-parameter spectral/hp shell finite element for large deformation analysis of composite shells. Composite Structures 151:183-196, DOI 10.1016/j.compstruct.2016.02.068
[20] Kapania RK, Raciti S (1989) Recent advances in analysis of laminated beams and plates. Part I: Shear effects and buckling. AIAA Journal 27(7):923-935, DOI 10.2514/3.10202
[21] Kapania RK, Raciti S. (1989) Recent advances in analysis of laminated beams and plates. Part II: Vibrations and Wave Propagation. AIAA Journal 27(7):935-946, DOI 10.2514/3.59909
[22] Maji A, Mahato PK (2022) Development and applications of shear deformation theories for laminated composite plates: An overview. Journal of Thermoplastic Composite Materials 35(12):2576-2619, DOI 10.1177/0892705720930765
[23] Mindlin RD (1951) Influence of rotatory inertia and shear on flexural motions of isotropic, elastic plates. Journal of Applied Mechanics 18(1):31-38, DOI 10.1115/1.4010217
[24] Petrolo M, Augello R, Carrera E, Scano D, Pagani A (2023) Evaluation of transverse shear stresses in layered beams/plates/shells via stress recovery accounting for various CUF-based theories. Composite Structrure 307():116625, DOI 10.1016/j.compstruct.2022.116625
[25] Pagani A, Carrera E (2017) Large-deflection and post-buckling analyses of laminated composite beams by Carrera Unified Formulation. Composite Structures 170:40-52, DOI 10.1016/j.compstruct.2017.03.008
[26] Pagani A, Carrera E (2018) Unified formulation of geometrically nonlinear refined beam theories. Mechanics of Advanced Materials and Structures 25(1):15-31, DOI 10.1080/15376494.2016.1232458
[27] Pagani A, Carrera E, Augello R, Scano D (2021) Use of Lagrange polynomials to build refined theories for laminated beams, plates and shells. Composite Structures 276:114505, DOI 10.1016/j.compstruct.2021.114505
[28] Petrolo M, Iannotti P (2023) Best theory diagrams for laminated composite shells based on failure indexes. Aerotecnica Missili & Spazio 102(3):199-218, DOI 10.1007/s42496-023-00158-5
[29] Reddy JN (2014) Introduction to Nonlinear Finite Element Analysis with Applications to Heat Transfer, Fluid Mechanics, and Solid Mechanics. Oxford University Press, New York, USA
[30] Reddy JN (1997) Mechanics of Laminated Composite Plates and Shells: Theory and Analysis. CRC press, New York, USA, DOI 10.1201/b12409

[31] Reissner E (1994) The effect of transverse shear deformation on the bending of elastic plates. Journal Applied Mechanics 12(2):A69-A77, DOI 10.1115/1.4009435
[32] Scano D, Carrera E, Zappino E (2025) Plate finite elements with arbitrary displacement fields along the thickness. Finite Elements in Analysis and Design 244:104296, DOI 10.1016/j.finel.2024.104296
[33] Scano D, Carrera E, Zappino E (2025) Shell finite elements with arbitrary displacement fields along the thickness. International Journal of Solids and Structures 321:113544, DOI 10.1016/j.ijsolstr.2025.113544
[34] Urthaler Y, Reddy JN (2008) A mixed finite element for the nonlinear bending analysis of laminated composite plates based on FSDT. Mechanics of Advanced Materials and Structures 15(5):335-354, DOI 10.1080/15376490802045671
[35] Wu B, Pagani A, Filippi M, Chen WQ, Carrera E (2019) Large-deflection and post-buckling analyses of isotropic rectangular plates by Carrera Unified Formulation. International Journal of Non-Linear Mechanics 116:18-31, DOI 10.1016/j.ijnonlinmec.2019.05.004
[36] Zhang G, Hu Y, Yan B, Tong M, Wang F (2024) Buckling and post-buckling analysis of composite stiffened panels: A ten-year review (2014–2023). Thin-Walled Structures 205:112525, DOI 10.1016/j.tws.2024.112525

Chapter 2
Free Vibration Analysis of Doubly-Curved Laminated Shells by Legendre Spectral Method and a Higher-Order Shell Theory

Susana C.F. Fernandes, Jesús Cuartero, António J.M. Ferreira, and Ana M.A. Neves

Abstract A highly efficient pseudospectral approach is developed for the free vibration analysis of laminated shells based on Reddy's higher-order shear deformation shell theory. Kronecker products are employed to construct two-dimensional differentiation matrices, while univariate differentiation matrices are generated using Legendre polynomials.

2.1 Introduction

In this work, it is demonstrated for the first time that the free vibration analysis of laminated composite shells using a third-order formulation by Reddy and Liu [6], can be successfully carried out using a pseudospectral method based on Legendre polynomials [4, 7]. The use of Chebyshev pseudospectrals was considered in [3]. The computed natural frequencies exhibit excellent agreement with the analytical results reported by Reddy [6].

2.2 Legendre Spectral Collocation Method

The Legendre Spectral Collocation Method (LSCM) [1, 2, 7] is a powerful numerical technique for solving differential equations using global polynomial approximations. It is based on the use of Legendre polynomials as basis functions and the selection

Susana C. F. Fernandes · Jesús Cuartero
Department of Mechanical Engineering, University of Zaragoza, 50009 Zaragoza, Spain, e-mail: sfernandes@umaia.pt, jcs@unizar.es

António J.M. Ferreira (corresponding author) · Ana M.A. Neves
Faculty of Engineering, University of Porto, Rua Dr. Roberto Frias, s/n 4200-465 Porto, Portugal, e-mail: ferreira@fe.up.pt, ananeves@fe.up.pt

H. Altenbach et al. (eds.), *Composite Structures and Technology*,
Advanced Structured Materials 255,
https://doi.org/10.1007/978-3-032-24656-1_2

of collocation points at the roots of these polynomials or their derivatives. Legendre polynomials $\{P_n(x)\}_{n=0}^{\infty}$ are a sequence of orthogonal polynomials on the interval $[-1, 1]$ with respect to the standard L^2 inner product:

$$\langle P_m, P_n \rangle = \int_{-1}^{1} P_m(x)P_n(x)\,dx = \frac{2}{2n+1}\delta_{mn},$$

where δ_{mn} is the Kronecker delta. They satisfy the recurrence relation:

$$(n+1)P_{n+1}(x) = (2n+1)xP_n(x) - nP_{n-1}(x), \quad n \geq 1,$$

with $P_0(x) = 1$ and $P_1(x) = x$.

A smooth function $u(x)$ defined on $[-1, 1]$ can be approximated by a truncated series of Legendre polynomials:

$$u_N(x) = \sum_{n=0}^{N} \hat{u}_n P_n(x),$$

where $\hat{u}_n$ are the spectral coefficients, typically computed via projection:

$$\hat{u}_n = \frac{2n+1}{2} \int_{-1}^{1} u(x)P_n(x)\,dx.$$

In the spectral collocation method, instead of computing the coefficients via integrals, we enforce the differential equation at specific collocation points. For Legendre spectral collocation, these are usually the Legendre-Gauss-Lobatto (LGL) points:

$$x_j, \quad j = 0, 1, \ldots, N,$$

which include the endpoints $x_0 = -1$ and $x_N = 1$ and the roots of $(1-x^2)P_N'(x) = 0$. These points are advantageous because they minimize Runge's phenomenon and allow exact integration of polynomials up to degree $2N - 1$ via quadrature.

To apply the method to differential equations, we define a differentiation matrix D that approximates derivatives at the collocation points. For $u(x)$ interpolated at LGL points as

$$u_N(x) = \sum_{j=0}^{N} u(x_j)\ell_j(x),$$

where $\ell_j(x)$ are Lagrange interpolating polynomials:

$$\ell_j(x) = \prod_{\substack{k=0 \\ k \neq j}}^{N} \frac{x - x_k}{x_j - x_k},$$

the derivative at the collocation points is approximated by

$$u'_N(x_i) = \sum_{j=0}^{N} D_{ij} u(x_j), \quad i = 0, 1, \ldots, N,$$

where the entries of D are

$$D_{ij} = \begin{cases} \dfrac{P_N(x_i)}{P_N(x_j)} \dfrac{1}{x_i - x_j}, & i \neq j, \\ -\dfrac{N(N+1)}{4}, & i = j = 0, \\ \dfrac{N(N+1)}{4}, & i = j = N, \\ 0, & \text{otherwise.} \end{cases}$$

Unlike finite elements, the current method presents spectral accuracy, as shown in the examples below.

Consider a boundary value problem

$$\mathcal{L}u(x) = f(x), \quad x \in [-1, 1], \quad u(-1) = u(-1), \, u(1) = u_1.$$

The Legendre spectral collocation method seeks $u_N(x)$ such that

$$\mathcal{L}u_N(x_j) = f(x_j), \quad j = 1, \ldots, N - 1,$$

while enforcing boundary conditions at x_0 and x_N. Using the differentiation matrix D, this leads to a linear system

$$\mathbf{Lu} = \mathbf{f},$$

where $\mathbf{u} = [u(x_0), \ldots, u(x_N)]^\top$, $\mathbf{f} = [f(x_0), \ldots, f(x_N)]^\top$, and $\mathbf{L}$ is constructed from D according to the operator $\mathcal{L}$.

2.3 Third-Order Shear Theory

In Reddy's third-order theory for laminated shells [6] of constant thickness h, the orthogonal curvilinear coordinates (x, y, z) are defined such that the z-curves are lines perpendicular to the middle surface $z = 0$, and the x- and y-curves are lines of curvature on the middle surface $z = 0$, and where R_x and R_y are the radii of curvature. The transverse displacement is set as constant, $w = w_0$, and the in-plane displacements are expressed as

$$
\begin{aligned}
u &= \left(1 + \frac{z}{R_x}\right) u_0 + z\theta_x - \frac{4}{3h^2} z^3 \left(\theta_x + \frac{1}{\alpha_x}\frac{\partial w_0}{\partial x}\right); \\
v &= \left(1 + \frac{z}{R_y}\right) v_0 + z\theta_y - \frac{4}{3h^2} z^3 \left(\theta_y + \frac{1}{\alpha_y}\frac{\partial w_0}{\partial y}\right)
\end{aligned}
\tag{2.1}
$$

where α_i are the surface metrics, u_0, v_0, and w_0 are the displacements of the shell's middle surface, and θ_x and θ_y are the normal rotations to the x and y-axes, respectively.

Deformations can be obtained by derivatives of the displacements as

$$
\begin{aligned}
&\epsilon_1 = \epsilon_1^{(0)} + z(k_1^{(0)} + z^2 k_1^{(2)});\ \epsilon_2 = \epsilon_2^{(0)} + z(k_2^{(0)} + z^2 k_2^{(2)}); \\
&\epsilon_3^{(0)} = 0;\ \epsilon_4 = \epsilon_4^{(0)} + z^2 k_4^{(1)} \\
&\epsilon_5 = \epsilon_5^{(0)} + z^2 k_5^{(1)};\ \epsilon_6 = \epsilon_6^{(0)} + z(k_6^{(0)} + z^2 k_6^{(2)})
\end{aligned}
\tag{2.2}
$$

where

$$
\begin{aligned}
\epsilon_1^{(0)} &= \frac{\partial u_0}{\partial x} + \frac{w}{R_x};\ \epsilon_2^{(0)} = \frac{\partial v_0}{\partial y} + \frac{w}{R_y};\ \epsilon_6^{(0)} = \frac{\partial u_0}{\partial y} + \frac{\partial v_0}{\partial x};\ \epsilon_4^{(0)} = \frac{\partial w_0}{\partial y} + \theta_y; \\
\epsilon_5^{(0)} &= \frac{\partial w_0}{\partial x} + \theta_x
\end{aligned}
\tag{2.3}
$$

$$
\begin{aligned}
k_1^{(0)} &= \frac{\partial \theta_x}{\partial x};\ k_2^{(0)} = \frac{\partial \theta_y}{\partial y};\ k_6^{(0)} = \frac{\partial \theta_y}{\partial x} + \frac{\partial \theta_x}{\partial y};\ k_4^{(1)} = -\frac{4}{h^2}\left(\theta_y + \frac{\partial w_0}{\partial y}\right); \\
k_5^{(1)} &= -\frac{4}{h^2}\left(\theta_x + \frac{\partial w_0}{\partial x}\right) \\
k_1^{(2)} &= -\frac{4}{3h^2}\left(\frac{\partial \theta_x}{\partial x} + \frac{\partial^2 w_0}{\partial x^2}\right);\ k_2^{(2)} = -\frac{4}{3h^2}\left(\frac{\partial \theta_y}{\partial y} + \frac{\partial^2 w_0}{\partial y^2}\right) \\
k_6^{(2)} &= -\frac{4}{3h^2}\left(\frac{\partial \theta_y}{\partial x} + \frac{\partial \theta_x}{\partial y} + 2\frac{\partial^2 w_0}{\partial x \partial y}\right)
\end{aligned}
\tag{2.4}
$$

The stress-strain relation for a k_{th} layer are

$$
\boldsymbol{\sigma}_k = \boldsymbol{D}_k \boldsymbol{\epsilon}_k
\tag{2.5}
$$

where

$$
\begin{bmatrix} \sigma_1^{(k)} \\ \sigma_2^{(k)} \\ \sigma_6^{(k)} \\ \sigma_4^{(k)} \\ \sigma_5^{(k)} \end{bmatrix}
=
\begin{bmatrix}
Q_{11}^{(k)} & Q_{12}^{(k)} & Q_{16}^{(k)} & 0 & 0 \\
 & Q_{22}^{(k)} & Q_{26}^{(k)} & 0 & 0 \\
symm. & & Q_{66}^{(k)} & 0 & 0 \\
 & & & Q_{44}^{(k)} & 0 \\
 & & & & Q_{55}^{(k)}
\end{bmatrix}
\begin{bmatrix} \epsilon_1^{(k)} \\ \epsilon_2^{(k)} \\ \epsilon_6^{(k)} \\ \epsilon_4^{(k)} \\ \epsilon_5^{(k)} \end{bmatrix}
\tag{2.6}
$$

being $Q_{ij}^{(k)}$ material constants for a k_{th} layer in the laminate coordinate system [5]. The equations of motion are obtained by Hamilton's principle as [5]

$$\frac{\partial N_{xx}}{\partial x}+\frac{\partial N_{xy}}{\partial y}=I_0\ddot{u}_0+J_1\ddot{\theta}_x-c_1I_3\frac{\partial \ddot{w}_0}{\partial x} \tag{2.7}$$

$$\frac{\partial N_{xy}}{\partial x}+\frac{\partial N_{yy}}{\partial y}=I_0\ddot{v}_0+J_1\ddot{\theta}_y-c_1I_3\frac{\partial \ddot{w}_0}{\partial y} \tag{2.8}$$

$$\frac{\partial Q_x}{\partial x}+\frac{\partial Q_y}{\partial y}-\frac{4}{h^2}\left(\frac{\partial K_x}{\partial x}+\frac{\partial K_y}{\partial y}\right)+\frac{4}{3h^2}\left(\frac{\partial^2 P_{xx}}{\partial x^2}+\frac{\partial^2 P_{yy}}{\partial y^2}+2\frac{\partial^2 P_{xy}}{\partial x\partial y}\right)-\frac{N_{xx}}{R_x}-\frac{N_{yy}}{R_y}+q=$$
$$I_0\ddot{w}_0-c_1^2I_6\left(\frac{\partial^2 \ddot{w}_0}{\partial x^2}+\frac{\partial^2 \ddot{w}_0}{\partial y^2}\right)+c_1\left[I_3\left(\frac{\partial \ddot{u}_0}{\partial x}+\frac{\partial \ddot{v}_0}{\partial y}\right)+J_4\left(\frac{\partial \ddot{\theta}_x}{\partial x}+\frac{\partial \ddot{\theta}_y}{\partial y}\right)\right] \tag{2.9}$$

$$\frac{\partial M_{xx}}{\partial x}+\frac{\partial M_{xy}}{\partial y}-Q_x+\frac{4}{h^2}K_x-\frac{4}{3h^2}\left(\frac{\partial P_{xx}}{\partial x}+\frac{\partial P_{xy}}{\partial y}\right)=J_1\ddot{u}_0+K_2\ddot{\theta}_x-c_1J_4\frac{\partial \ddot{w}_0}{\partial x} \tag{2.10}$$

$$\frac{\partial M_{xy}}{\partial x}+\frac{\partial M_{yy}}{\partial y}-Q_y+\frac{4}{h^2}K_y-\frac{4}{3h^2}\left(\frac{\partial P_{xy}}{\partial x}+\frac{\partial P_{yy}}{\partial y}\right)=J_1\ddot{v}_0+K_2\ddot{\theta}_y-c_1J_4\frac{\partial \ddot{w}_0}{\partial y} \tag{2.11}$$

where q is the distributed transverse load, N_i, M_i, etc. are the stress resultants, given by

$$(N_i, M_i, P_i)=\sum_{k=1}^{NL}\int_{z_{k-1}}^{z_k}\sigma_i^{(k)}(1,z,z^3)dz \qquad (i=xx,yy,xy)$$
$$(Q_x,K_x)=\sum_{k=1}^{NL}\int_{z_{k-1}}^{z_k}\sigma_{xz}^{(k)}(1,z^2)dz;\quad (Q_y,K_y)=\sum_{k=1}^{NL}\int_{z_{k-1}}^{z_k}\sigma_{yz}^{(k)}(1,z^2)dz \tag{2.12}$$

where NL is the total number of layers of the shell and z_k, z_{k-1} the top and bottom z- coordinates of the k_{th} layer.

The in-plane stress resultants are obtained as

$$\boldsymbol{N}=\boldsymbol{A}\boldsymbol{\epsilon}^{(0)}+\boldsymbol{B}\boldsymbol{k}^{(0)}+\boldsymbol{E}\boldsymbol{k}^{(2)};\quad \boldsymbol{M}=\boldsymbol{B}\boldsymbol{\epsilon}^{(0)}+\boldsymbol{D}\boldsymbol{k}^{(0)}+\boldsymbol{F}\boldsymbol{k}^{(2)};\quad \boldsymbol{P}=\boldsymbol{E}\boldsymbol{\epsilon}^{(0)}+\boldsymbol{F}\boldsymbol{k}^{(0)}+\boldsymbol{H}\boldsymbol{k}^{(2)} \tag{2.13}$$

and the out-of-plane terms as

$$\boldsymbol{Q}=\boldsymbol{A}\boldsymbol{\epsilon}^{(0)}+\boldsymbol{D}\boldsymbol{k}^{(0)};\quad \boldsymbol{K}=\boldsymbol{D}\boldsymbol{\epsilon}^{(0)}+\boldsymbol{F}\boldsymbol{k}^{(0)} \tag{2.14}$$

being A_{ij}, B_{ij}, etc. the laminate stiffness components, obtained by integration in the thickness direction

$$(A_{ij}, B_{ij}, D_{ij}, E_{ij}, F_{ij}, H_{ij}) = \sum_{k=1}^{NL} \int_{z_{k-1}}^{z_k} Q_{ij}^{(k)} (1, z, z^2, z^3, z^4, z^6) dz \tag{2.15}$$

and where

$$I_i = \sum_{k=1}^{NL} \int_{z_{k-1}}^{z_k} \rho^{(k)} z^i dz, \, (i = 0, 1, 2, ..., 6) \tag{2.16}$$

$$J_i = I_i - c_i I_{i+2}, \; K_2 = I_2 - 2c_1 I_4 + c_1^2 I_6, \; c_1 = \frac{4}{3h^2}, \; c_2 = 3c_1 \tag{2.17}$$

being $\rho^{(k)}$ the density of the material of layer k.

2.4 Free Vibrations of Spherical and Cylindrical Laminated Shells

We investigate the free vibration behavior of spherical and cylindrical laminated shells, where he material properties of each lamina are given by $E_1 = 25.0E_2, G_{12} = G_{13} = 0.5E_2, G_{23} = 0.2E_2, \nu_{12} = \nu_{13} = 0.25$. Two cross-ply laminates are considered: $[0°/90°/90°/0°]$ and $[0°/90°/0°]$. Adimensional natural frequencies are displayed in Tables 2.1 and 2.2, for spherical and cylindrical shells, respectively. When compared to analytical solutions by Reddy and Liu [6], excellent agreement are observed even with a small number of collocation points.

Figure 2.1 illustrates the first four eigenmodes for cross-ply laminated spherical shells, computed on a 13 × 13 grid with $R/a = 10$ and $a/h = 100$.

Table 2.1: Nondimensionalized fundamental frequencies of cross-ply laminated spherical shells, $\overline{\omega} = \omega(a^2/h)\sqrt{\rho/E_2}$, laminate ($[0°/90°/90°/0°]$).

	a/h	Method	R/a					
			5	10	20	50	100	10^9
$[0°/90°/90°/0°]$	10	present (13 × 13)	12.198	12.019	11.974	11.961	11.959	11.958
		present (17 × 17)	12.198	12.019	11.974	11.961	11.959	11.958
		present (21 × 21)	12.191	12.024	11.968	11.958	11.963	11.959
		HSDT [6]	12.040	11.840	11.790	11.780	11.780	11.780
	100	present (13 × 13)	31.088	20.379	16.631	15.417	15.235	15.175
		present (17 × 17)	31.088	20.379	16.631	15.417	15.235	15.175
		present (21 × 21)	31.088	20.379	16.631	15.417	15.235	15.175
		HSDT [6]	31.100	20.380	16.630	15.420	15.230	15.170

Table 2.2: Nondimensionalized fundamental frequencies of cross-ply cylindrical shells, $\overline{\omega} = \omega(a^2/h)\sqrt{\rho/E_2}$.

		[0/90/0]		[0/90/90/0]	
R/a	Method	$a/h = 100$	$a/h = 10$	$a/h = 100$	$a/h = 10$
5	present (13 × 13)	20.310	11.890	20.345	11.996
	present (17 × 17)	20.310	11.890	20.345	11.996
	present (21 × 21)	20.310	11.890	20.345	11.996
	FSDT [6]	20.332	12.207	20.361	12.267
	HSDT [6]	20.330	11.850	20.360	11.830
10	present (13 × 13)	16.612	11.864	16.623	11.968
	present (17 × 17)	16.612	11.864	16.623	11.968
	present (21 × 21)	16.612	11.864	16.623	11.968
	FSDT [6]	16.625	12.173	16.634	12.236
	HSDT [6]	16.620	11.800	16.630	11.790
100	present (13 × 13)	15.190	11.855	15.190	11.959
	present (17 × 17)	15.190	11.855	15.190	11.959
	present (21 × 21)	15.190	11.855	15.190	11.959
	FSDT [6]	15.198	12.163	15.199	12.227
	HSDT [6]	15.190	11.790	15.190	11.780
Plate	present (N = 13 × 13)	15.175	11.855	15.175	11.959
	present (17 × 17)	15.175	11.855	15.175	11.959
	present (21 × 21)	15.175	11.855	15.175	11.959
	FSDT [6]	15.183	12.162	15.184	12.226
	HSDT [6]	15.170	11.790	15.170	11.780

2.5 Conclusions

We showed that the Legendre spectral collocation method is a very accurate method for the free vibration analysis of laminated shells, as shown in the examples. The method is implemented on the equations of motion obtained by Reddy's shell theory, and via differentiation matrices that are derived by Kronecker products.

Acknowledgements The authors dedicate this book chapter to the memory of Professor Cristóvão Mota Soares and his contributions to the field of laminated plates and shells. Cristóvão has a true friend, we will miss him.

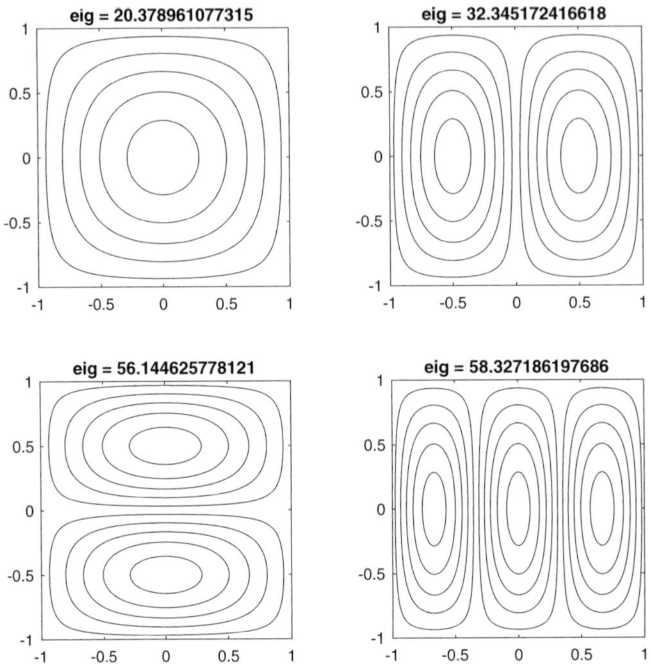

Fig. 2.1: First 4 eigenmodes for cross-ply laminated spherical shells, 13×13 grid, $R/a = 10$, $a/h = 100$.

References

[1] Boyd JP: (2001) Chebyshev and Fourier Spectral Methods, second edn. Dover Books on Mathematics. Dover Publications, Mineola, NY

[2] Canuto C, Hussaini M, Quarteroni A, Zang T (1988) Spectral Methods in Fluid Dynamics. Springer, Berlin, Heidelberg, DOI 10.1007/978-3-642-84108-8

[3] Fernandes SCF, Cuartero J, Ferreira AJM (2025) Free vibration analysis of doubly-curved laminated shells by pseudospectrals and Reddy's higher-order shell theory. Mechanics Research Communications 149:104547, DOI 10.1016/j.mechrescom.2025.104547

[4] Fornberg B (1996) A Practical Guide to Pseudospectral Methods. Cambridge Monographs on Applied and Computational Mathematics. Cambridge University Press

[5] Reddy JN (1997) Mechanics of Laminated Composite Plates. CRC Press, New York

[6] Reddy JN, Liu CF (1985) A higher-order shear deformation theory of laminated elastic shells. International Journal of Engineering Science 23(3):319-330, DOI 10.1016/0020-7225(85)90051-5 (1985)

[7] Trefethen LN (2000) Spectral Methods in MatLab. Society for Industrial and Applied Mathematics, USA

Chapter 3
The Dual Mesh Control Domain Method: Theory and Applications to Functionally-Graded Beams and Orthotropic Plates

Tanmaye Y. Heblekar, Junuthula N. Reddy, and Arun R. Srinivasa

Abstract We present a succinct yet self-contained treatment of the *Dual Mesh Control Domain Method*, a novel numerical technique that hybridizes the finite element and finite volume methods. The theoretical as well as implementation aspects of this method are addressed, and its applications to the linear and nonlinear (in the von Kármán sense) bending analysis of functionally graded beams and orthotropic plates are demonstrated. While the chapter is intended for a broad audience, including readers new to numerical methods, it is specifically devoted to those interested in implementing the Dual Mesh Control Domain Method. To that end, computational aspects related to assembly, generation of shape functions and their derivatives, numerical integration, and solution techniques for nonlinear systems of equations are discussed in reasonable depth. Additionally, examples replete with numerical data and plots are provided to assist interested readers in reproducing and extending the presented results. Although the emphasis of the chapter leans toward structural applications, the method has also been successfully applied to problems in heat transfer and fluid mechanics, making it valuable to anyone seeking to learn and adopt this novel and versatile numerical technique.

Key words: Dual Mesh Control Domain Method, Functionally graded beams, Orthotropic plates, von Kármán nonlinearity, Higher-order elements

Tanmaye Y. Heblekar · Junuthula N. Reddy (corresponding author) · Arun R. Srinivasa
Texas A&M University, MS 3123, College Station, TX 77843, USA, e-mail: tanmayeyashodan.h@tamu.edu, jnreddy@tamu.edu, asrinivasa@tamu.edu

H. Altenbach et al. (eds.), *Composite Structures and Technology*,
Advanced Structured Materials 255,
https://doi.org/10.1007/978-3-032-24656-1_3

3.1 Introduction

3.1.1 Background

The dual mesh control domain method (DMCDM) is a numerical technique for solving boundary value problems, similar to the finite element method (FEM) or the finite volume method (FVM). In fact, the DMCDM is a hybrid of the FEM and FVM, as we will see shortly. It is worthwhile to begin with a qualitative understanding before delving into the mathematical intricacies.

The Dual Mesh Control Domain Method (DMCDM), introduced by Reddy [13], is a blend of several desirable features of the finite volume and finite element methods. As in the finite volume method, the integral form of the governing differential equation(s) is satisfied (unlike the finite element method, which employs the weighted integral form of the governing differential equation). However, departing from traditional finite volume approaches that replace derivatives in the integral statement with their Taylor series approximations, the DMCDM utilizes systematic finite element interpolation of the dependent variables that facilitates the straightforward computation of the derivatives.

The main motivation for introducing DMCDM is threefold:

1. The satisfaction of the integral form of the governing principle represents a physically more meaningful approach; [1]
2. The approximation of the dependent variables using finite element interpolation is not only systematic and modular, but also naturally accommodates domains with curved geometries and unstructured meshes, avoiding the excruciating and ad hoc approaches typical of conventional finite volume schemes;
3. The DMCDM naturally respects the duality between the primary variables and the secondary (or flux-like) variables, and the imposition of boundary conditions is elegantly handled, as in the finite element method.

As in the finite element method (FEM), we begin by discretizing the computational domain into a mesh of finite elements, as shown in Fig. 3.1(a). In the DMCDM nomenclature, this mesh is referred to as the primal mesh. It is over this mesh that the dependent variables of the boundary value problem are approximated using finite element interpolations; that is, within each element of the primal mesh, a dependent variable is expressed as a linear combination of its nodal values and the element interpolation functions.

Next, we construct the dual mesh, as shown in Fig. 3.1(b). For the quadrilateral primal mesh shown, the dual mesh consists of disjoint polygonal units, each surrounding a node. We refer to these polygonal units as control domains. *There are as many control domains in the dual mesh as there are nodes in the primal mesh.*

[1] The weak-form Galerkin finite element formulations of problems in solid and structural mechanics are equivalent to the principle of virtual work, and thus retains physical meaning. However, in areas such as fluid mechanics, heat transfer, and mass transfer, the weak form lacks a clear physical interpretation and serves primarily as a mathematical construct.

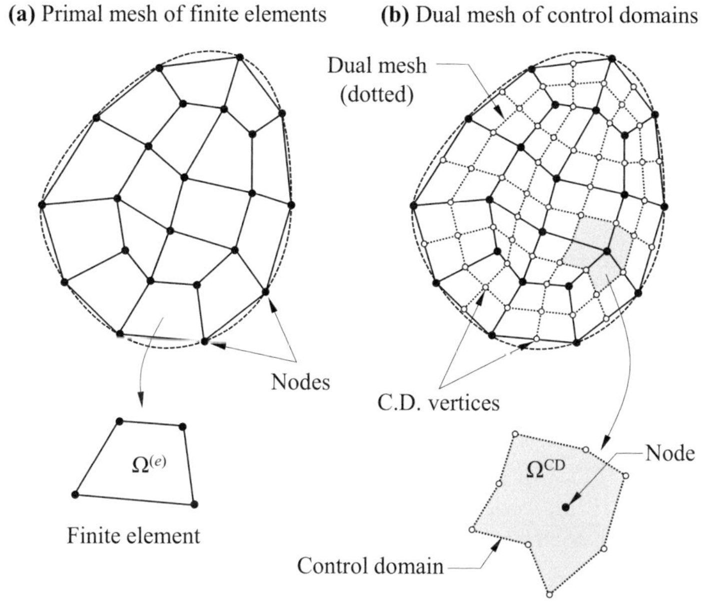

Fig. 3.1: Discretization of the computational domain into the primal and dual meshes.

The essence of the DMCDM lies in satisfying the integral form of the boundary value problem over each control domain in the dual mesh, thereby yielding as many equations as there are unknowns. Thus, the DMCDM may be viewed as an evolved variant of the finite volume method (FVM), wherein the local conservation property, arising from exactly satisfying the integral balance law, is retained, and the cumbersome Taylor series replacement of derivatives is supplanted by systematic finite element interpolation.

3.1.2 The Node-Based Procedure

For the purpose of illustrating this method, let us consider a simple one-dimensional model boundary value problem of finding u that satisfies the differential equation

$$-\frac{\mathrm{d}}{\mathrm{d}x}\left(a\frac{\mathrm{d}u}{\mathrm{d}x}\right) + b\frac{\mathrm{d}u}{\mathrm{d}x} + cu = f \tag{3.1}$$

in $\Omega \equiv [0, L]$, subject to the Dirichlet boundary conditions

$$u(0) = 0 \qquad u(L) = 1 \tag{3.2}$$

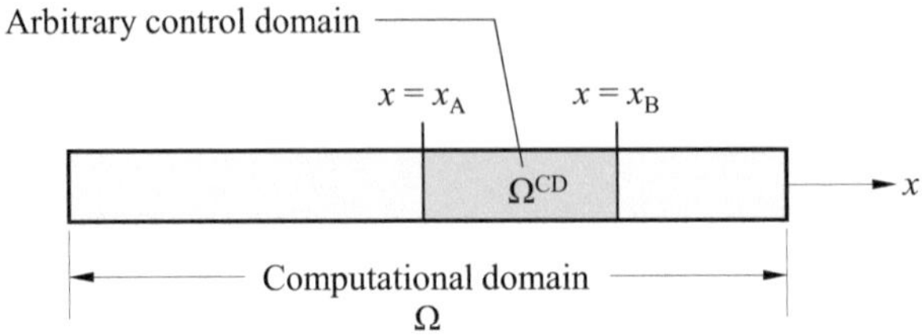

Fig. 3.2: One-dimensional computational domain Ω with an arbitrary control domain $\Omega^{CD} \subset \Omega$.

We now describe a step-by-step procedure for solving this boundary value problem using the DMCDM. The differential equation (3.1) arises in a wide range of engineering applications, including heat transfer, solid mechanics, and fluid flow.

Step 1: Obtain the integral form of the governing equation

The DMCDM, like the FVM, works with the integral form of the governing differential equation. This is in contrast to the FEM, where a weighted integral statement is used. Referring to Fig. 3.2, let us denote $\Omega^{CD} \equiv [x_A, x_B]$ as an arbitrary domain inside the computational domain Ω (i.e., $\Omega^{CD} \subset \Omega$). In the DMCDM nomenclature, we refer to Ω^{CD} as a control domain, which may be thought of along the same lines as control volumes in the FVM literature. Next, we integrate Eq. (3.1) over the control domain Ω^{CD}:

$$\int_{x_A}^{x_B} \left[-\frac{\mathrm{d}}{\mathrm{d}x}\left(a\frac{\mathrm{d}u}{\mathrm{d}x} \right) + b\frac{\mathrm{d}u}{\mathrm{d}x} + cu \right] \mathrm{d}x = \int_{x_A}^{x_B} f \; \mathrm{d}x \tag{3.3}$$

As will be shown shortly, the DMCDM relies on the use of interpolation functions that are C^0 continuous. However, an inspection of the integral statement in (3.3) reveals that the dependent function u is differentiated twice (in the first term). Therefore, we must integrate the first term so that only the first derivative of u appears in the integral statement. Integrating the first term accordingly yields

$$\underbrace{-\left[a\frac{\mathrm{d}u}{\mathrm{d}x} \right]_{x_B} + \left[a\frac{\mathrm{d}u}{\mathrm{d}x} \right]_{x_A}}_{\text{Boundary terms}} + \int_{x_A}^{x_B} \left(b\frac{\mathrm{d}u}{\mathrm{d}x} + cu \right) \mathrm{d}x = \int_{x_A}^{x_B} f \; \mathrm{d}x \tag{3.4}$$

Equation (3.4) will be our starting point for deriving the discretized equations. In many situations, the governing equation is already available in the form shown in Eq. (3.4), and Step 1 can be completely bypassed. However, there are two important requirements to ensure. First, as mentioned earlier, the integral statement must contain at most the first derivative of the dependent variable (and no higher derivatives). Second, the boundary terms must represent physically meaningful quantities that are dual to the dependent variable.

For example, in the present case, if u represents the temperature at a point in a fin of area A and thermal conductivity k, and if we set $a = kA$ (thermal resistance), then Eq. (3.4) becomes an integral statement of energy conservation in the fin, and the term

$$a\frac{\mathrm{d}u}{\mathrm{d}x} = kA\frac{\mathrm{d}u}{\mathrm{d}x}$$

represents the heat flow (dual to the temperature gradient). Similarly, if u represents the axial displacement of a one-dimensional bar of area A and elastic modulus E, and if we set $a = EA$ (axial stiffness), then Eq. (3.4) becomes an integral statement of the principle of the balance of linear momentum for the bar, and the term

$$a\frac{\mathrm{d}u}{\mathrm{d}x} = EA\frac{\mathrm{d}u}{\mathrm{d}x}$$

represents the axial force (dual to the axial displacement). Thus, we must ensure that our integral statement contains boundary terms that are physically meaningful. The importance of this requirement will become clear when we discuss the imposition of boundary conditions in Subsect. 3.1.4.

Step 2: Discretize the domain

As in all numerical methods, we begin by discretizing, or meshing, the computational domain Ω [see Fig. 3.3(a)]. In the DMCDM, we first divide Ω into a mesh of finite elements, as shown in Fig. 3.3(b). Let us denote the set of finite elements as

$$\left\{\Omega^{(e)}\right\}_{e=1}^{\mathrm{NEM}}$$

where NEM denotes the number of elements in the mesh. In the DMCDM nomenclature, this finite element mesh is referred to as the primary or primal mesh. It is named so because we use this mesh to approximate the primary variable u of the boundary value problem. As shown in Fig. 3.3(b), the primal mesh consists of nodes and elements. For simplicity, we will consider a primal mesh of two-node (i.e., linear) elements with Lagrange interpolation functions:

$$\psi_1^{(e)}(x) = 1 - \frac{x - x_0^{(e)}}{h_e}, \quad \psi_2^{(e)}(x) = \frac{x - x_0^{(e)}}{h_e} \qquad x \in \Omega^{(e)} \tag{3.5}$$

where $\psi_1^{(e)}$ and $\psi_2^{(e)}$ are the interpolation (or shape) functions of an arbitrary element $\Omega^{(e)}$, $x_0^{(e)}$ is the x-coordinate of the first (i.e., left) node of $\Omega^{(e)}$, and h_e is its length. Once the primal mesh is established, we next construct the dual mesh. To that end, as shown in Fig. 3.3(c), we create a control domain around every node of the primal mesh, ensuring that no two control domains overlap. The set of these disjoint nodal control domains is referred to as the dual mesh. As the reader may recall, it is over the control domains of the dual mesh that we will satisfy the integral form of the governing equation. To summarize, *the primal mesh serves the purpose of allowing us to approximate the unknown function u (using finite element interpolation), while the dual mesh is used to satisfy the governing equation in an integral sense.*

Although the DMCDM features two meshes, we will show later that, from a computational standpoint, only the primal mesh needs to be generated and stored. Also we note that there is no unique dual-mesh construction corresponding to a given primal mesh. In all our discussions, however, we will construct the dual mesh such that the nodes lie at the center of their respective control domains (i.e., there is no "stagger").

Step 3: Derive the discrete equations

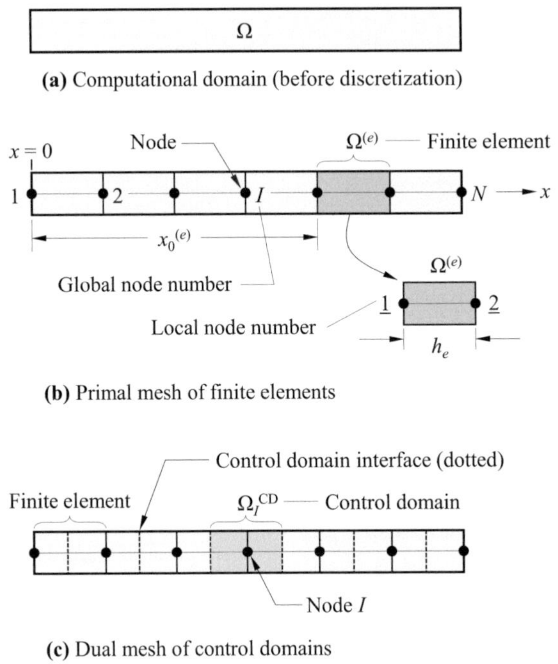

Fig. 3.3: Discretization procedure in the dual mesh control domain method.

With the integral governing equation (3.4) and the primal and dual meshes in hand, we are now ready to derive the discrete version of the continuous problem. In the DMCDM, this is accomplished using finite element interpolation. Referring to Fig. 3.4, inside an arbitrary element $\Omega^{(e)}$ of the primal mesh, u can be approximated as

$$u(x) \approx u^h(x) = \sum_{j=1}^{2} u_j^{(e)} \psi_j^{(e)}(x) = u_1^{(e)} \psi_1^{(e)}(x) + u_2^{(e)} \psi_2^{(e)}(x), \quad x \in \Omega^{(e)} \tag{3.6}$$

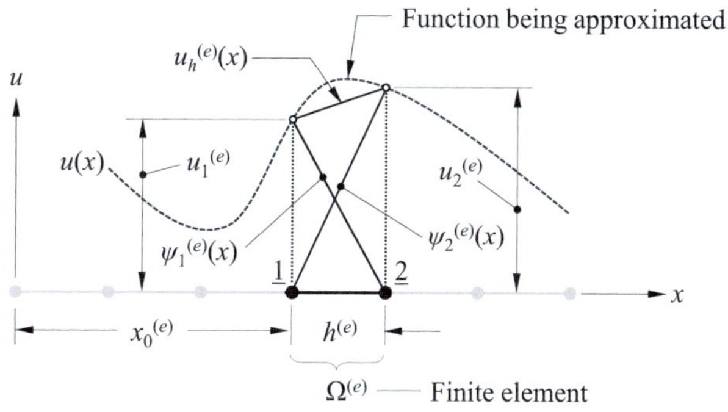

Fig. 3.4: Approximation of a function over a finite element using linear shape functions.

where $u_1^{(e)}$ and $u_2^{(e)}$ are the *a priori* unknown values of the function u at the two nodes of the element $\Omega^{(e)}$. Our objective is to determine the unknown nodal values of u at all nodes in the primal mesh. To this end, we return to the integral statement (3.4), but now written for the control domain surrounding a specific node I, as shown in Fig. 3.5:

$$-\left[a\frac{\mathrm{d}u}{\mathrm{d}x}\right]_{x_B^I} + \left[a\frac{\mathrm{d}u}{\mathrm{d}x}\right]_{x_A^I} + \int_{x_A^I}^{x_B^I}\left(b\frac{\mathrm{d}u}{\mathrm{d}x} + cu\right)\mathrm{d}x = \int_{x_A^I}^{x_B^I} f\,\mathrm{d}x \tag{3.7}$$

For simplicity, let us assume that $b = 0$ and that the coefficients (a, c, f) are constants. For the mesh system shown in Fig. 3.5, we observe that one half of the control domain around node I lies inside element $\Omega^{(I-1)}$, and the other half lies inside element $\Omega^{(I)}$. Thus, the control domain equation (3.7) can be rewritten (after splitting the integrals) as

$$\underbrace{\left[a\frac{\mathrm{d}u}{\mathrm{d}x}\right]_{x_A^I} + \int_{x_A^I}^{x_I} cu\,\mathrm{d}x}_{\Omega^{(I-1)}} \underbrace{- \left[a\frac{\mathrm{d}u}{\mathrm{d}x}\right]_{x_B^I} + \int_{x_I}^{x_B^I} cu\,\mathrm{d}x}_{\Omega^{(I)}} = \int_{x_A^I}^{x_I} f\,\mathrm{d}x + \int_{x_I}^{x_B^I} f\,\mathrm{d}x \tag{3.8}$$

where x_I denotes the x-coordinate of the Ith node of the primal mesh. The terms indicated over $\Omega^{(I-1)}$ in the underbrace must be evaluated using the shape functions of element $\Omega^{(I-1)}$, while those indicated over $\Omega^{(I)}$ must be evaluated using the shape functions of element $\Omega^{(I)}$.

Also, owing to our assumption that the nodes are located at the geometric centers of their respective control domains, we have

$$x_A^I = \frac{x_{I-1} + x_I}{2}\,, \quad x_B^I = \frac{x_I + x_{I+1}}{2} \tag{3.9}$$

Keeping the above points in mind, we can now introduce the finite element interpolation functions into the integral equation (3.8), which leads to a linear algebraic equation relating the nodal unknowns U_{I-1}, U_I, and U_{I+1}, as

$$K_0^I U_{I-1} + K_1^I U_I + K_2^I U_{I+1} = F_I \tag{3.10}$$

where

$$\begin{aligned} K_0^I &= \frac{ch_{I-1}}{2} - \frac{a}{h_{I-1}}\,, \quad K_1^I = \left(\frac{c}{2} + \frac{a}{h_{I-1}\,h_I}\right)(h_{I-1} + h_I)\,, \\ K_2^I &= \frac{ch_{I+1}}{2} - \frac{a}{h_{I+1}}\,, \quad F_I = \frac{f}{2}\,(h_{I-1} + h_I) \end{aligned} \tag{3.11}$$

In Eq. (3.11), we have introduced $h_{I-1} = x_I - x_{I-1}$ and $h_I = x_{I+1} - x_I$, which are,

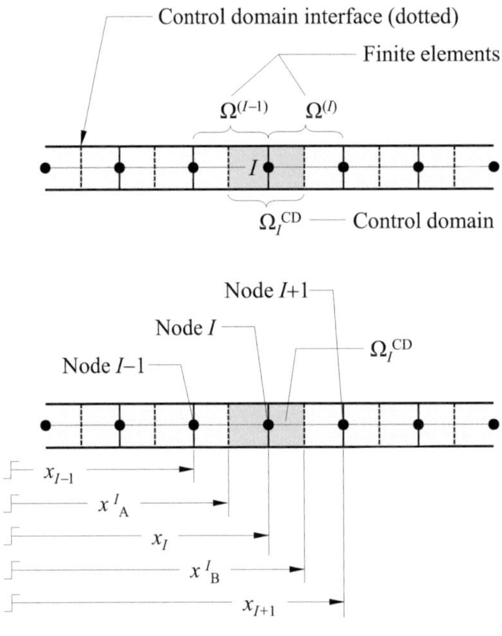

Fig. 3.5: Portions of control domain Ω_I^{CD} around node I lying in elements $\Omega^{(I-1)}$ and Ω^I.

respectively, the lengths of the elements $\Omega^{(I-1)}$ and $\Omega^{(I)}$. The detailed algebraic steps leading up to Eqs. (3.10) and (3.11) are straightforward and have been omitted for conciseness. However, for the convenience of the reader, we briefly show the computation for the first set of terms in Eq. (3.8) that originate from element $\Omega^{(I-1)}$:

$$\left[a\frac{\mathrm{d}u}{\mathrm{d}x}\right]_{x_A^I} + \int_{x_A^I}^{x_I} cu\,\mathrm{d}x = \left\{a\frac{\mathrm{d}}{\mathrm{d}x}\left[U_{I-1}\psi_1^{(I-1)} + U_I\psi_2^{(I-1)}\right]\right\}_{x_A^I}$$

$$+\int_{x_A^I}^{x_I} c\left[U_{I-1}\psi_1^{(I-1)} + U_I\psi_2^{(I-1)}\right] \mathrm{d}x$$

$$= \left\{a\frac{\mathrm{d}}{\mathrm{d}x}\left[U_{I-1}\left(1 - \frac{x - x_{I-1}}{x_I - x_{I-1}}\right) + U_I\left(\frac{x - x_{I-1}}{x_I - x_{I-1}}\right)\right]\right\}_{x_A^I}$$

$$+\int_{x_A^I}^{x_I} c\left[U_{I-1}\left(1 - \frac{x - x_{I-1}}{x_I - x_{I-1}}\right) + U_I\left(\frac{x - x_{I-1}}{x_I - x_{I-1}}\right)\right] \mathrm{d}x$$

$$= \left(\frac{c\,h_{I-1}}{2} + \frac{a}{h_{I-1}}\right)U_I + \left(\frac{c\,h_{I-1}}{2} - \frac{a}{h_{I-1}}\right)U_{I-1} \tag{3.12}$$

The integral was evaluated exactly in the above computation since the integrand is linear. In situations (as we will see later) where the coefficients are arbitrary functions of x and the interpolation functions are not linear, we must resort to numerical integration.

Equation (3.10) is called a stencil equation and relates the value of u at a node to the values at its neighboring nodes. Since we assumed linear interpolation, only the immediate left and right neighbors of a node appear in its stencil equation. The higher the order of interpolation, the larger the resulting "zone of influence" of a node (i.e., the stencil equation will involve more nodes beyond its immediate neighbors). We now apply the stencil to every node of the primal mesh, which yields a system of algebraic equations, also called the global system of equations, of the form:

$$\begin{aligned}
K_1^1 U_1 + K_2^1 U_2 &= F_1\\
K_0^2 U_1 + K_1^2 U_2 + K_2^2 U_3 &- F_2\\
&\vdots\\
K_0^I U_{I-1} + K_1^I U_I + K_2^I U_{I+1} &= F_I\\
&\vdots\\
K_0^{N_g-1} U_{N_g-2} + K_1^{N_g-1} U_{N_g-1} + K_2^{N_g-1} U_{N_g} &= F_{N_g-1}\\
K_0^{N_g} U_{N_g-1} + K_1^{N_g} U_{N_g} &= F_{N_g}
\end{aligned} \tag{3.13}$$

where N_g is the number of nodes in the primal mesh. Note that the control domains of the first and last nodes in the primal mesh lie entirely inside the first element $\Omega^{(1)}$ and the last element $\Omega^{(N_g-1)}$. Consequently, the first and last equations in the system (3.13) contain only two nodal unknowns.

Step 4: Apply boundary conditions and solve

All that remains to be done now is to impose the prescribed boundary conditions and solve the system of equations (3.13). In the present illustration, we have simple

Dirichlet boundary conditions

$$u(0) = 0 \qquad u(L) = 1$$

which translate in the discrete system as

$$U_1 = 0 \qquad U_{N_g} = 1 \tag{3.14}$$

Thus, after incorporating the boundary conditions, Eqs. (3.13) reduce to

$$\begin{aligned} K_1^2 U_2 + K_2^2 U_3 &= F_2 \\ &\vdots \\ K_0^I U_{I-1} + K_1^I U_I + K_2^I U_{I+1} &= F_I \\ &\vdots \\ K_0^{N_g-1} U_{N_g-2} + K_1^{N_g-1} U_{N_g-1} &= F_{N_g-1} - K_2^{N_g-1} U_{N_g} \end{aligned} \tag{3.15}$$

The reduced system of equations (3.15) involves $N_g - 2$ unknowns and can be solved using any numerical linear algebra software package.

The approach presented in this section is popularly referred to as the node-by-node or stencil-based approach because we write equations by parsing through the nodes of the primal mesh. This was the approach presented in [13] and was also adopted in much of the initial work on the DMCDM; see, for instance, [9, 18–20]. While the node-by-node construction of the global system of equations is intuitive and a good starting point, it quickly becomes cumbersome from an algebraic standpoint and lacks modularity. The complexity grows substantially when dealing with higher order elements or with problems involving multiple dependent variables. In addition, the node-by-node approach is not easily integrated within conventional FEM frameworks. In such frameworks, element-level information, such as the connectivity that identifies the nodes belonging to a given element, is readily available. However, the reverse information, namely which elements share a given node, is not directly stored and must be precomputed. Furthermore, conventional FEM implementations loop over elements rather than nodes, which makes it preferable to work with element-level equations instead of node-level stencil equations.

The following section introduces an element-by-element formulation of the DMCDM, as presented in [4]. By formulating the discrete equations at the element level, the implementation becomes more straightforward and more compatible with existing FEM tools. This enables us to leverage standard FEM capabilities for assembly, application of boundary conditions, and related operations. The element-by-element perspective therefore provides a practical, clean, and robust framework for extending the DMCDM to a wide range of problems.

3.1.3 Element-Based Procedure

We will understand the element-based approach of the DMCDM by considering the example of a one-dimensional bar shown in Fig. 3.6. Suppose we are interested in computing the displacement field $u(x)$ of a one-dimensional bar of axial stiffness EA and length L, embedded in an elastic medium of stiffness k_f and subjected to a uniformly distributed load f. The strong form of the governing equation is given by

$$-\frac{\mathrm{d}}{\mathrm{d}x}\left(EA\frac{\mathrm{d}u}{\mathrm{d}x}\right) + k_f\, u = f \tag{3.16}$$

Assume that the ends of the bar at $x = 0$ and $x = L$ are fixed. Following the usual procedure, we first mesh the domain into a set of finite elements, as shown in Fig. 3.7. To keep the discussion general, we will assume elements with an arbitrary number of nodes $N \geq 2$ (the construction of shape functions for such elements is discussed in Sect. 3.3.1). To understand the element-based approach, we now isolate a finite element $\Omega^{(e)}$ from its surroundings, as shown in Fig. 3.7. It is important to recognize that the Qs are in fact equal to

$$EA\,\frac{\mathrm{d}u}{\mathrm{d}x}$$

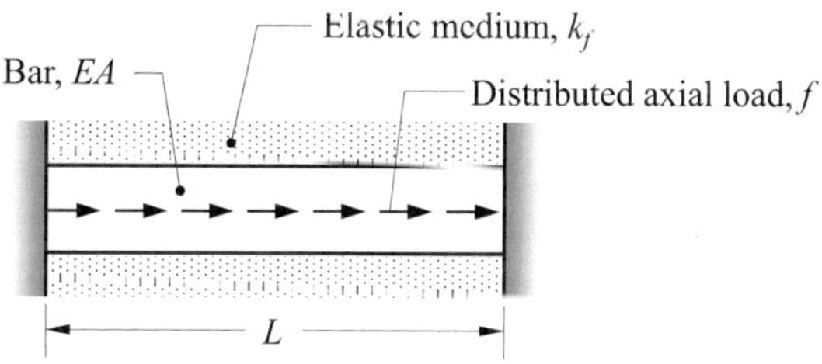

Fig. 3.6 One-dimensional elastic bar embedded in an elastic medium and subjected to a distributed axial load.

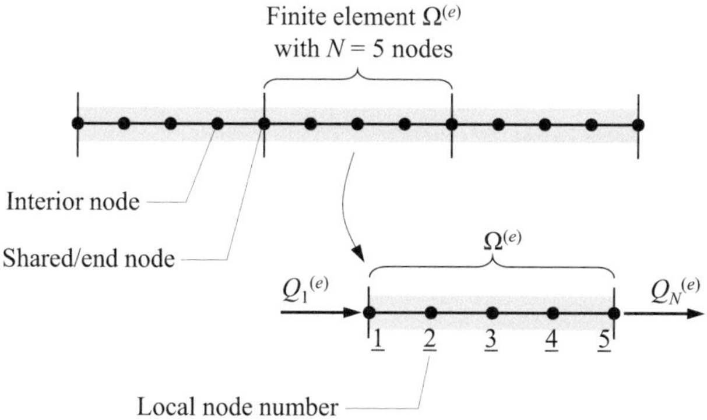

Fig. 3.7: An arbitrary element $\Omega^{(e)}$ isolated from the primal mesh and shown with the end forces $Q_1^{(e)}$ and $Q_N^{(e)}$.

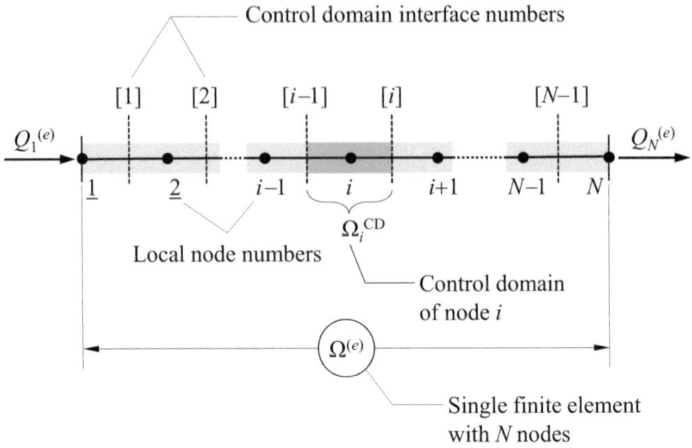

Fig. 3.8: Numbering convention for the control domain interfaces for an element with N nodes.

evaluated at their respective locations. However, we intentionally avoid expressing them this way, because later we will replace u by its finite element approximation everywhere except in the Qs. This is because we want the Qs to correspond to the true values of the secondary variable at the end nodes, not to their interpolated or approximate values. More precisely, the Qs represent the natural boundary conditions at the end nodes of the element.

In Fig. 3.8, we have marked and numbered the control domain interfaces inside the element, and for clarity we enclose the interface numbers and the indices used to refer to them within square brackets. Following from our previous discussion, we can write the integral form of equilibrium for each nodal control subdomain as

For Node 1:

$$-\left[EA\frac{\mathrm{d}u}{\mathrm{d}x}\right]\left(x_{[1]}\right)+\int_{\Omega_1^{\mathrm{CD}}} k_f\,u\,\mathrm{d}x=\int_{\Omega_1^{\mathrm{CD}}} f\,\mathrm{d}x+Q_1^{(e)}$$

For Nodes $i \in [2, N-1]$:

$$-\left[EA\frac{\mathrm{d}u}{\mathrm{d}x}\right]\left(x_{[i]}\right)+\left[EA\frac{\mathrm{d}u}{\mathrm{d}x}\right]\left(x_{[i-1]}\right)+\int_{\Omega_i^{\mathrm{CD}}} k_f\,u\,\mathrm{d}x=\int_{\Omega_i^{\mathrm{CD}}} f\,\mathrm{d}x$$

For Node N:

$$\left[EA\frac{\mathrm{d}u}{\mathrm{d}x}\right]\left(x_{[N-1]}\right)+\int_{\Omega_N^{\mathrm{CD}}} k_f\,u\,\mathrm{d}x=\int_{\Omega_N^{\mathrm{CD}}} f\,\mathrm{d}x+Q_N^{(e)}$$

Next, we replace the dependent variable u in each of the above integral equations with its finite element approximation given by

$$u(x) \approx u_h(x) = \sum_{j=1}^{N} u_j \psi_j(x) \quad x \in \Omega^{(e)} \tag{3.17}$$

which yields the following system of N algebraic equations for the element under consideration

$$\begin{aligned}
i = 1: \quad & \sum_{j=1}^{N} \left\{ -\left[EA\frac{\mathrm{d}\psi_j}{\mathrm{d}x} \right]\left(x_{[1]}\right) + \int_{\Omega_1^{\mathrm{CD}}} k_f \psi_j \,\mathrm{d}x \right\} U_j = \int_{\Omega_1^{\mathrm{CD}}} f \,\mathrm{d}x + Q_1^{(e)} \\
2 \le i \le N-1: \quad & \sum_{j=1}^{N} \left\{ \left[EA\frac{\mathrm{d}\psi_j}{\mathrm{d}x} \right]\left(x_{[i-1]}\right) - \left[EA\frac{\mathrm{d}\psi_j}{\mathrm{d}x} \right]\left(x_{[i]}\right) + \int_{\Omega_i^{\mathrm{CD}}} k_f \psi_j \,\mathrm{d}x \right\} U_j \\
& = \int_{\Omega_i^{\mathrm{CD}}} f \,\mathrm{d}x \\
i = N: \quad & \sum_{j=1}^{N} \left\{ \left[EA\frac{\mathrm{d}\psi_j}{\mathrm{d}x} \right]\left(x_{[N-1]}\right) + \int_{\Omega_N^{\mathrm{CD}}} k_f \psi_j \,\mathrm{d}x \right\} U_j = \int_{\Omega_N^{\mathrm{CD}}} f \,\mathrm{d}x + Q_N^{(e)}
\end{aligned} \tag{3.18}$$

It is possible to represent the element-level system of equations (3.18) in the familiar form

$$\sum_{j=1}^{N} K_{ij} U_j = F_i + Q_i \quad i = 1, \ldots, N \quad \Longleftrightarrow \quad \boldsymbol{K} \cdot \boldsymbol{U} = \boldsymbol{F} + \boldsymbol{Q} \tag{3.19}$$

where $\boldsymbol{K}$ and $\boldsymbol{F}$ are, respectively, the dual mesh control domain stiffness matrix and force vector of element $\Omega^{(e)}$. In the discussion that follows, we will derive concise expressions for $\boldsymbol{K}$ and $\boldsymbol{F}$ that can be easily implemented in a computer program. Once again, recall that $Q_1^{(e)}$ and $Q_2^{(e)}$ are the true values of the secondary variable (axial force in this case) at the end nodes, and not their approximations. They correspond to the natural boundary conditions at the end points of $\Omega^{(e)}$.

While the representation in Eqs. (3.18) serves to convey that the DMCDM also leads to an element-level system of equations, it is not convenient from a computational standpoint. We therefore go a step further and develop a clean and efficient scheme to compute $\boldsymbol{K}$. To that end, let us first define a matrix $\boldsymbol{R}$ as

$$R_{[m]j} = -EA\frac{\mathrm{d}\psi_j}{\mathrm{d}x}\left(x_{[m]}\right) \quad \begin{cases} [m] = 1, \ldots, N-1 \\ \;\; j = 1, \ldots, N \end{cases} \tag{3.20}$$

We are essentially computing and storing the value of the boundary term $EA\psi_j'$ at every control domain interface. Recall that the subscript m is intentionally enclosed in square brackets to emphasize that it refers to a control domain interface. Once $\boldsymbol{R}$

is computed using Eq. (3.20), the stiffness coefficients can be computed as

$$
\begin{aligned}
K_{1j} &= R_{[1]j} + \int_{\Omega_1^{\mathrm{CD}}} k_f \psi_j \, \mathrm{d}x \,, \\
K_{2j} &= R_{[2]j} - R_{[1]j} + \int_{\Omega_2^{\mathrm{CD}}} k_f \psi_j \, \mathrm{d}x \,, \\
&\vdots \\
K_{ij} &= R_{[i]j} - R_{[i-1]j} + \int_{\Omega_i^{\mathrm{CD}}} k_f \psi_j \, \mathrm{d}x \,, \\
&\vdots \\
K_{Nj} &= -R_{[N-1]j} + \int_{\Omega_N^{\mathrm{CD}}} k_f \psi_j \, \mathrm{d}x
\end{aligned}
\tag{3.21}
$$

We can condense the representation in Eq. (3.21) by introducing a correspondence matrix $\boldsymbol{H}$ that identifies the interfaces adjacent to each node. This mapping ensures that, when evaluating the boundary term associated with a given node, contributions from all relevant interfaces are assembled correctly, including their appropriate signs. More precisely, we define a matrix $\boldsymbol{H}$ such that the product $\boldsymbol{H} \cdot \boldsymbol{R}$ extracts the boundary-term contributions needed to construct the element stiffness matrix $\boldsymbol{K}$.

For the node and interface numbering convention shown in Fig. 3.8, it is easy to see that the correspondence matrix is given by

$$
\boldsymbol{H} = \begin{bmatrix}
+1 & 0 & 0 & 0 & \dots & 0 & 0 \\
-1 & +1 & 0 & 0 & \dots & 0 & 0 \\
0 & -1 & +1 & 0 & \dots & 0 & 0 \\
\vdots & \vdots & \vdots & \vdots & \ddots & \vdots & \vdots \\
0 & 0 & 0 & 0 & \dots & -1 & +1 \\
0 & 0 & 0 & 0 & \dots & 0 & -1
\end{bmatrix}_{N \times (N-1)}
\tag{3.22}
$$

We can now write a succinct and computationally friendly expression for the components of the element stiffness matrix $\boldsymbol{K}$ and force vector $\boldsymbol{F}$ as

$$
\begin{aligned}
K_{ij} &= \sum_{[m]=1}^{N-1} H_{i[m]} R_{[m]j} + \int_{\Omega_i^{\mathrm{CD}}} k_f \psi_j \, \mathrm{d}x \,, \quad i, j = 1, \dots, N \\
F_i &= \int_{\Omega_i^{\mathrm{CD}}} f \, \mathrm{d}x \,, \quad i = 1, \dots, N
\end{aligned}
\tag{3.23}
$$

We compute the element stiffness matrix $\boldsymbol{K}$ and force vector $\boldsymbol{F}$ for every element of the primal mesh using Eq. (3.23), and then assemble these arrays, exactly as in the finite element method, to form the global system of equations. This global system can then be solved after applying the boundary conditions.

For completeness, we briefly review the assembly algorithm used in finite element programs, which applies directly to the DMCDM as well. From the meshing software, we obtain two important data structures: the array of nodal coordinates $\boldsymbol{X}$ and the element connectivity array $\boldsymbol{C}$. The latter is central to the assembly procedure. The connectivity array is a data structure that specifies which global nodes belong to each element. In other words, it maps the local numbering of nodes within an element to their corresponding global node numbers[2]. This mapping defines the mesh topology. For the three–element mesh of four-node elements shown in Fig. 3.9, the connectivity array $\boldsymbol{C}$ would read as follows:

```
Element     Node 1     Node 2     Node 3     Node 4
1           1          2          3          4
2           4          5          6          7
3           7          8          9          10
```

With this information in hand, the procedure for computing and assembling the element equations is presented in Algorithm 1. The notation is consistent with modern linear algebra languages such as MATLAB and Julia so that the algorithm

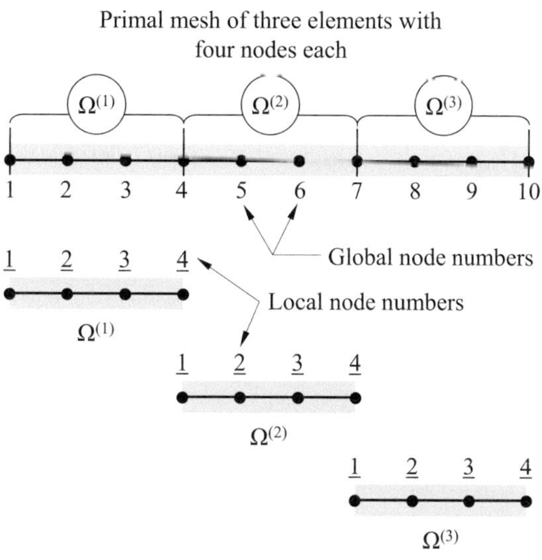

Fig. 3.9 Global and local node numbering convention for one-dimensional elements.

[2] By a global node label we mean a unique identifier assigned to each node in the entire mesh. It determines the node's position in the global coordinate system and is shared across all elements. Local node labels, on the other hand, refer to the numbering of nodes within a specific element, used for defining shape functions and connectivity; these map to the global node labels through the element connectivity array.

Algorithm 1 Finite Element Assembly Procedure

Inputs: a, b, c, f ▹ Problem data
Inputs: $\boldsymbol{X}, \boldsymbol{C}$ ▹ Nodal coordinates, Connectivity matrix
Outputs: $\boldsymbol{K}^G, \boldsymbol{F}^G$ ▹ Global stiffness matrix and force vector

Initialize $\boldsymbol{K}^G \leftarrow \boldsymbol{0}_{N\times N}$, $\boldsymbol{F}^G \leftarrow \boldsymbol{0}_{N\times 1}$
for $E = 1$ to NEM **do** ▹ Loop over all elements
  $\boldsymbol{C}_e \leftarrow \boldsymbol{C}(E,:)$ ▹ Nodes of the current element
  $\boldsymbol{X}_e \leftarrow \boldsymbol{X}(\boldsymbol{C}_e,:)$ ▹ Extract element nodal coordinates
  Compute $\boldsymbol{K}^{(e)}, \boldsymbol{F}^{(e)}$ using Eq. (3.23) [3] ▹ Element stiffness and force
  Assemble Stiffness: $\boldsymbol{K}^G(\boldsymbol{C}_e, \boldsymbol{C}_e) \leftarrow \boldsymbol{K}^G(\boldsymbol{C}_e, \boldsymbol{C}_e) + \boldsymbol{K}^{(e)}$
  Assemble Force: $\boldsymbol{F}^G(\boldsymbol{C}_e) \leftarrow \boldsymbol{F}^G(\boldsymbol{C}_e) + \boldsymbol{F}^{(e)}$
end for

can be readily translated into code. The algorithm shown here applies to problems with a single unknown function and therefore one degree of freedom per node.

In the assembly algorithm, the element coefficients are simply placed in the correct locations of the global system, with the positions determined by the mapping array $\boldsymbol{C}_e$, which is just the corresponding row of the connectivity array $\boldsymbol{C}$ for the element being assembled. The computation and assembly of element equations can also be parallelized, since the calculations for different elements are independent of one another.

We will now try to gain a deeper insight into the physical implication of assembly in the context of the DMCDM. For this purpose, we return to the one-dimensional bar problem shown in Fig. 3.6. For simplicity, we use a uniform primal mesh of two linear elements, as shown in Fig. 3.10(a), and we also assume that $k_f = 0$ (without any loss of generality).

The assembly of element equations is based on (a) the continuity of the primary variable u and (b) equilibrium at the nodes. As an example, for node 2 of the mesh with two linear elements shown in Fig. 3.10(a), the first requirement is enforced simply by assigning $u_2^{(e_1)} = u_1^{(e_2)} = U_2$, where $u_2^{(e_1)}$ and $u_1^{(e_2)}$ are the local nodal values of u at nodes 2 and 1 of elements $\Omega^{(e_1)}$ and $\Omega^{(e_2)}$, respectively. The common value U_2 is the global degree of freedom associated with that node. In general, when two or more elements share a node, continuity of the primary variable is enforced by setting all corresponding local nodal unknowns equal to a single global degree of freedom.

To understand the second aspect of assembly, namely nodal equilibrium, consider the control domain Ω_2^{CD} of node 2, as shown in Fig. 3.10(b). One half of Ω_2^{CD} lies in element $\Omega^{(e_1)}$, denoted $\Omega_{2\mathrm{L}}$, and the other half lies in element $\Omega^{(e_2)}$, denoted $\Omega_{2\mathrm{R}}$. Referring to the free-body diagrams of $\Omega_{2\mathrm{L}}$ and $\Omega_{2\mathrm{R}}$ in Fig. 3.10(c), the equilibrium equations for these two subdomains can be written as

[3] Note that to compute the element stiffness matrix $\boldsymbol{K}^{(e)}$ and force vector $\boldsymbol{F}^{(e)}$ using Eq. (3.23), we need, in addition to the functions a, b, c, f, the element nodal coordinates $\boldsymbol{X}_e$ and hence it has been shown on the preceding line.

$$\begin{aligned} \Omega_{2\mathrm{L}}: &\quad Q_{\mathrm{L}} + Q_2^{(e_1)} + \int_{\Omega_{2\mathrm{L}}} f \,\mathrm{d}x = 0 \\ \Omega_{2\mathrm{R}}: &\quad Q_2^{(e_2)} + Q_{\mathrm{R}} + \int_{\Omega_{2\mathrm{R}}} f \,\mathrm{d}x = 0 \end{aligned} \tag{3.24}$$

Recall that the internal force Q (the secondary variable dual to u) is given by

$$Q = EA\frac{\mathrm{d}u}{\mathrm{d}x}n \tag{3.25}$$

where n is the unit normal of the face on which Q is evaluated. We replace the forces Q_{L} and Q_{R} in Eqs. (3.24) using the expression in Eq. (3.25) so that u can be discretized through its finite element approximation. However, the quantities $Q_1^{(e_1)}$ and $Q_1^{(e_2)}$ are left unchanged, because they represent the true values of the secondary variable at the nodes rather than their approximations. With this substitution, Eqs. (3.24) become

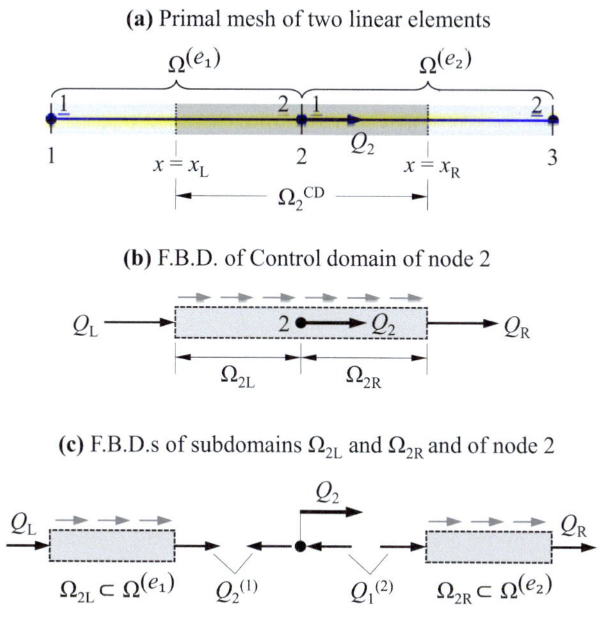

Fig. 3.10: Graphical illustration of the assembly process.

$$\begin{aligned}
\Omega_{2\mathrm{L}}: &\quad -\left[EA\frac{\mathrm{d}u}{\mathrm{d}x}\right]_L + Q_2^{(e_1)} + \int_{\Omega_{2\mathrm{L}}} f\,\mathrm{d}x = 0 \\
\Omega_{2\mathrm{R}}: &\quad Q_2^{(e_2)} + \left[EA\frac{\mathrm{d}u}{\mathrm{d}x}\right]_R + \int_{\Omega_{2\mathrm{R}}} f\,\mathrm{d}x = 0
\end{aligned} \tag{3.26}$$

Equation $(3.26)_1$ is the second control domain equation ($i = 2$) for element $\Omega^{(e_1)}$, and Eq. $(3.26)_2$ is the first control domain equation ($i = 1$) for element $\Omega^{(e_2)}$. To obtain the discrete equations, one would normally substitute the finite element approximation for u, but that step is not needed here because our goal is simply to understand the physical motivation behind the assembly process. Now that we have the control domain equations associated with node 2 from both elements that share it, we are ready to assemble them. The assembly step is simply the addition of these equations:

$$-\left[EA\frac{\mathrm{d}u}{\mathrm{d}x}\right]_L + Q_2^{(e_1)} + \int_{\Omega_{2\mathrm{L}}} f\,\mathrm{d}x + Q_2^{(e_2)} + \left[EA\frac{\mathrm{d}u}{\mathrm{d}x}\right]_R + \int_{\Omega_{2\mathrm{R}}} f\,\mathrm{d}x = 0 \tag{3.27}$$

Our objective now is to determine whether Eq. (3.27), obtained through assembly, truly represents the equilibrium of the entire control domain Ω_2^{CD}. This is straightforward to verify. Referring to the free–body diagram of node 2 in Fig. 3.10(c), we can write the force balance at node 2 as

$$Q_2 = Q_2^{(e_1)} + Q_1^{(e_2)} \tag{3.28}$$

which simplifies Eq. (3.27) to

$$\begin{aligned}
&-\left[EA\frac{\mathrm{d}u}{\mathrm{d}x}\right]_L + \int_{\Omega_{2\mathrm{L}}} f\,\mathrm{d}x + \left[EA\frac{\mathrm{d}u}{\mathrm{d}x}\right]_R + \int_{\Omega_{2\mathrm{R}}} f\,\mathrm{d}x + Q_2 = 0 \\
\Longrightarrow &\left[EA\frac{\mathrm{d}u}{\mathrm{d}x}\right]_L - \left[EA\frac{\mathrm{d}u}{\mathrm{d}x}\right]_R = \int_{\Omega_2^{\mathrm{CD}}} f\,\mathrm{d}x + Q_2 \quad (\text{since } \Omega_2^{\mathrm{CD}} = \Omega_{2\mathrm{L}} \cup \Omega_{2\mathrm{R}})
\end{aligned} \tag{3.29}$$

Recognize that Eq. (3.29) indeed represents the equilibrium of the entire control domain Ω_2^{CD} [see Fig. 3.10(b)]. In the node based approach, we would start directly from Eq. (3.29) and discretize it. In the element based approach, we instead compute the contributions from each element and then add them through the assembly operation.

3.1.4 Neumann and Robin Boundary Conditions

Recall that the illustration we considered in Sect. 3.1.2 was a Dirichlet problem, that is, both boundary conditions specified the primary variable u. However, this is not always the case. We may instead have a mixed prescription of the form

$$u(0) = 0 \qquad Q(x = L) = F_L \quad \text{where} \quad Q = EA\frac{\mathrm{d}u}{\mathrm{d}x} \tag{3.30}$$

The second boundary condition is a Neumann (or natural) boundary condition, which entails specifying the secondary variable Q. The imposition of the Dirichlet boundary condition is straightforward and has already been addressed in Subsect. 3.1.2. We now examine how the natural boundary condition can be imposed in the DMCDM. Referring to Fig. 3.11, consider the discretized control domain equation of the last node, where the secondary variable is prescribed [see Eq. $(3.18)_3$]:

$$\sum_{j=1}^{N}\left\{\left[EA\frac{\mathrm{d}\psi_j}{\mathrm{d}x}\right](x_{[N-1]}) + \int_{\Omega_N^{\mathrm{CD}}} k_f\,\psi_j\,\mathrm{d}x\right\}U_j = \int_{\Omega_N^{\mathrm{CD}}} f\,\mathrm{d}x + Q_N^{(e)} \tag{3.31}$$

where we have repeatedly emphasized that $Q_N^{(e)}$ represents the true value of the secondary variable at the end node and not an approximation. To impose the natural boundary condition, we simply set $Q_N^{(e)} = F_L$, so that the discretized control domain equation becomes

$$\sum_{j=1}^{N}\left\{\left[EA\frac{\mathrm{d}\psi_j}{\mathrm{d}x}\right](x_{[N-1]}) + \int_{\Omega_N^{\mathrm{CD}}} k_f\,\psi_j\,\mathrm{d}x\right\}U_j = \int_{\Omega_N^{\mathrm{CD}}} f\,\mathrm{d}x + F_L \tag{3.32}$$

From an implementation standpoint, we compute the element equations as described in Subsect. 3.1.3 and simply add F_L to the right-hand side of the last equation.

Let us next consider the case where a Robin boundary condition is prescribed at the right boundary ($x = L$):

$$Q + Hu = G \text{ at } x = L \tag{3.33}$$

where H and G are known, and $Q = EA\mathrm{d}u/\mathrm{d}x$. To impose this boundary condition, we replace u in Eq. (3.33) with its finite element approximation and compute the secondary variable Q as

$$Q = Q_N^{(e)} = G - H\sum_{j=1}^{N} u_j\psi_j(x_N) = G - H\sum_{j=1}^{N} u_j\delta_{jN} = G - H\,u_N \tag{3.34}$$

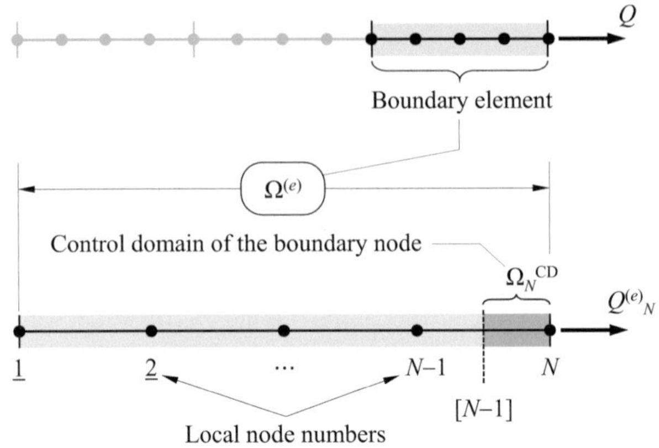

Fig. 3.11: Control domain for a boundary node.

where we have used the interpolation property[4] of the functions $\psi_i(x_j) = \delta_{ij}$. We now substitute $Q_N^{(e)}$ from Eq. (3.34) into Eq. (3.31), which yields

$$\sum_{j=1}^{N}\left\{\left[EA\frac{\mathrm{d}\psi_j}{\mathrm{d}x}\right](x_{[N-1]}) + \int_{\Omega_N^{\mathrm{CD}}} k_f\,\psi_j\,\mathrm{d}x\right\}U_j = \int_{\Omega_N^{\mathrm{CD}}} f\,\mathrm{d}x + G - H\,u_N$$

$$\Longrightarrow \sum_{j=1}^{N}\left\{\left[EA\frac{\mathrm{d}\psi_j}{\mathrm{d}x}\right](x_{[N-1]}) + \int_{\Omega_N^{\mathrm{CD}}} k_f\,\psi_j\,\mathrm{d}x + H\,\delta_{jN}\right\}U_j = \int_{\Omega_N^{\mathrm{CD}}} f\,\mathrm{d}x + G \tag{3.35}$$

The key takeaway from Eq. (3.35) is that, in imposing the Robin boundary condition in Eq. (3.33), the quantity H simply adds to the last diagonal entry of the element stiffness matrix, while the term G adds to the last row of the element force vector.

3.1.5 Complete Picture

With the major concepts described in the preceding sections, we will now gather our thoughts and lay out a systematic procedure for solving any boundary value problem using the DMCDM.

1. **Governing Integral Equation.** The starting point of the DMCDM is the integral form of the governing principle (for example, conservation of linear momentum or

[4] When we say the approximation functions satisfy the interpolation property, we mean that they equal unity at its own node and zero at all others.

conservation of energy). The boundary terms in this integral form must represent physically meaningful secondary variables so that natural boundary conditions can be imposed. Since the DMCDM uses C^0 continuous Lagrange interpolation functions to approximate the dependent variable, the integral equation must involve at most the first derivative of the dependent variable.

2. **Construction of the Primal Mesh.** Next, just as in the FEM, we discretize the computational domain into a mesh of finite elements. This mesh is called the primal mesh and is used to approximate the dependent variable using Lagrange interpolation functions. Within each element of the primal mesh, the dependent variable is approximated as a linear combination of the element shape functions, with the (a priori unknown) nodal values as the coefficients.
3. **Compute Element Equations and Assemble.** As described in Sect. 3.1.3, we satisfy the integral equations over the control domains of the dual mesh[5] and then discretize using finite element interpolation. This process produces element-level equations of the form given in Eq. (3.19), which are computed for every element and subsequently assembled (see Algorithm 1) to produce the global system of equations.
4. **Boundary Conditions and Solution.** We modify the global system of equations to incorporate the prescribed boundary conditions. Subsection 3.1.4 details the procedure for imposing Neumann and Robin boundary conditions. Once the boundary conditions are imposed, we solve the global system of equations $\boldsymbol{K}^G \cdot \boldsymbol{U}^G = \boldsymbol{F}^G$ (using a linear algebra package) to obtain the vector of nodal unknowns $\boldsymbol{U}^G$.
5. **Post-processing.** Almost always, we are interested in computing derived quantities such as fluxes (for heat transfer problems) or stresses (for solid mechanics problems). The procedure for computing such quantities is essentially the same as in the FEM, and a detailed explanation is therefore omitted here. Interested readers are referred to [12, 14] for a comprehensive discussion on post-processing.

3.2 Extension to 2D

Having developed the procedure for one-dimensional problems, extending it to two dimensions (and even to three dimensions) requires only bookkeeping and programming considerations, and the core ideas remain essentially the same. This section is organized as follows: in Sect. 3.2.1, we review the generalized Poisson equation and use it as a model boundary value problem to illustrate the dual mesh control domain discretization procedure in 2D. In Sect. 3.2.2, we derive the discretized equations

[5] By now it should be clear that the only mesh that must be generated and stored is the primal (finite element) mesh, in the form of $\boldsymbol{X}$ (nodal coordinates) and $\boldsymbol{C}$ (element connectivity). The dual mesh is used only during the derivation of the element-level equations (3.23), and its "information" is already embedded in those equations together with the assumption that the nodes are located at the geometric centers of their control domains. Thus, no explicit data structure for the dual mesh needs to be created or stored.

using the DMCDM. Finally, in Sect. 3.2.3, we introduce a vectorized notation system that makes the representation of element equations more compact and easily programmable.

3.2.1 Model Boundary-Value Problem

Consider the problem of finding the function $u(x, y)$ that satisfies the following:

Differential equation

$$-\frac{\partial}{\partial x}\left(a_{11}\frac{\partial u}{\partial x}+a_{12}\frac{\partial u}{\partial y}\right)-\frac{\partial}{\partial y}\left(a_{21}\frac{\partial u}{\partial x}+a_{22}\frac{\partial u}{\partial y}\right)+a_{00}\,u=f \quad \text{in } \Omega \tag{3.36}$$

Boundary conditions:

$$\begin{aligned} u &= \hat{u} \quad \text{on } \Gamma_u \\ q_n &= \hat{q}_n \quad \text{on } \Gamma_q \end{aligned} \qquad \overline{\Gamma}_u \cup \overline{\Gamma}_q = \partial\Omega \quad \text{and} \quad \Gamma_u \cap \Gamma_q = \varnothing \tag{3.37}$$

where the coefficients $a_{11}, a_{12}, a_{21}, a_{22}$, and a_{00} are known and may depend on (x, y) as well as on the dependent variable u and its first derivatives (which makes the problem nonlinear). The function f is also known and is assumed to depend only on (x, y). The computational domain is shown in Fig. 3.12. The boundary value problem defined by Eqs. (3.36) and (3.37) is commonly referred to as the generalized Poisson problem [12].

As we saw, the dual mesh control domain discretization procedure is based on the integral form of Eq. (3.36). As shown in Fig. 3.12, let us consider an arbitrary control domain inside the computational domain (i.e., $\Omega^{CD} \subset \Omega$) and integrate the

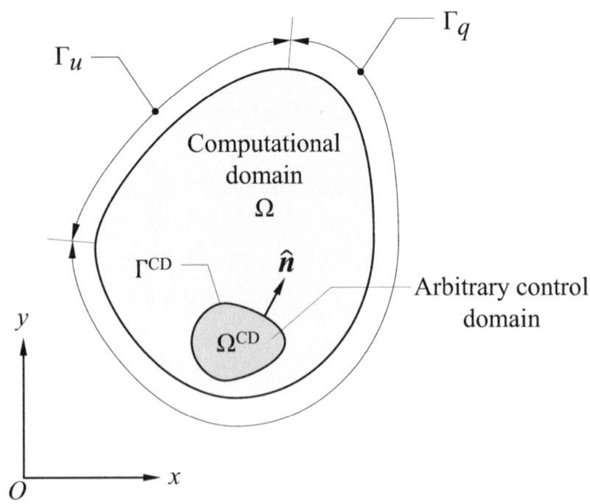

Fig. 3.12 Typical two-dimensional computational domain Ω with an arbitrary control domain $\Omega^{CD} \subset \Omega$

differential equation (3.36) over Ω^{CD} as follows:

$$-\iint_{\Omega^{\text{CD}}} \left[\frac{\partial}{\partial x}\left(a_{11}\frac{\partial u}{\partial x} + a_{12}\frac{\partial u}{\partial y} \right) + \frac{\partial}{\partial y}\left(a_{21}\frac{\partial u}{\partial x} + a_{22}\frac{\partial u}{\partial y} \right) \right] \mathrm{d}\Omega + \iint_{\Omega^{\text{CD}}} a_{00}\, u \,\mathrm{d}\Omega = \iint_{\Omega^{\text{CD}}} f \,\mathrm{d}\Omega \tag{3.38}$$

Notice that in Eq. (3.38), the dependent variable u is differentiated twice. To enable the use of C^0 continuous interpolation functions to approximate u, we must reduce the differentiability requirements. To achieve this, we apply the Gauss theorem to rewrite the first term in Eq. (3.38), ensuring that u is differentiated at most once in the resulting integral equation:

$$-\oint_{\Gamma^{\text{CD}}} \left[\left(a_{11}\frac{\partial u}{\partial x} + a_{12}\frac{\partial u}{\partial y} \right) n_x + \left(a_{21}\frac{\partial u}{\partial x} + a_{22}\frac{\partial u}{\partial y} \right) n_y \right] \mathrm{d}\Gamma + \iint_{\Omega^{\text{CD}}} a_{00}\, u \,\mathrm{d}\Omega = \iint_{\Omega^{\text{CD}}} f \,\mathrm{d}\Omega \tag{3.39}$$

Equation (3.39) is the form that we will work with in order to obtain the discretized equations. An important point to make at this juncture is that the integrand of the path integral must represent a physically meaningful quantity. It corresponds to the secondary variable of the problem that is dual to the primary variable u. In this case, as seen in Eq. (3.39), the secondary variable is the flux normal to the control domain boundary:

$$q_n = \left(a_{11}\frac{\partial u}{\partial x} + a_{12}\frac{\partial u}{\partial y} \right) n_x + \left(a_{21}\frac{\partial u}{\partial x} + a_{22}\frac{\partial u}{\partial y} \right) n_y \tag{3.40}$$

Once again, note that it is not always necessary to start from the strong form (3.36), integrate, and then apply the Gauss theorem to arrive at the integral equation (3.39). In many cases, the governing equations can be formulated directly in their integral form. For example, consider the problem of determining the temperature field $T(x, y)$ in an isotropic medium of thermal conductivity k. Under steady-state conditions, the application of the principle of conservation of energy to a control domain Ω^{CD} (which is a fixed portion of the material) yields

$$\text{Heat outflow from } \Omega^{\text{CD}} = \text{Internal heat generation within } \Omega^{\text{CD}}$$

$$\oint_{\Gamma^{\text{CD}}} \boldsymbol{q}\cdot\hat{\boldsymbol{n}}\,\mathrm{d}\Gamma = \iint_{\Omega^{\text{CD}}} f\,\mathrm{d}\Omega \tag{3.41}$$

where $\boldsymbol{q}$ is the heat flux vector and f is the internal rate of heat generation per unit volume. Introducing Fourier's law of conduction, which relates the heat flux to the temperature gradient,

$$\boldsymbol{q} = -k\,\boldsymbol{\nabla} T \tag{3.42}$$

into the integral energy balance statement (3.41) yields an integral equation in terms of the unknown $T(x, y)$:

$$-\oint_{\Gamma^{\mathrm{CD}}} k\,\boldsymbol{\nabla} T \cdot \hat{\boldsymbol{n}}\,\mathrm{d}\Gamma = \iint_{\Omega^{\mathrm{CD}}} f\,\mathrm{d}\Omega \iff -\oint_{\Gamma^{\mathrm{CD}}} k\left(\frac{\partial T}{\partial x} n_x + \frac{\partial T}{\partial y} n_y\right)\mathrm{d}\Gamma = \iint_{\Omega^{\mathrm{CD}}} f\,\mathrm{d}\Omega \tag{3.43}$$

which is in the form presented in Eq. (3.39). In the next section, we will discuss the discretization of Eq. (3.39) using the dual mesh control domain method.

3.2.2 Discretized model

To use the DMCDM, we first discretize the computational domain into finite elements, as shown in Fig. 3.13. This mesh is called the primal mesh. Throughout this chapter, we will restrict our attention to quadrilateral elements (readers are referred to [5, 16] for a detailed discussion of primal meshes constructed from triangular elements). For generality, we develop the formulation for elements of arbitrarily high order p. In Fig. 3.13, the primal mesh consists of biquadratic ($p = 2$) elements, and

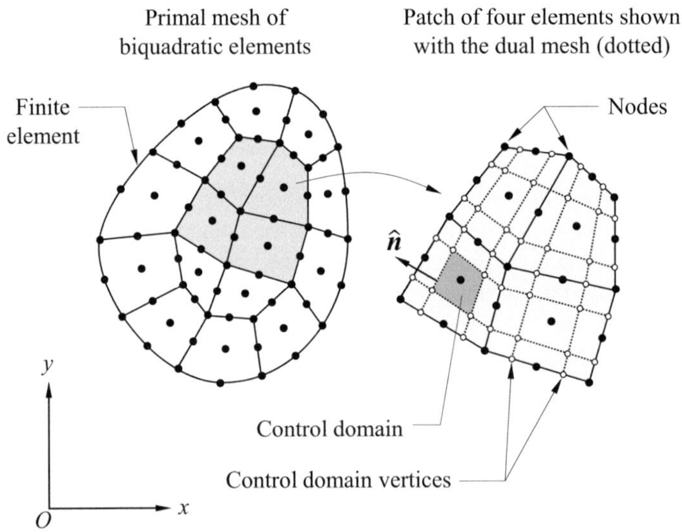

Fig. 3.13: Discretization of the computational domain into a primal mesh of biquadratic elements. Also shown in a patch of four elements with the dual mesh overlaid (dotted).

each element therefore has $N = (p+1)^2 = 9$ nodes. The details of how the shape functions for such elements are generated are deferred to Subsubsect. 3.3.1.2.

Next, we construct the dual mesh by forming a control domain around every node of the primal mesh, as shown in Fig. 3.13. The control domains must be disjoint, meaning they must not overlap one another. Recall that, just as in the one-dimensional case, the construction of the dual mesh is not unique. In our analysis, for simplicity, we will construct the control domains such that their centroids coincide with their respective nodes.

We will use the element-based approach to develop the discretized equations. An element $\Omega^{(e)}$ is shown in Fig. 3.14, with the dual mesh overlaid. The numbers labeling the nodes are the local node numbers, and their correspondence to global nodes is contained in the element connectivity matrix discussed in Subsect. 3.1.3. Let Ω_i^{CD} denote the control domain associated with the ith local node of $\Omega^{(e)}$. Writing Eq. (3.39) for Ω_i^{CD} yields

$$-\oint_{\Gamma_i^{\mathrm{CD}}} \left[\left(a_{11}\frac{\partial u}{\partial x} + a_{12}\frac{\partial u}{\partial y}\right) n_x + \left(a_{21}\frac{\partial u}{\partial x} + a_{22}\frac{\partial u}{\partial y}\right) n_y\right] \mathrm{d}\Gamma + \iint_{\Omega_i^{\mathrm{CD}}} a_{00}\, u \,\mathrm{d}\Omega = \iint_{\Omega_i^{\mathrm{CD}}} f\,\mathrm{d}\Omega \tag{3.44}$$

It is important to point out that in Eq. (3.44), it is technically correct to include the circle on the integral symbol only for the control domains of internal nodes, since the integration paths for such nodes are closed curves, as seen in Fig. 3.14. For nodes lying on the element boundary (that is, corner and edge nodes), the path integration must be performed only over those portions of the control domain boundary that lie inside $\Omega^{(e)}$. Consequently, these are not closed path integrals. We must also recognize that, for such boundary nodes, the equality in Eq. (3.44) does not strictly hold; nonzero flux terms $Q_i^{(e)}$ will remain in the control domain equations for these nodes. However, these terms either vanish after assembly (because contributions from adjacent elements cancel) or are imposed as natural boundary conditions when the corresponding element edge coincides with a portion of the physical boundary where the secondary variable is prescribed. Before proceeding to discretize, we rewrite Eq. (3.44) in a form that applies uniformly to all nodes of the element:

$$-\int_{\Gamma_i^{\mathrm{CD}}} \left[\left(a_{11}\frac{\partial u}{\partial x} + a_{12}\frac{\partial u}{\partial y}\right) n_x + \left(a_{21}\frac{\partial u}{\partial x} + a_{22}\frac{\partial u}{\partial y}\right) n_y\right] \mathrm{d}\Gamma + \iint_{\Omega_i^{\mathrm{CD}}} a_{00}\, u \,\mathrm{d}\Omega = \iint_{\Omega_i^{\mathrm{CD}}} f\,\mathrm{d}\Omega + Q_i^{(e)} \tag{3.45}$$

where $Q_i^{(e)} = 0$ for all internal nodes, and for the boundary nodes they represent the true values of the integrated fluxes. To discretize Eq. (3.45), we replace u with its

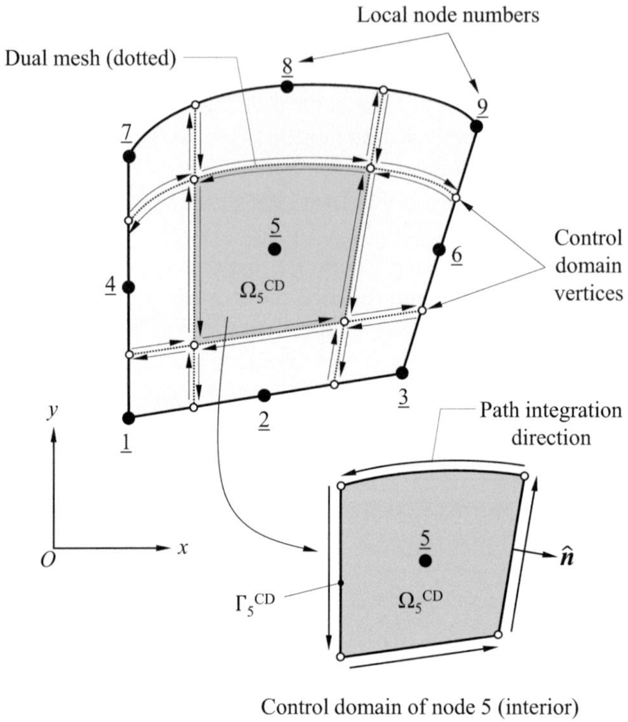

Fig. 3.14: Control domains and integration paths inside a biquadratic element. Note that only for node 5, the integration path is closed (since the node is an interior node).

finite element approximation given by

$$u \approx u_h(x) = \sum_{j=1}^{N} u_j^{(e)} \psi_j^{(e)}(x, y) \quad (x, y) \in \Omega^{(e)} \tag{3.46}$$

where $u_j^{(e)}$ are the unknown values of u at the nodes of $\Omega^{(e)}$, $\psi_j^{(e)}$ are the element shape functions, and N denotes the number of nodes of the element. For quadrilateral elements, $N = (p+1)^2$, where p is the polynomial order of the shape functions. Substituting Eq. (3.46) into the integral equation (3.45) yields

$$-\sum_{j=1}^{N} u_j \int_{\Gamma_i^{CD}} \left[\left(a_{11} \frac{\partial \psi_j}{\partial x} + a_{12} \frac{\partial \psi_j}{\partial y} \right) n_x + \left(a_{21} \frac{\partial \psi_j}{\partial x} + a_{22} \frac{\partial \psi_j}{\partial y} \right) n_y \right] \mathrm{d}\Gamma$$

$$+ \sum_{j=1}^{N} u_j \iint_{\Omega_i^{CD}} a_{00}\, \psi_j \,\mathrm{d}\Omega = \iint_{\Omega_i^{CD}} f \,\mathrm{d}\Omega + Q_i^{(e)} \tag{3.47}$$

We have omitted the superscript (e) over ψ_j and u_j with the tacit understanding that the discussion pertains to the element $\Omega^{(e)}$, and that the associated shape functions and nodal variables are implicitly defined with respect to this element. Equation (3.47) represents the discrete control domain equation of the ith local node of $\Omega^{(e)}$. Writing Eq. (3.47) for every nodal control domain of $\Omega^{(e)}$, we arrive at the element-level system of equations

$$\boldsymbol{K}^{(e)} \cdot \boldsymbol{\Delta}^{(e)} = \boldsymbol{F}^{(e)} + \boldsymbol{Q}^{(e)} \tag{3.48}$$

where the components of the element stiffness matrix $\boldsymbol{K}^{(e)}$ and force vector $\boldsymbol{F}^{(e)}$ are given by

$$\begin{aligned} K_{ij}^{(e)} &= -\int_{\Gamma_i^{CD}} \left[\left(a_{11} \frac{\partial \psi_j}{\partial x} + a_{12} \frac{\partial \psi_j}{\partial y} \right) n_x + \left(a_{21} \frac{\partial \psi_j}{\partial x} + a_{22} \frac{\partial \psi_j}{\partial y} \right) n_y \right] \mathrm{d}\Gamma \\ &\quad + \iint_{\Omega_i^{CD}} a_{00}\, \psi_j \, \mathrm{d}\Omega \,, \quad i, j = 1, \ldots, N \\ F_i^{(e)} &= \iint_{\Omega_i^{CD}} f \, \mathrm{d}\Omega \,, \quad i = 1, \ldots, N \end{aligned} \tag{3.49}$$

and $\boldsymbol{\Delta}^{(e)}$ is the vector of nodal unknowns of element $\Omega^{(e)}$.

Once again, as in the one-dimensional case, we write the element equations for each element in the mesh and then assemble them into the global system of equations using Algorithm 1 to obtain the global equation system

$$\boldsymbol{K} \cdot \boldsymbol{U} = \boldsymbol{F}$$

This system can then be solved after applying the prescribed boundary conditions. Before closing this subsection, we will briefly discuss the imposition of boundary conditions. Recall that we had the following boundary conditions:

$$\begin{aligned} u &= \hat{u} \quad \text{on } \Gamma_u \\ q_n &= \hat{q}_n \quad \text{on } \Gamma_q \end{aligned} \qquad \overline{\Gamma}_u \cup \overline{\Gamma}_q = \partial\Omega \quad \text{and} \quad \Gamma_u \cap \Gamma_q = \emptyset$$

The imposition of the Dirichlet (or essential) boundary condition $u = \hat{u}$ on Γ_u is straightforward. We simply set the value of u at all nodes lying on Γ_u to $\hat{u}$ and eliminate the corresponding equations from the global system, thereby obtaining a reduced system of equations that must be of full rank to ensure a unique solution.

The Neumann (or natural) boundary condition involves specifying the secondary variable (flux)

$$q_n = \left(a_{11} \frac{\partial u}{\partial x} + a_{12} \frac{\partial u}{\partial y} \right) n_x + \left(a_{21} \frac{\partial u}{\partial x} + a_{22} \frac{\partial u}{\partial y} \right) n_y \tag{3.50}$$

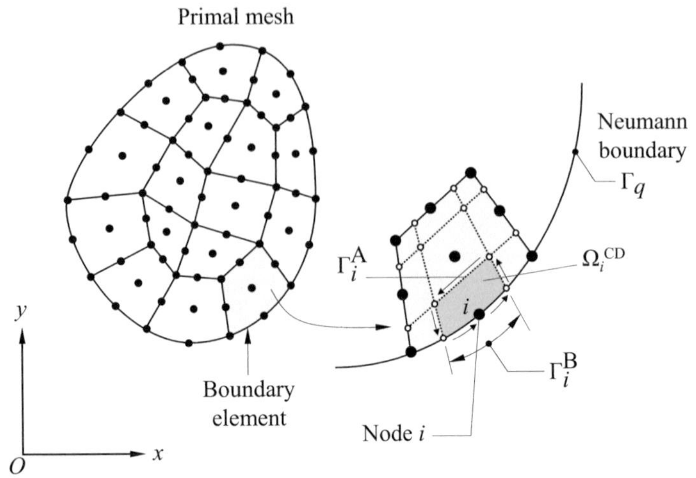

Fig. 3.15: Control domain with edge common with the Neumann boundary.

on Γ_q. This condition can be imposed easily in the DMCDM. In Fig. 3.15, we schematically depict an element $\Omega^{(e)}$ that shares an edge with the Neumann boundary Γ_q. Writing the integral equation (3.44) for the control domain Ω_i^{CD} of node i, which lies on Γ_q, we obtain (assuming $a_{00} = 0$ without loss of generality)

$$
\begin{aligned}
&-\int_{\Gamma_i^A} \left[\left(a_{11} \frac{\partial u}{\partial x} + a_{12} \frac{\partial u}{\partial y} \right) n_x + \left(a_{21} \frac{\partial u}{\partial x} + a_{22} \frac{\partial u}{\partial y} \right) n_y \right] d\Gamma \\
&- \int_{\Gamma_i^B} \left[\left(a_{11} \frac{\partial u}{\partial x} + a_{12} \frac{\partial u}{\partial y} \right) n_x + \left(a_{21} \frac{\partial u}{\partial x} + a_{22} \frac{\partial u}{\partial y} \right) n_y \right] d\Gamma = \iint_{\Omega_i^{CD}} f \, d\Omega
\end{aligned}
\tag{3.51}
$$

where we have decomposed the control domain boundary Γ_i^{CD} into two segments: Γ_i^A, which lies inside the element, and Γ_i^B, which (approximately[6]) lies on Γ_q, as shown in Fig. 3.15. The second integral in Eq. (3.51) is known, since the secondary variable q_n defined in Eq. (3.50) is prescribed on $\Gamma_i^B \subset \Gamma_q$. Thus, Eq. (3.51) can be rewritten as

[6] We use the term "approximately" to indicate that the element boundaries are inexact representations of the actual boundary, as the geometry is interpolated in the same manner as the dependent variable. In our analyses, we use the same interpolation functions for both the geometry and the dependent unknowns. Such formulations are called isoparametric, and more details will be discussed in Subsect. 3.3.1.

$$-\int_{\Gamma_i^A} \left[\left(a_{11} \frac{\partial u}{\partial x} + a_{12} \frac{\partial u}{\partial y} \right) n_x + \left(a_{21} \frac{\partial u}{\partial x} + a_{22} \frac{\partial u}{\partial y} \right) n_y \right] \mathrm{d}\Gamma = \iint_{\Omega_i^{CD}} f \, \mathrm{d}\Omega + \int_{\Gamma_i^B} \hat{q}_n \, \mathrm{d}\Gamma \tag{3.52}$$

The left-hand side of Eq. (3.52) can be discretized in the usual manner by replacing u with its finite element approximation. From a computational standpoint, implementing the Neumann boundary condition simply entails computing the element-level equations as usual and then adding the integrated flux contribution to the right-hand side for each node lying on the Neumann boundary Γ_q.

3.2.3 Vectorized Representation

In order to make the notation more compact, as well as to make the implementation more convenient and efficient, it is worthwhile to vectorize the computations instead of using traditional nested-loop approaches when evaluating the stiffness coefficients in Eq. (3.49). Most languages such as MATLAB and Julia not only support but also encourage the use of vectorized statements, since operations such as matrix multiplication are highly optimized internally and execute much faster than their loop-based counterparts.

Let us repeat the stiffness coefficients in Eq. (3.49) for ease of reference:

$$\begin{aligned} K_{ij}^{(e)} = &-\int_{\Gamma_i^{CD}} \left[\left(a_{11} \frac{\partial \psi_j}{\partial x} + a_{12} \frac{\partial \psi_j}{\partial y} \right) n_x + \left(a_{21} \frac{\partial \psi_j}{\partial x} + a_{22} \frac{\partial \psi_j}{\partial y} \right) n_y \right] \mathrm{d}\Gamma \\ &+ \iint_{\Omega_i^{CD}} a_{00} \, \psi_j \, \mathrm{d}\Omega \, , \quad i, j = 1, \ldots, N \end{aligned} \tag{3.53}$$

In order to convert the above equation from an indicial notation to a vectorized notation, we define the matrix $\boldsymbol{C}$ and the unit vector $\hat{\boldsymbol{n}}$ as

$$\boldsymbol{C} = \begin{bmatrix} a_{11} & a_{12} \\ a_{21} & a_{22} \end{bmatrix} , \quad \hat{\boldsymbol{n}} = \begin{Bmatrix} n_x \\ n_y \end{Bmatrix} \tag{3.54}$$

Let us also define an array of shape functions and their derivatives in the following manner

$$\boldsymbol{\psi} = \begin{Bmatrix} \psi_1 \\ \psi_2 \\ \vdots \\ \psi_N \end{Bmatrix} \quad \frac{\partial \boldsymbol{\psi}}{\partial \boldsymbol{x}} = \begin{bmatrix} \partial \psi_1 / \partial x & \partial \psi_1 / \partial y \\ \partial \psi_2 / \partial x & \partial \psi_2 / \partial y \\ \vdots & \vdots \\ \partial \psi_N / \partial x & \partial \psi_N / \partial y \end{bmatrix} \tag{3.55}$$

Then Eq. (3.49) can be rewritten as follows (the reader may verify):

$$K_{ij}^{(e)} = -\int\limits_{\Gamma_i^{\mathrm{CD}}} \frac{\partial \psi_j}{\partial \boldsymbol{x}} \cdot \boldsymbol{C}^{\mathrm{T}} \cdot \hat{\boldsymbol{n}}\, \mathrm{d}\Gamma + \iint\limits_{\Omega_i^{\mathrm{CD}}} a_{00}\, \psi_j \, \mathrm{d}\Omega \tag{3.56}$$

where $\partial \psi_j / \partial \boldsymbol{x}$ denotes the jth row of $\partial \boldsymbol{\psi} / \partial \boldsymbol{x}$ defined in Eq. (3.55). We could go a step further and write

$$\boldsymbol{K}^{(e)}(i,:) = \left\{ -\int\limits_{\Gamma_i^{\mathrm{CD}}} \frac{\partial \boldsymbol{\psi}}{\partial \boldsymbol{x}} \cdot \boldsymbol{C}^{\mathrm{T}} \cdot \hat{\boldsymbol{n}}\, \mathrm{d}\Gamma + \iint\limits_{\Omega_i^{\mathrm{CD}}} a_{00}\, \boldsymbol{\psi} \, \mathrm{d}\Omega \right\}^{\mathrm{T}} \tag{3.57}$$

While vectorization brings efficiency and compact notation, it must not come at the cost of losing clarity or making the equations difficult to interpret. To that end, we will retain the index i in Eq. (3.57), which computationally corresponds to looping over the control domains.

3.3 Computational Aspects

In this section, we discuss the computational aspects of the dual mesh control domain method so that the formulations presented in the preceding sections can be readily translated into a usable computer program. The DMCDM shares many similarities with the FEM in terms of computational blocks and overall program architecture, and readers familiar with the FEM will recognize these parallels in the discussion that follows. We begin in Subsect. 3.3.1 with the generation of shape functions and their derivatives (using automatic differentiation) for both one- and two-dimensional elements. In Subsect. 3.3.2, we elaborate on numerical integration in the context of the DMCDM. Finally, in Subsect. 3.3.3, we present a novel rail-system approach for computing the line integrals that appear in the stiffness components of the DMCDM, an approach that significantly accelerates the implementation and avoids duplicate evaluations at control domain interfaces.

3.3.1 Shape Functions

In Subsect. 3.1.2, where we discussed the DMCDM using linear two-node elements, notice that the shape functions in Eq. (3.5) were defined in terms of the physical coordinate x. This is not a convenient approach, especially when dealing with 2D and 3D formulations. More precisely, expressing shape functions in physical coordinates becomes inconvenient for two important computational tasks:

1. the computation of the shape functions and their derivatives, and
2. numerical integration.

Both tasks become unnecessarily cumbersome if performed directly on the physical element. For this reason, we resort to the use of the natural or master element coordinate $\xi \in [-1, 1]$. A master element is a reference element of fixed geometry, which we take to be the interval $[-1, 1]$ in one dimension, where the element shape functions are defined in terms of the natural coordinate ξ.

In two dimensions, the idea extends naturally to a master element defined in the coordinate system (ξ, η), where each coordinate varies within the range $[-1, 1]$. Accordingly, the master element in 2D is taken as the square domain $[-1, 1]\times[-1, 1]$, which serves as a reference (or parent) for all physical elements, regardless of their actual shape or size. The shape functions are then expressed in terms of the natural coordinates (ξ, η), allowing for a systematic and consistent formulation of interpolation and integration procedures over arbitrary physical elements.

3.3.1.1 Shape Functions in One Dimension

Figure 3.16(a) shows the geometry of a typical linear master element. The two nodes are located at $\xi = -1$ and $\xi = +1$. The Lagrange shape functions for this element are given by

$$\psi_1(\xi) = \frac{1-\xi}{2}, \quad \psi_2(\xi) = \frac{1+\xi}{2} \tag{3.58}$$

As mentioned earlier, we are interested in using higher order elements, that is, ele-

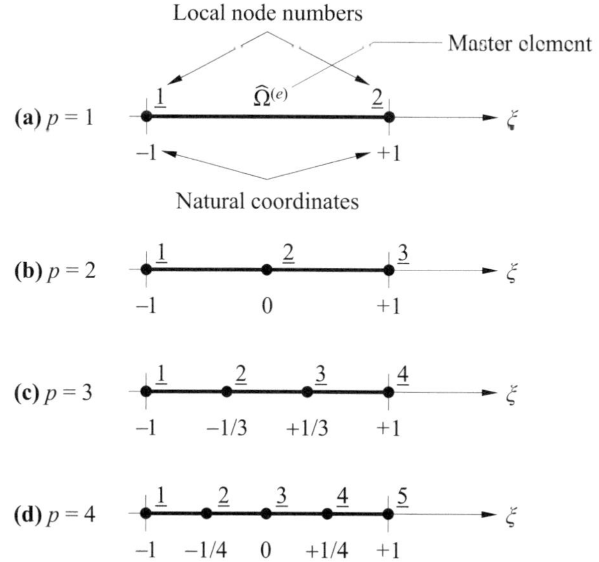

Fig. 3.16: One-dimensional master (or parent) finite elements of different orders and equispaced nodes.

ments equipped with shape functions of polynomial order greater than one. Figures 3.16(b)–(d) show the geometry of elements of arbitrary order $p \geq 2$. A pth-order element has $p+1$ nodes. We refer to the nodes at $\xi = -1$ and $\xi = +1$ as the primary, end, or shared nodes, because these nodes are typically shared between adjacent elements. The remaining nodes in the interior of the element are called internal nodes. In all our analyses, we will use elements with equally spaced nodes[7].

A pth-order element has $(p+1)$ Lagrange shape functions, each associated with a particular node. The Lagrange shape functions of a pth-order element are given by the general formula

$$\psi_i(\xi) = \prod_{\substack{k=1 \\ k \neq i}}^{p+1} \frac{\xi - \xi_k}{\xi_i - \xi_k} \qquad i = 1, \ldots, p+1 \tag{3.59}$$

where ξ_k, $k = 1, \ldots, p+1$ are the locations of the nodes in the natural coordinate system.

With this background, we will now see how a physical element (that is, an element in the mesh of the computational domain) can be obtained from the master element. Let us denote the master element as $\hat{\Omega}^{(e)} \equiv [-1, 1]$. Let $\Omega^{(e)}$ be a physical element and let $\boldsymbol{X}_e$ be the vector of nodal coordinates (i.e., physical coordinates) of $\Omega^{(e)}$. We will approximate the location x of any point in $\Omega^{(e)}$ as

$$x = \sum_{j=1}^{p+1} X_j^{(e)} \psi_j(\xi) \qquad \xi \in \hat{\Omega}^{(e)} \tag{3.60}$$

where $X_j^{(e)}$ is the jth entry in $\boldsymbol{X}_e$. We have thus used the same interpolation functions that were used to interpolate the unknown field to now interpolate the geometry. Such a formulation is referred to as an isoparametric formulation. We may visualize the master element $\hat{\Omega}^{(e)}$ being deformed into $\Omega^{(e)}$, with the deformation mapping $x = \chi(\xi)$ given by Eq. (3.60). The natural coordinate ξ may therefore be viewed as a dimensionless variable that parametrizes the position within the master element, and any physical element in the mesh is obtained from the master element through the mapping given in Eq. (3.60).

We are also interested in computing the derivatives of the shape functions defined in Eq. (3.59). While there are several ways to obtain these derivatives, a simple and tidy approach is to use automatic differentiation (specifically, forward-mode auto-

---

[7] It is generally recommended that for elements of order $p \geq 2$, the nodes be placed at the zeros of a special class of orthogonal polynomials. For equally spaced nodes, the interpolation functions tend to exhibit pronounced oscillations, especially near the element boundaries, leading to a degradation of interpolation accuracy. This behavior is known as the Runge phenomenon. The oscillations can be minimized by using node distributions that cluster toward the ends of the interval, such as Chebyshev nodes or Gauss–Lobatto–Legendre (GLL) nodes. However, in the present text, we will not be dealing with very high values of p, and equally spaced nodes perform just as well as spectral nodes.

matic differentiation). This approach is motivated by the concept of dual numbers, which we will briefly review next.

Dual numbers are expressions of the form $\xi_V + \xi_D\,\varepsilon$, where ξ_V and ξ_D are real numbers and ε is a symbol defined to satisfy $\varepsilon^2 = 0$ with $\varepsilon \neq 0$. They can be added or subtracted component-wise, that is,

$$(\xi_V + \xi_D\varepsilon) \pm (\zeta_V + \zeta_D\varepsilon) = (\xi_V \pm \zeta_V) + (\xi_D \pm \zeta_D)\,\varepsilon \tag{3.61}$$

while the multiplication of two dual numbers is defined by

$$\begin{aligned}(\xi_V + \xi_D\varepsilon)\cdot(\zeta_V + \zeta_D\varepsilon) &= \xi_V\zeta_V + (\xi_V\zeta_D + \xi_D\zeta_V)\,\varepsilon + \xi_D\zeta_D\underbrace{\varepsilon^2}_{=0}\\ &= \xi_V\zeta_V + (\xi_V\zeta_D + \xi_D\zeta_V)\,\varepsilon\end{aligned} \tag{3.62}$$

To divide two dual numbers, we multiply the denominator by its conjugate (just as we do when dividing complex numbers):

$$\begin{aligned}\frac{\xi_V + \xi_D\varepsilon}{\zeta_V + \zeta_D\varepsilon} &= \frac{\xi_V + \xi_D\varepsilon}{\zeta_V + \zeta_D\varepsilon}\cdot\frac{\zeta_V - \zeta_D\varepsilon}{\zeta_V - \zeta_D\varepsilon}\\ &= \frac{\xi_V\zeta_V + (\xi_D\zeta_V - \xi_V\zeta_D)\,\varepsilon - \xi_D\zeta_D\varepsilon^2}{\zeta_V^2 - \zeta_D^2\varepsilon^2}\\ &= \frac{\xi_V}{\zeta_V} + \frac{\xi_D\zeta_V - \xi_V\zeta_D}{\zeta_D^2}\,\varepsilon\end{aligned} \tag{3.63}$$

where once again we have used the stipulation $\varepsilon^2 = 0$.

Let us now see how this notion of dual numbers is useful in computing derivatives. If, in Eqs. (3.61) through (3.63), ξ_V and ζ_V represent the values of some functions f and g, while ξ_D and ζ_D represent the derivatives f' and g' of these functions (with respect to a common independent variable), then notice that the outcome of each binary operation in Eqs. (3.61) through (3.63) yields not only the function value but also the derivative. The derivative appears as the coefficient of ε.

We will describe how this computational procedure can be implemented in a program. The steps are listed systematically in Algorithm 2. We are interested in computing the derivatives of the shape functions $\psi_i(\xi)$ defined in Eq. (3.59). In our implementation, we define a data structure called the dual number data structure, which has two attributes: (1) the value attribute ($\cdot$val) and (2) the derivative attribute ($\cdot$der). Since our goal is to compute the derivatives of the polynomials $\psi_i(\xi)$ with respect to the independent variable ξ, we declare ξ to be an instance (or object) of this dual number data structure, as shown in Algorithm 2. The independent variable is initialized as follows[8]:

$$\begin{aligned}\underline{\xi}.\text{val} &= \xi_V\\ \underline{\xi}.\text{der} &= 1\end{aligned} \tag{3.64}$$

[8] Variables that are declared as dual numbers are represented with an underline to distinguish them from other variables that are real (or integers).

Algorithm 2 Computation of Shape functions and their derivatives

```
Inputs: p, ξ_V                                        ▷ Element order, Value of ξ
Inputs: ξ_e                                           ▷ Array of natural coordinates of nodes
Outputs: ψ, dψ/dξ                                     ▷ Shape functions and their derivatives at ξ = ξ_V
ξ ← dual(val = ξ_V, der = 0)                          ▷ Initialize ξ as a dual number
for i = 1 to p + 1 do                                 ▷ Loop over nodes
    B = dual(val = 1, der = 0)                        ▷ Temporary variable
    for k = 1 to p + 1 do
        if k = i then
            skip
        end if
        A_0 ← dual(val = ξ_e(i) − ξ_e(k), der = 0)    ▷ Temporary variable
        A_1 ← dual(val = −ξ_e(k) , der = 0)           ▷ Temporary variable
        A_2 ← add(ξ , A_1)                            ▷ Temporary variable
        A_3 ← divd(A_2 , A_0)                         ▷ Temporary variable
        B ← mult(B , A_3)                             ▷ Refer to Eq. (3.59)
    end for
    ψ(i) ← B.val                                      ▷ Extract and store ith shape function value
    dψ/dξ(i) ← B.der                                  ▷ Extract and store ith shape function derivative value
end for
______________________________________________________________
function a = dual(val, der)                           ▷ Dual number constructor
    a · val ← val
    a · der ← der
end function
function a = add( b , c )                             ▷ Dual number addition
    a · val ← b · val + c · val
    a · der ← b · der + c · der
end function
function a = mult( b , c )                            ▷ Dual number multiplication
    a · val ← b · val × c · val
    a · der ← b · der × c · val + b · val × c · der
end function
function a = divd( b , c )                            ▷ Dual number division
    a · val ← b · val/c · val
    a · der ← (b · der × c · val − b · val × c · der) / (c · val)^2
end function
```

where ξ_V is the value of the master element coordinate at which we want to compute the shape functions and their derivatives. We have underlined the variable to explicitly indicate that it is a dual number. Notice that the derivative attribute of $\underline{\xi}$ is equal to one, because the derivative of a variable with respect to itself is one. We then follow Algorithm 2 to simultaneously compute ψ_i and ψ_i'. The algorithm is based on Eq. (3.59).

3.3.1.2 Shape Functions in Two Dimensions

Figure 3.17 shows two-dimensional master elements of different orders. By convention, we number the nodes starting from the bottom-left corner and then traverse rightward and upward. The shape functions for a two-dimensional element can be conveniently generated by taking tensor products of one-dimensional shape functions. We describe the procedure below, with the explanation tailored for the implementer or programmer.

Step 1 Say we are interested in computing the shape functions ψ_i for $i = 1, \ldots, N$ and their derivatives at a given point (ξ, η) in the master element. We first begin by computing the following vectors

$$
\begin{aligned}
\boldsymbol{\psi}_\xi &= \left\{\psi_1^{1\mathrm{D}}(\xi)\ \psi_2^{1\mathrm{D}}(\xi)\ \cdots\ \psi_M^{1\mathrm{D}}(\xi)\right\}^{\mathrm{T}} \\
\boldsymbol{\psi}_\eta &= \left\{\psi_1^{1\mathrm{D}}(\eta)\ \psi_2^{1\mathrm{D}}(\eta)\ \cdots\ \psi_M^{1\mathrm{D}}(\eta)\right\}^{\mathrm{T}}
\end{aligned}
\tag{3.65}
$$

where $M = p + 1$ (the number of nodes per direction), and the superscript "1D" indicates that these shape functions are computed using the procedure explained in Subsubsect. 3.3.1.1 (or Algorithm 1) for one-dimensional shape functions. The vectors $\boldsymbol{\psi}_\xi$ and $\boldsymbol{\psi}_\eta$ are simply the one-dimensional shape functions evaluated separately at the coordinates ξ and η.

Step 2 Next, we compute the tensor product of the vectors computed in the previous step, which yields an $M \times M$ matrix as follows

$$
\boldsymbol{\psi}_{\xi\eta} = \boldsymbol{\psi}_\xi \cdot \boldsymbol{\psi}_\eta^{\mathrm{T}} = \begin{bmatrix}
\psi_1^{1\mathrm{D}}(\xi)\psi_1^{1\mathrm{D}}(\eta) & \psi_1^{1\mathrm{D}}(\xi)\psi_2^{1\mathrm{D}}(\eta) & \cdots & \psi_1^{1\mathrm{D}}(\xi)\psi_M^{1\mathrm{D}}(\eta) \\
\psi_2^{1\mathrm{D}}(\xi)\psi_1^{1\mathrm{D}}(\eta) & \psi_2^{1\mathrm{D}}(\xi)\psi_2^{1\mathrm{D}}(\eta) & \cdots & \psi_2^{1\mathrm{D}}(\xi)\psi_M^{1\mathrm{D}}(\eta) \\
\vdots & \vdots & \ddots & \vdots \\
\psi_M^{1\mathrm{D}}(\xi)\psi_1^{1\mathrm{D}}(\eta) & \psi_M^{1\mathrm{D}}(\xi)\psi_2^{1\mathrm{D}}(\eta) & \cdots & \psi_M^{1\mathrm{D}}(\xi)\psi_M^{1\mathrm{D}}(\eta)
\end{bmatrix}
\tag{3.66}
$$

Step 3 The matrix generated in the previous step contains the two-dimensional shape functions of interest. However, it is more convenient to store them in a vector rather than a matrix. Therefore, we flatten[9] the matrix into a vector in column-major order (this ensures that the shape functions are consistent with the node numbering shown in Fig. 3.17), yielding the two-dimensional shape functions

[9] Flattening is the process of converting a multi-dimensional matrix into a one-dimensional array.

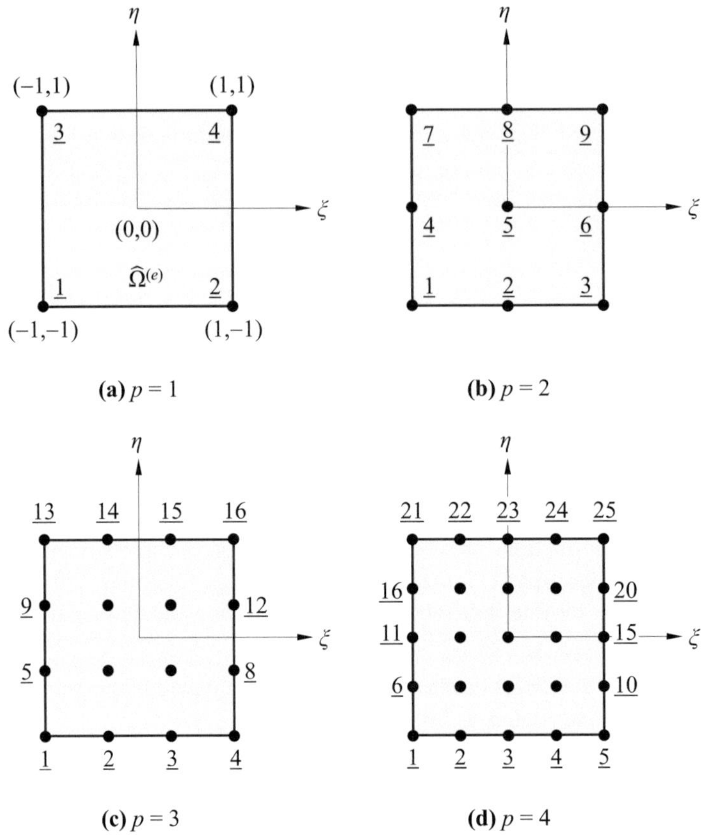

Fig. 3.17: Two-dimensional master (or parent) finite elements of different orders and equispaced nodes.

$$
\begin{array}{lll}
\psi_1 = \psi_1^{1D}(\xi)\psi_1^{1D}(\eta) & \psi_{M+1} = \psi_1^{1D}(\xi)\psi_2^{1D}(\eta) & \psi_{M^2-M+1} = \psi_1^{1D}(\xi)\psi_M^{1D}(\eta) \\
\psi_2 = \psi_2^{1D}(\xi)\psi_1^{1D}(\eta) & \psi_{M+2} = \psi_2^{1D}(\xi)\psi_2^{1D}(\eta) & \psi_{M^2-M+2} = \psi_2^{1D}(\xi)\psi_M^{1D}(\eta) \\
\vdots & \vdots & \ldots \vdots \\
\psi_M = \psi_M^{1D}(\xi)\psi_1^{1D}(\eta) & \psi_{2M} = \psi_M^{1D}(\xi)\psi_2^{1D}(\eta) & \psi_N = \psi_M^{1D}(\xi)\psi_M^{1D}(\eta)
\end{array}
\tag{3.67}
$$

where $N = M^2$ (which is equal to the number of nodes of the two-dimensional element). We can succinctly condense the entire procedure discussed above into the following single equation

$$
\psi_i(\xi,\eta) = \psi_j^{1D}(\xi)\,\psi_k^{1D}(\eta) \quad \begin{cases} 1 \le j \le M \\ 1 \le k \le M \\ i = j + (k-1) \times M \end{cases} \tag{3.68}
$$

This approach is known as the tensor product generation of shape functions and is extremely convenient from a computational standpoint. Any point $\boldsymbol{x}$ in the physical element $\Omega^{(e)}$ is obtained by mapping a corresponding point $\boldsymbol{\xi} = (\xi, \eta)$ in the master element $\hat{\Omega} = [-1, 1] \times [-1, 1]$ through the isoparametric transformation

$$\boldsymbol{x}(\boldsymbol{\xi}) = \sum_{j=1}^{N} \boldsymbol{x}_j \, \psi_j(\boldsymbol{\xi}), \tag{3.69}$$

where $\boldsymbol{x}_j = (x_j, y_j)$ are the coordinates of the jth node of $\Omega^{(e)}$. Equivalently,

$$\begin{aligned} x(\xi, \eta) &= \sum_{j=1}^{N} x_j \, \psi_j(\xi, \eta), \\ y(\xi, \eta) &= \sum_{j=1}^{N} y_j \, \psi_j(\xi, \eta). \end{aligned} \tag{3.70}$$

The partial derivatives of the shape functions with respect to the master element coordinates (ξ, η) can be computed as:

$$\frac{\partial \psi_i}{\partial \xi} = \frac{\partial \psi_j^{\mathrm{1D}}(\xi)}{\partial \xi} \psi_k^{\mathrm{1D}}(\eta) \,, \quad \frac{\partial \psi_i}{\partial \eta} = \psi_j^{\mathrm{1D}}(\xi) \frac{\psi_k^{\mathrm{1D}}(\eta)}{\partial \eta} \quad \begin{cases} 1 \le j \le M \\ 1 \le k \le M \\ i = j + (k-1) \times M \end{cases} \tag{3.71}$$

or, writing vectorially as

$$\frac{\partial \boldsymbol{\psi}}{\partial \xi} = \text{flatten}\left\{ \left(\frac{\mathrm{d}\boldsymbol{\psi}_\xi}{\mathrm{d}\xi} \right)^{\mathrm{T}} \cdot \boldsymbol{\psi}_\eta \right\}, \quad \frac{\partial \boldsymbol{\psi}}{\partial \eta} = \text{flatten}\left\{ \boldsymbol{\psi}_\xi \cdot \left(\frac{\mathrm{d}\boldsymbol{\psi}_\eta}{\mathrm{d}\eta} \right)^{\mathrm{T}} \right\} \tag{3.72}$$

where the flattening is done in column-major order. Also, note that while computing $\boldsymbol{\psi}_\xi$ and $\boldsymbol{\psi}_\eta$ defined in Eq. (3.65) using Algorithm 2, the derivatives $\mathrm{d}\boldsymbol{\psi}_\xi/\mathrm{d}\xi$ and $\mathrm{d}\boldsymbol{\psi}_\eta/\mathrm{d}\eta$ appearing in Eq. (3.72) are generated simultaneously. Therefore, computing the partial derivatives defined in Eq. (3.72) should be extremely straightforward.

3.3.1.3 Global Derivatives

In the preceding sections, notice that we computed the shape function derivatives with respect to the master element coordinates, while in the stiffness integrals such as Eq. (3.21) (for 1D) and Eq. (3.49) (for 2D), the shape functions appear differentiated with respect to the physical coordinates (i.e., x and y). Thus, in this section we will briefly review how these derivatives, often referred to as the global derivatives, can be computed. We will not provide detailed derivations or explanations in order to keep the discussion concise, and interested readers are referred to [12, 14] for a more comprehensive treatment. We will list the formulas for the two-dimensional case and then show the reduction to one dimension.

To begin with, we compute the Jacobian matrix of the transformation, which is a 2×2 matrix given by

$$\boldsymbol{J} \equiv \frac{\partial \boldsymbol{x}}{\partial \boldsymbol{\xi}} = \boldsymbol{X}_e^{\mathrm{T}} \cdot \frac{\partial \boldsymbol{\psi}}{\partial \boldsymbol{\xi}} \quad \text{where} \quad \frac{\partial \boldsymbol{\psi}}{\partial \boldsymbol{\xi}} \equiv \begin{bmatrix} \frac{\partial \boldsymbol{\psi}}{\partial \xi} & \frac{\partial \boldsymbol{\psi}}{\partial \eta} \end{bmatrix} \tag{3.73}$$

where $\boldsymbol{X}_e$ is a $N \times 2$ matrix of nodal coordinates (the first column contains the x coordinates and the second contains the y coordinates). The global derivatives are then computed using

$$\frac{\partial \boldsymbol{\psi}}{\partial \boldsymbol{x}} \equiv \begin{bmatrix} \frac{\partial \boldsymbol{\psi}}{\partial x} & \frac{\partial \boldsymbol{\psi}}{\partial y} \end{bmatrix} = \frac{\partial \boldsymbol{\psi}}{\partial \boldsymbol{\xi}} \cdot \boldsymbol{J}^{-1} \tag{3.74}$$

Once again, we note that for brevity we have not provided the detailed steps leading up to the formulas listed, and interested readers are encouraged to verify them on their own.

For the one-dimensional case, Eqs. (3.73) and (3.74) simply reduce, respectively, to the following (the Jacobian becomes a scalar):

$$J = \boldsymbol{X}_e^{\mathrm{T}} \cdot \frac{\mathrm{d}\boldsymbol{\psi}}{\mathrm{d}\xi} \quad \text{and} \quad \frac{\mathrm{d}\boldsymbol{\psi}}{\mathrm{d}x} = \frac{1}{J} \frac{\mathrm{d}\boldsymbol{\psi}}{\mathrm{d}\xi} \tag{3.75}$$

3.3.2 *Quadrature*

In this section, we will shed light on the numerical evaluation of the integrals that appear in the element stiffness and force coefficients. A detailed treatise on numerical integration can be found in [14, 17], and here we present only the final results that can be directly applied.

3.3.2.1 In One Dimension

In one-dimensional problems, the integrals we encounter are of the form

$$I = \int_{\Omega^{\mathrm{CD}}} f(x)\, \mathrm{d}x = \int_{x_a}^{x_b} f(x)\, \mathrm{d}x \tag{3.76}$$

where, as we already know, $\Omega^{\mathrm{CD}} \equiv [x_a, x_b]$ lies inside the element $\Omega^{(e)}$ in which we are performing the integration. The first step is to transform the integral from the physical element to the master element (that is, change the variable from x to ξ). Recall from elementary calculus that this can be done as follows

$$\int_{x_a}^{x_b} f(x)\,\mathrm{d}x = \int_{\xi_a}^{\xi_b} \hat{f}(\xi)\,J\,\mathrm{d}\xi \tag{3.77}$$

where ξ_a and ξ_b are the control domain endpoints in the master element coordinate system (corresponding, respectively, to x_a and x_b), J is the Jacobian of transformation defined in Eq. (3.75), and $\hat{f}(\xi)$ is simply $f(x(\xi))$.

Next, we may choose any numerical integration scheme to evaluate the integral in Eq. (3.77). A convenient choice is the Gauss–Legendre (GL) quadrature rule:

$$\int_{\xi_a}^{\xi_b} \hat{f}(\xi)\,J\,\mathrm{d}\xi \approx \frac{\xi_b - \xi_a}{2} \sum_{I=1}^{N_{\mathrm{GP}}} W_I\,\hat{f}\left(\frac{\xi_b - \xi_a}{2}\zeta_I + \frac{\xi_b + \xi_a}{2}\right) \tag{3.78}$$

where W_I and ζ_I are the quadrature weights and points, respectively, for integration over the standard interval $[-1, 1]$, and can be found tabulated in [8] for up to $N_{\mathrm{GP}} = 64$ (although we would never need so many integration points in practice!). The number of integration points N_{GP} is a user-selected integer. It is helpful to note that a GL quadrature rule with N_{GP} integration points exactly integrates polynomials of degree $2N_{\mathrm{GP}} - 1$ or lower (this is not true in general for other numerical integration schemes, such as the Newton–Cotes rules). For the dual mesh control domain method, it is sufficient to use $N_{\mathrm{GP}} = 2$ even for problems with higher-order elements.

The curious reader may wonder why we did not directly apply Eq. (3.78) to the integral in Eq. (3.76), and what the point was in changing variables from x to ξ as done in Eq. (3.77). The reason is that, when evaluating the integrals appearing in the stiffness coefficients, the function f will involve shape functions (and/or their derivatives), which are defined in terms of the master element coordinate ξ and not the physical coordinate x. Thus, what we can actually compute is $\hat{f}(\xi)$ and not $f(x)$.

However, for the force coefficients, we can integrate directly without transforming to the master coordinate system, since the integrands in that case do not involve the shape functions or their derivatives.

3.3.2.2 In Two Dimensions

In two dimensions, we encounter expressions of the form

$$I = \int_{\Gamma^{\mathrm{CD}}} \frac{\partial f}{\partial \boldsymbol{x}} \cdot \boldsymbol{C}^{\mathrm{T}} \cdot \hat{\boldsymbol{n}}\,\mathrm{d}\Gamma + \int_{\Omega^{\mathrm{CD}}} g\,\mathrm{d}\Omega \tag{3.79}$$

where f, $\boldsymbol{C}$, and g can, in general, be functions of (x, y). Once again, we perform a change of variables and transform the integral to one over the master element. To do so for the first term, we use Nanson's formula [11]

$$\hat{\boldsymbol{n}}\,\mathrm{d}\Gamma = \boldsymbol{J}^{-\mathrm{T}} \cdot \hat{\boldsymbol{N}}\,\mathcal{J}\,\mathrm{d}\hat{\Gamma} \quad \text{where } \mathcal{J} = \det(\boldsymbol{J}) \tag{3.80}$$

while for the second term we will use the well-known relation [11]

$$d\Omega = \mathcal{J}\, d\hat{\Omega} \tag{3.81}$$

Substituting Eqs. (3.80) and (3.81) into Eq. (3.79), we obtain

$$I = \int_{\hat{\Gamma}^{\mathrm{CD}}} \frac{\partial f}{\partial \boldsymbol{x}} \cdot \boldsymbol{C}^{\mathrm{T}} \cdot \boldsymbol{J}^{-\mathrm{T}} \cdot \hat{\boldsymbol{N}}\, \mathcal{J}\, d\hat{\Gamma} + \int_{\hat{\Omega}^{\mathrm{CD}}} g\, \mathcal{J}\, d\hat{\Omega} \tag{3.82}$$

In the master element system, the paths $\hat{\Gamma}^{\mathrm{CD}}$ are always either vertical or horizontal, as seen in Fig. 3.18, and the first integral can be easily computed using the procedure described in Subsect. 3.3.2 for one-dimensional integration. To compute the second integral over the rectangular control domain $\hat{\Omega}^{\mathrm{CD}} \equiv [\xi_a, \xi_b] \times [\eta_a, \eta_b]$ shown in Fig. 3.18, we may use the 2D Gauss quadrature rule (which is a tensor product of two 1D Gauss rules, one in each coordinate direction)

$$\int_{\hat{\Omega}^{\mathrm{CD}}} g\, J\, d\hat{\Omega} = \int_{\eta_a}^{\eta_b}\int_{\xi_a}^{\xi_b} g\, J\, d\xi\, d\eta \approx \frac{\eta_b - \eta_a}{2} \cdot \frac{\xi_b - \xi_a}{2} \sum_{I=1}^{N_{\mathrm{GP}}}\sum_{J=1}^{N_{\mathrm{GP}}} W_I\, W_J\, \mathcal{G}\,(\xi_I, \eta_J) \tag{3.83}$$

where $\mathcal{G}(\boldsymbol{\xi}) = g(\boldsymbol{x}(\boldsymbol{\xi}))\, \mathcal{J}(\boldsymbol{\xi})$ and

$$\xi_I = \frac{\xi_b - \xi_a}{2}\zeta_I + \frac{\xi_b + \xi_a}{2}, \quad \eta_J = \frac{\eta_b - \eta_a}{2}\zeta_J + \frac{\eta_b + \eta_a}{2} \tag{3.84}$$

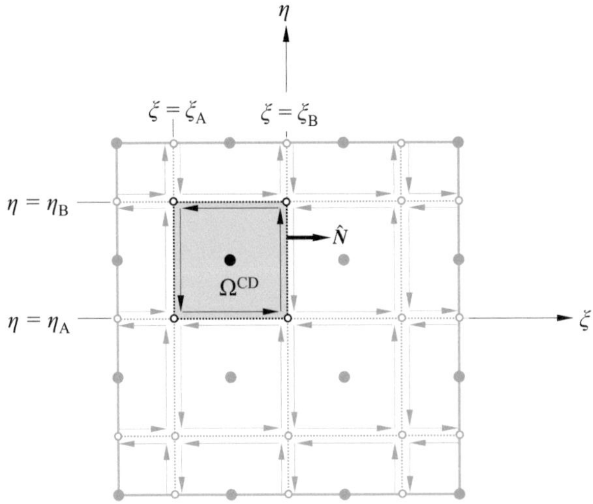

Fig. 3.18: Integration over an arbitrary control domain in the master element.

In Eq. (3.84), ζ_I, ζ_J and W_I, W_J denote the standard Gauss points and weights described earlier and can be found tabulated in [8].

3.3.3 Rail-System Approach

In this section, we will discuss an efficient method to compute the stiffness coefficients in Eq. (3.56) for 2D problems, specifically the part that arises from evaluating the path integrals. To illustrate this procedure, we will consider the stiffness coefficients without the a_{00} term:

$$K_{ij}^{(e)} = -\int_{\Gamma_i^{\mathrm{CD}}} \frac{\partial \psi_j}{\partial \boldsymbol{x}} \cdot \boldsymbol{C}^{\mathrm{T}} \cdot \boldsymbol{J}^{-\mathrm{T}} \cdot \boldsymbol{N} \mathcal{J} \, \mathrm{d}\hat{\Gamma} \tag{3.85}$$

If we were to evaluate the coefficients in Eq. (3.85) by moving from node to node and computing the path integrals each time, it becomes evident, as shown in Fig. 3.18, that a significant amount of duplicate computation would occur, since every edge is shared by two control domains. The rail-system procedure discussed here avoids this issue.

We begin by numbering every path in the dual mesh, as illustrated in Fig. 3.19(a) and (b). For ease of bookkeeping, the vertical and horizontal paths are numbered separately. Each set of paths (vertical and horizontal) is referred to as a rail system. Let N_p denote the number of paths per system. For the element shown in Fig. 3.19, $N_p = 12$. For clarity, we will enclose all path numbers within square brackets.

Next, we define two matrices $\boldsymbol{R}$ and $\boldsymbol{S}$, each of order $N_p \times N$ and belonging respectively to the horizontal and vertical rail systems, as

$$\begin{aligned} R_{[m]j} &= -\int_{\Gamma_{[m]}} \frac{\partial \psi_j}{\partial \boldsymbol{x}} \cdot \boldsymbol{C}^{\mathrm{T}} \cdot \boldsymbol{J}^{-\mathrm{T}} \cdot \boldsymbol{N}_{\mathrm{H}} \mathcal{J} \, \mathrm{d}\xi \\ S_{[m]j} &= -\int_{\Gamma_{[m]}} \frac{\partial \psi_j}{\partial \boldsymbol{x}} \cdot \boldsymbol{C}^{\mathrm{T}} \cdot \boldsymbol{J}^{-\mathrm{T}} \cdot \boldsymbol{N}_{\mathrm{V}} \mathcal{J} \, \mathrm{d}\eta \end{aligned} \tag{3.86}$$

where, as indicated in Fig. 3.19, $\boldsymbol{N}_{\mathrm{H}}$ and $\boldsymbol{N}_{\mathrm{V}}$ are the normals to the horizontal and vertical path segments, respectively. Note that in Eq. (3.86) we can avoid the index j by vectorization, as demonstrated in going from Eq. (3.56) to Eq. (3.57). In that case, we would have only one index $[m]$, which corresponds to looping over all the path segments in a rail system. That is,

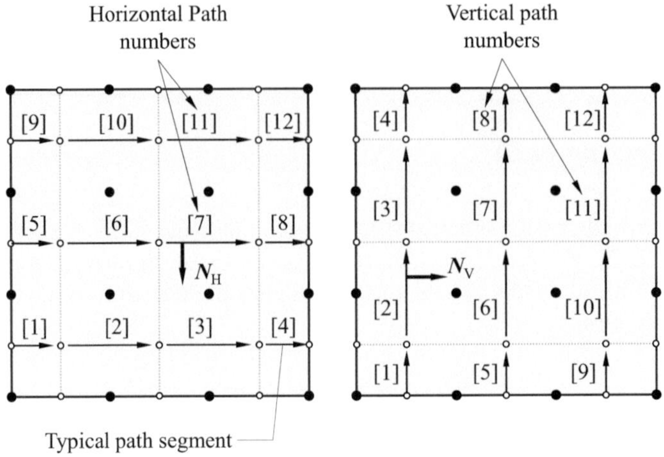

(a) Horizontal rail system **(b)** Vertical rail system

Fig. 3.19: Numbering convention for the paths of the rail systems in the master element.

$$
\begin{aligned}
\boldsymbol{R}([m],:) &= -\left\{ \int\limits_{\Gamma_{[m]}} \frac{\partial \psi}{\partial \boldsymbol{x}} \cdot \boldsymbol{C}^{\mathrm{T}} \cdot \boldsymbol{J}^{-\mathrm{T}} \cdot \boldsymbol{N}_{\mathrm{H}} \, \mathcal{J} \, \mathrm{d}\xi \right\}^{\mathrm{T}} \\
\boldsymbol{S}([m],:) &= -\left\{ \int\limits_{\Gamma_{[m]}} \frac{\partial \psi}{\partial \boldsymbol{x}} \cdot \boldsymbol{C}^{\mathrm{T}} \cdot \boldsymbol{J}^{-\mathrm{T}} \cdot \boldsymbol{N}_{\mathrm{V}} \, \mathcal{J} \, \mathrm{d}\eta \right\}^{\mathrm{T}}
\end{aligned}
\qquad 1 \le [m] \le N_p \tag{3.87}
$$

What we are essentially doing is computing and storing the stiffness integrals for every path segment in each rail system. Next, in order to construct the element stiffness matrix $\boldsymbol{K}^{(e)}$ from the matrices $\boldsymbol{R}$ and $\boldsymbol{S}$, we need a mapping matrix that identifies which paths correspond to which nodal control domain so that the contributions can be assembled correctly, including their proper signs. In other words, we seek two matrices, $\boldsymbol{H}^{\mathrm{H}}$ and $\boldsymbol{H}^{\mathrm{V}}$, such that we can write

$$
\boldsymbol{K}^{(e)} = \boldsymbol{H}^{\mathrm{H}} \cdot \boldsymbol{R} + \boldsymbol{H}^{\mathrm{V}} \cdot \boldsymbol{S} \implies K_{ij}^{(e)} = \sum_{m=1}^{N_p} \left(H_{i[m]}^{\mathrm{H}} R_{[m]j} + H_{i[m]}^{\mathrm{V}} S_{[m]j} \right) \tag{3.88}
$$

The matrices $\boldsymbol{H}^{\mathrm{H}}$ and $\boldsymbol{H}^{\mathrm{V}}$ are each of order $N \times N_p$ and depend only on the element order p. Hence, they can be precomputed and supplied to the function that evaluates the element coefficients. Although we do not detail the procedure for constructing these matrices, since it is purely a bookkeeping task, we list the entries of $\boldsymbol{H}^{\mathrm{H}}$ and $\boldsymbol{H}^{\mathrm{V}}$ for different element orders in Appendix 1 for the reader's convenience. These matrices are sparse (that is, mostly filled with zeros). As expected, the nonzero entries are either +1 or −1, depending on the orientation of a path relative to the corresponding control domain.

3.3.4 Solution Procedures for Nonlinear Equations

We had not explicitly addressed nonlinearity in the previous sections, but we will discuss it now. If one or more of the coefficients (a, b, c) in Eq. (3.1) (for the one–dimensional model problem) or the coefficients a_{ij} in Eq. (3.36) (for the two–dimensional model problem) depend on u (and/or its first derivatives), then the problem becomes nonlinear, and the discretized system of equations takes the form

$$\boldsymbol{K}(\boldsymbol{\Delta}) \cdot \boldsymbol{\Delta} = \boldsymbol{F} \tag{3.89}$$

where the stiffness matrix $\boldsymbol{K}$ depends on the nodal unknowns in $\boldsymbol{\Delta}$. Consequently, $\boldsymbol{K}$ cannot be computed until $\boldsymbol{\Delta}$ is known, and we must resort to special iterative techniques to solve such nonlinear systems of equations. We will present two commonly used approaches in practice, namely, the fixed-point (Picard) iteration and Newton's method.

3.3.4.1 Fixed-Point Iteration

This method is known as the direct iteration method or the method of successive substitutions. Since $\boldsymbol{K}(\boldsymbol{\Delta})$ cannot be evaluated until $\boldsymbol{\Delta}$ is known, we must employ an iterative process. In this process, we assume that the solution at the end of the $(p-1)$st iteration is known and seek the solution at the end of the pth iteration. This idea can be expressed mathematically as:

$$\boldsymbol{K}(\boldsymbol{\Delta}^{(p-1)}) \cdot \boldsymbol{\Delta}^{(p)} = \boldsymbol{F}(\boldsymbol{\Delta}^{(p-1)}). \tag{3.90}$$

Here, $\boldsymbol{\Delta}^{(p-1)}$ represents the solution from the $(p-1)$st iteration, and $\boldsymbol{\Delta}^{(p)}$ is the solution corresponding to the pth iteration, which is obtained from Eq. (3.90). That is,

$$\boldsymbol{\Delta}^{(p)} = -\left[\boldsymbol{K}(\boldsymbol{\Delta}^{(p-1)})\right]^{-1} \cdot \boldsymbol{F}(\boldsymbol{\Delta}^{(p-1)}). \tag{3.91}$$

Next, we require a termination condition to determine when the iterations should cease, essentially a measure to gauge how close we are to the actual solution of the system of equations. One approach is to compute the normalized difference between the solutions of two successive iterations using the Euclidean norm and compare it against a user-specified error tolerance ε_{tol}. Thus, the termination condition can be expressed as follows:

$$\varepsilon = \sqrt{\frac{\langle \delta\boldsymbol{\Delta}, \delta\boldsymbol{\Delta} \rangle}{\langle \boldsymbol{\Delta}^{(p)}, \boldsymbol{\Delta}^{(p)} \rangle}} \leq \varepsilon_{\text{tol}} \tag{3.92}$$

where $\delta\boldsymbol{\Delta} = \boldsymbol{\Delta}^{(p)} - \boldsymbol{\Delta}^{(p-1)}$.

3.3.4.2 Newton's Method

A comprehensive discussion of the Newton method in the context of nonlinear finite element analysis can be found in [12]. In this method, we compute the residual $\boldsymbol{R} = \boldsymbol{K}(\boldsymbol{\Delta}) \cdot \boldsymbol{\Delta} - \boldsymbol{F}$ and the tangent matrix $\boldsymbol{T}$, whose components are given by

$$\begin{aligned} T_{ij} \equiv \frac{\partial R_i}{\partial \Delta_j} &= \frac{\partial}{\partial \Delta_j}\left(\sum_{m=1}^{N_{\text{eq}}} K_{im}\Delta_m - F_i\right) \\ &= \sum_{m=1}^{N_{\text{eq}}}\left(K_{im}\delta_{mj} + \Delta_m \frac{\partial K_{im}}{\partial \Delta_j}\right) \\ &= K_{ij} + \sum_{m=1}^{N_{\text{eq}}} \Delta_m \frac{\partial K_{im}}{\partial \Delta_j} \end{aligned} \tag{3.93}$$

where N_{eq} denotes the number of equations in the nonlinear system. Once the tangent matrix $\boldsymbol{T}$ is known, we can calculate the solution increment $\delta\boldsymbol{\Delta}$ using:

$$\delta\boldsymbol{\Delta} = -\boldsymbol{T}^{-1} \cdot \boldsymbol{R} \tag{3.94}$$

At the end of each iteration, the solution is updated as follows:

$$\boldsymbol{\Delta}^{(p)} = \boldsymbol{\Delta}^{(p-1)} + \delta\boldsymbol{\Delta}. \tag{3.95}$$

where it is important to note that $\boldsymbol{T}$ and $\boldsymbol{R}$ are computed using $\boldsymbol{\Delta}^{(p-1)}$. A few remarks are in order:

1. We can apply the same termination condition in Eq. (3.92), and it is generally observed that the Newton iteration method converges with far fewer iterations than the direct iteration method for a given tolerance ε_{tol}.
2. We presented the approach for a generic system of equations $\boldsymbol{K}(\boldsymbol{\Delta}) \cdot \boldsymbol{\Delta} = \boldsymbol{F}$. In practice, we compute the tangent matrix and the residual vector at the element level and then assemble them exactly as we would assemble the stiffness matrix and force vector. After assembly and the imposition of boundary conditions, we solve for the global solution increment and update the solution accordingly.
3. For problems such as those we will see in the applications, we may have multiple unknown functions and thus multiple degrees of freedom per node. In such cases, the element system of equations assumes the form

$$\begin{bmatrix} \boldsymbol{K}^{11} & \cdots & \boldsymbol{K}^{1\mathcal{P}} \\ \vdots & \ddots & \vdots \\ \boldsymbol{K}^{\mathcal{P}1} & \cdots & \boldsymbol{K}^{\mathcal{P}\mathcal{P}} \end{bmatrix} \begin{Bmatrix} \boldsymbol{\Delta}^1 \\ \vdots \\ \boldsymbol{\Delta}^{\mathcal{P}} \end{Bmatrix} = \begin{Bmatrix} \boldsymbol{F}^1 \\ \vdots \\ \boldsymbol{F}^{\mathcal{P}} \end{Bmatrix} \text{ or } \sum_{\beta=1}^{\mathcal{P}} \boldsymbol{K}^{\alpha\beta} \cdot \boldsymbol{\Delta}^{\beta} = \boldsymbol{F}^{\alpha}\,,\ \alpha = 1, \ldots, \mathcal{P} \tag{3.96}$$

where $\mathcal{P}$ is the number of unknown functions (and hence the number of degrees of freedom per node). The element tangent matrix coefficients are then given by

$$T_{ij}^{\alpha\beta} = K_{ij}^{\alpha\beta} + \hat{K}_{ij}^{\alpha\beta} = K_{ij}^{\alpha\beta} + \sum_{\gamma=1}^{\mathcal{P}} \sum_{m=1}^{N} \frac{\partial K_{im}^{\alpha\gamma}}{\partial \Delta_j^{\beta}} \Delta_m^{\gamma} \quad \begin{cases} i, j = 1, \ldots, N \\ \alpha, \beta = 1, \ldots, \mathcal{P} \end{cases} \tag{3.97}$$

where N denotes the number of nodes per element. The components of the element residual vector are given by

$$R_i^{\alpha} = \sum_{\gamma=1}^{\mathcal{P}} \sum_{m=1}^{N} K_{im}^{\alpha\gamma} \Delta_m^{\gamma} - F_i^{\alpha} \quad \begin{cases} i = 1, \ldots, N \\ \alpha = 1, \ldots, \mathcal{P} \end{cases} \tag{3.98}$$

4. In many situations, applying the entire load all at once may lead to a failure in convergence even with Newton's method, and it becomes necessary to split the load into small chunks called increments. For each load step, we solve the governing equations iteratively (using either fixed-point iteration or Newton's method), check for convergence, and then move to the next load step. Such a solution procedure is referred to as load stepping or incremental loading.
5. In the implementation, it is important to guard against the program entering an infinite loop if the convergence criterion is not met. This can be done by terminating the iterative process once a specified maximum number of iterations has been reached.

3.3.5 *Computer Programs*

All of the formulations presented in this chapter have been implemented in an efficient and modular MATLAB program that is freely available and hosted at the following GitHub repository.

`https://github.com/Tanmaye-Y-Heblekar/DMSOL_Solids_PubRepo.git`

On this site, the reader can find the source code (which is thoroughly commented and organized to match the presentation of this book chapter, making it easy to follow side by side), as well as the executable version for users without a MATLAB license and comprehensive documentation on the program and input-file usage. There are two main programs, namely, the beam analysis program and the plate analysis program. We have also provided the input files for the examples solved in this chapter for the convenience of the reader. Readers are encouraged to explore the program and extend it for their own applications.

3.4 Applications

In this section, we illustrate the application of the Dual Mesh Control Domain Method to the nonlinear bending analysis of beams and plates. For each case, we provide a brief description of the structural model along with the governing equilibrium equations in both strong and integral forms. The integral equations are then discretized, leading to a system of nonlinear algebraic equations (detailed derivations are omitted). We recommend solving these equations using Newton's method, as it offers a significantly faster rate of convergence compared to fixed-point iteration. To that end, we also present the tangent coefficients for each case. Finally, numerical examples are included, complete with all necessary data to ensure reproducibility and to enable readers to validate their own implementations.

3.4.1 Bending of Functionally-Graded Beams

3.4.1.1 Assumptions

We will use the Timoshenko beam theory (TBT) to model the functionally graded beam. In the TBT, the displacement field is postulated to be

$$\begin{aligned} u_1(x) &= u(x) + z\phi_x \\ u_2(x) &= 0 \\ u_3(x) &= w(x) \end{aligned} \tag{3.99}$$

where, as shown in Fig. 3.20, $u(x)$ and $w(x)$ are the axial and transverse displacements, respectively, of the beam's mid-plane $z = 0$, and ϕ_x is the section rotation. We refer to (u, w, ϕ_x) as the generalized displacements. These are the unknown functions that characterize the shape of the deformed beam and that we seek to determine.

We will assume that the beam has a rectangular cross-section of dimensions b and H, measured along the y and z directions, respectively. In formulating the model, we assume that the beam undergoes small strains and moderate rotations, and thus employ the von Kármán nonlinear strain–displacement relations. The reader is referred to [15] for a more detailed discussion of this topic.

In terms of the material gradation in the beam, we will assume that the elastic modulus varies only along the z-direction (i.e., through-thickness gradation) according to the following power-law formula

$$E(z) = (E_2 - E_1)\left(\frac{1}{2} + \frac{z}{H}\right)^n + E_2 \tag{3.100}$$

where E_1 and E_2 are the moduli of the constituent materials. We can clearly see that the top of the beam ($z = H/2$) is composed entirely of material 1, while the bottom

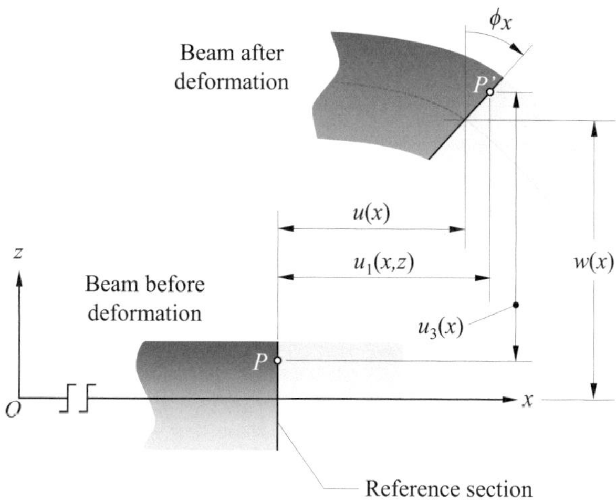

Fig. 3.20: Kinematics of beam deformation as per as the Timoshenko beam theory (TBT).

$(z = -H/2)$ is composed entirely of material 2. When $n = 0$, the beam becomes homogeneous with modulus E_1.

3.4.1.2 Strong-Form of Governing Equations

Strong form of the equilibrium equations are given by [15]

$$
\begin{aligned}
\frac{\mathrm{d}N_{xx}}{\mathrm{d}x} + f(x) = 0 \\
\frac{\mathrm{d}Q_x}{\mathrm{d}x} + \frac{\mathrm{d}}{\mathrm{d}x}\left(N_{xx}\frac{\mathrm{d}w}{\mathrm{d}x}\right) + q(x) = 0 \\
-\frac{\mathrm{d}M_{xx}}{\mathrm{d}x} + Q_x = 0
\end{aligned}
\tag{3.101}
$$

where N_{xx}, M_{xx}, and Q_x are called the stress resultants, and f and q are, respectively, the axial and transverse distributed loads applied to the beam.

For a functionally graded, isotropic, linearly elastic beam, and assuming von Kármán nonlinear strain–displacement relations along with the displacement field in Eqs. (3.99), the stress resultants appearing in Eq. (3.101) are given in terms of the generalized displacements (u, w, ϕ_x) by

$$
N_{xx} = A_{xx}\left[\frac{\mathrm{d}u}{\mathrm{d}x} + \frac{1}{2}\left(\frac{\mathrm{d}w}{\mathrm{d}x}\right)^2\right] + B_{xx}\frac{\mathrm{d}\phi_x}{\mathrm{d}x} \tag{102a}
$$

$$M_{xx} = B_{xx}\left[\frac{du}{dx} + \frac{1}{2}\left(\frac{dw}{dx}\right)^2\right] + D_{xx}\frac{d\phi_x}{dx} \tag{102b}$$

$$Q_x = S_{xx}\left(\frac{dw}{dx} + \phi_x\right) \tag{102c}$$

where closed-form expressions for the beam stiffnesses $(A_{xx}, B_{xx}, D_{xx}, S_{xx})$ are provided in Appendix 2.

3.4.1.3 Integral Form of Governing Equations

Integrating Eqs. (3.101) over the arbitrary control domain $\Omega^{\text{CD}} \equiv [x_1, x_2]$ shown in Fig. 3.21 yields the integral form of the equilibrium equations

$$\begin{aligned}
&\sum_{k=1}^{2} [N_{xx}](x_k)\, n_k + \int_{\Omega^{\text{CD}}} f\, dx = 0\\
&\sum_{k=1}^{2} \left[Q_x + N_{xx}\frac{dw}{dx}\right](x_k)\, n_k + \int_{\Omega^{\text{CD}}} q\, dx = 0\\
&\sum_{k=1}^{2} [M_{xx}](x_k)\, n_k - \int_{\Omega^{\text{CD}}} Q_x\, dx = 0
\end{aligned} \tag{3.103}$$

where $n_1 = -1$ and $n_2 = +1$ are the unit normals drawn at the left and right boundaries of a control subdomain Ω^{CD}. Substituting the constitutive equations 102(a)-(c) into Eqs. (3.101) yields the integral equilibrium equations in terms of the generalized displacements (u, w, ϕ_x) as shown below. While the detailed steps leading up to the presented equations have been omitted since they are purely algebraic, the reader may note that we have arranged the terms in each equation such that the terms containing each of the three dependent variables (u, w, ϕ_x) are segregated into separate groups.

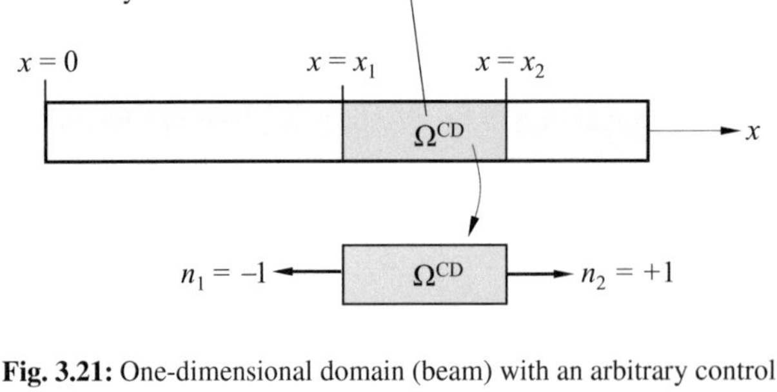

Fig. 3.21: One-dimensional domain (beam) with an arbitrary control domain.

Equation 1:

$$\sum_{k=1}^{2}\left[A_{xx}\frac{\mathrm{d}u}{\mathrm{d}x}\right](x_k)\,n_k+\sum_{k=1}^{2}\left[\frac{A_{xx}\,\tilde{W}_x}{2}\frac{\mathrm{d}w}{\mathrm{d}x}\right](x_k)\,n_k$$
$$+\sum_{k=1}^{2}\left[B_{xx}\frac{\mathrm{d}\phi_x}{\mathrm{d}x}\right](x_k)\,n_k+\int_{\Omega^{\mathrm{CD}}} f\,\mathrm{d}x=0 \tag{3.104}$$

Equation 2:

$$\sum_{k=1}^{2}\left[A_{xx}\,\tilde{W}_x\frac{\mathrm{d}u}{\mathrm{d}x}\right](x_k)\,n_k+\sum_{k=1}^{2}\left[\left(S_{xx}+\frac{A_{xx}\,\tilde{W}_x^2}{2}\right)\frac{\mathrm{d}w}{\mathrm{d}x}\right](x_k)\,n_k$$
$$+\sum_{k=1}^{2}\left[S_{xx}\,\phi_x+B_{xx}\,\tilde{W}_x\frac{\mathrm{d}\phi_x}{\mathrm{d}x}\right](x_k)\,n_k+\int_{\Omega^{\mathrm{CD}}} q\,\mathrm{d}x=0 \tag{3.105}$$

Equation 3:

$$\sum_{k=1}^{2}\left[B_{xx}\frac{\mathrm{d}u}{\mathrm{d}x}\right](x_k)\,n_k+\sum_{k=1}^{2}\left[\frac{B_{xx}\,\tilde{W}_x}{2}\frac{\mathrm{d}w}{\mathrm{d}x}\right](x_k)\,n_k$$
$$+\sum_{k=1}^{2}\left[D_{xx}\frac{\mathrm{d}\phi_x}{\mathrm{d}x}\right](x_k)\,n_k-\int_{\Omega^{\mathrm{CD}}} S_{xx}\left(\frac{\mathrm{d}w}{\mathrm{d}x}+\phi_x\right)\mathrm{d}x=0 \tag{3.106}$$

We have linearized the equations by writing the nonlinear terms (that is, products of the dependent variables or their derivatives) as

$$\left(\frac{\mathrm{d}u}{\mathrm{d}x},\frac{\mathrm{d}w}{\mathrm{d}x},\text{ or }\frac{\mathrm{d}\phi_x}{\mathrm{d}x}\right)\times\frac{\mathrm{d}w}{\mathrm{d}x}\quad\text{as}\quad\left(\frac{\mathrm{d}u}{\mathrm{d}x},\frac{\mathrm{d}w}{\mathrm{d}x},\text{ or }\frac{\mathrm{d}\phi_x}{\mathrm{d}x}\right)\times\tilde{W}_x$$

where $\tilde{W}_x$ denotes $\mathrm{d}w/\mathrm{d}x$ evaluated using the solution from the previous iteration (since we will be solving these equations iteratively, as outlined in Subsect. 3.3.4). We will use this notation consistently going forward for other variables as well. For example, $\tilde{U}_x$ denotes $\mathrm{d}u/\mathrm{d}x$ evaluated using the solution from the previous iteration and is therefore treated as known.

3.4.1.4 Dual Mesh Control Domain Model

To discretize the integral equations (3.104)–(3.106), we replace the dependents (u, w, ϕ_x) with their finite element approximations (we will assume that all three variables are interpolated using the same set of interpolation functions):

$$u(x) \approx \sum_{j=1}^{N} u_j \psi_j(x), \quad w(x) \approx \sum_{j=1}^{N} w_j \psi_j(x), \quad \phi_x(x) \approx \sum_{j=1}^{N} s_j \psi_j(x) \quad x \in \Omega^{(e)} \tag{3.107}$$

where (u_j, w_j, s_j), $j = 1, \ldots, N$ are the unknown nodal degrees of freedom of the element. Substitution of the finite element interpolations given in Eqs. (3.107) into Eqs. (3.104)–(3.106) results in a discrete system of $3N$ algebraic equations of the form

$$\begin{aligned}
&\sum_{j=1}^{N} K_{ij}^{11} u_j + \sum_{j=1}^{N} K_{ij}^{12} w_j + \sum_{j=1}^{N} K_{ij}^{13} s_j = F_i^1 \\
&\sum_{j=1}^{N} K_{ij}^{21} u_j + \sum_{j=1}^{N} K_{ij}^{22} w_j + \sum_{j=1}^{N} K_{ij}^{23} s_j = F_i^2 \qquad i = 1, 2, \ldots, N \\
&\sum_{j=1}^{N} K_{ij}^{31} u_j + \sum_{j=1}^{N} K_{ij}^{32} w_j + \sum_{j=1}^{N} K_{ij}^{33} s_j = F_i^3
\end{aligned} \tag{3.108}$$

which can be cast in matrix form as in the template shown in Eq. (3.96) (here $\mathcal{P} = 3$):

$$\begin{bmatrix} \boldsymbol{K}^{11} & \boldsymbol{K}^{12} & \boldsymbol{K}^{13} \\ \boldsymbol{K}^{21} & \boldsymbol{K}^{22} & \boldsymbol{K}^{23} \\ \boldsymbol{K}^{31} & \boldsymbol{K}^{32} & \boldsymbol{K}^{33} \end{bmatrix} \begin{Bmatrix} \boldsymbol{u} \\ \boldsymbol{w} \\ \boldsymbol{s} \end{Bmatrix} = \begin{Bmatrix} \boldsymbol{F}^1 \\ \boldsymbol{F}^2 \\ \boldsymbol{F}^3 \end{Bmatrix} \tag{3.109}$$

Each stiffness submatrix $\boldsymbol{K}^{\alpha\beta}$ in the above system is of order $N \times N$, while the force subvector $\boldsymbol{F}^\alpha$ is of dimension N. The components of the element stiffness submatrices are given by

$$K_{ij}^{11} = \sum_m H_{im} \left[A_{xx} \frac{\mathrm{d}\psi_j}{\mathrm{d}x} \right] (x_{[m]}), \quad K_{ij}^{12} = \sum_m H_{im} \left[\frac{A_{xx} \tilde{W}_x}{2} \frac{\mathrm{d}\psi_j}{\mathrm{d}x} \right] (x_{[m]}),$$

$$K_{ij}^{13} = \sum_m H_{im} \left[B_{xx} \frac{\mathrm{d}\psi_j}{\mathrm{d}x} \right] (x_{[m]}), \quad K_{ij}^{21} = \sum_m H_{im} \left[A_{xx} \tilde{W}_x \frac{\mathrm{d}\psi_j}{\mathrm{d}x} \right] (x_{[m]}),$$

$$K_{ij}^{22} = \sum_m H_{im} \left[\left(S_{xx} + \frac{A_{xx} \tilde{W}_x^2}{2} \right) \frac{\mathrm{d}\psi_j}{\mathrm{d}x} \right] (x_{[m]}),$$

$$K_{ij}^{23} = \sum_m H_{im} \left[S_{xx} \psi_j + B_{xx} \tilde{W}_x \frac{\mathrm{d}\psi_j}{\mathrm{d}x} \right] (x_{[m]}), \quad K_{ij}^{31} = \sum_m H_{im} \left[B_{xx} \frac{\mathrm{d}\psi_j}{\mathrm{d}x} \right] (x_{[m]}),$$

$$K_{ij}^{32} = \sum_m H_{im} \left[\frac{B_{xx} \tilde{W}_x}{2} \frac{\mathrm{d}\psi_j}{\mathrm{d}x} \right] (x_{[m]}) - \int_{\Omega_i^{\mathrm{CD}}} S_{xx} \frac{\mathrm{d}\psi_j}{\mathrm{d}x} \, \mathrm{d}x,$$

$$K_{ij}^{33} = \sum_m H_{im} \left[D_{xx} \frac{d\psi_j}{dx} \right] (x_{[m]}) - \int_{\Omega_i^{CD}} S_{xx} \psi_j \, dx \tag{3.110}$$

and the nodal force vector coefficients are given by

$$F_i^1 = - \int_{\Omega_i^{CD}} f \, dx \,, \quad F_i^2 = - \int_{\Omega_i^{CD}} q \, dx \,, \quad F_i^3 = 0 \tag{3.111}$$

3.4.1.5 Tangent Matrix Coefficients

Recall Eq. (3.97) to compute the tangent matrix coefficients for a nonlinear discrete model with multiple degrees of freedom per node (repeated below for ease of reference):

$$T_{ij}^{\alpha\beta} = K_{ij}^{\alpha\beta} + \hat{K}_{ij}^{\alpha\beta} = K_{ij}^{\alpha\beta} + \sum_{\gamma=1}^{\mathcal{P}} \sum_{k=1}^{N} \frac{\partial K_{ik}^{\alpha\gamma}}{\partial \Delta_j^{\beta}} \Delta_k^{\gamma} \quad \begin{cases} i, j = 1, \ldots, N \\ \alpha, \beta = 1, \ldots, \mathcal{P} \end{cases}$$

where $\Delta_j^1 = u_j$, $\Delta_j^2 = w_j$, and $\Delta_j^3 = s_j$. Since the nonlinear terms depend only on w_j, it is clear that $\boldsymbol{T}^{\alpha 1} = \boldsymbol{K}^{\alpha 1}$ and $\boldsymbol{T}^{\alpha 3} = \boldsymbol{K}^{\alpha 3}$ for all values of α. The tangent coefficients $\boldsymbol{T}^{\alpha 2}$, which do receive contributions from the second term in Eq. (3.97), are calculated as follows.

It is also helpful to pre-compute the following quantity in advance:

$$\frac{\partial \tilde{W}_x}{\partial w_j} = \frac{\partial}{\partial w_j} \left(\frac{dw}{dx} \right) = \frac{\partial}{\partial w_j} \left(\sum_{k=1}^{N} w_k \frac{d\psi_k}{dx} \right) = \sum_{k=1}^{N} \frac{d\psi_k}{dx} \frac{\partial w_k}{\partial w_j} = \sum_{k=1}^{N} \frac{d\psi_k}{dx} \delta_{jk}$$
$$= \frac{d\psi_j}{dx} \tag{3.112}$$

Term 1: $\alpha = 1, \beta = 2$

$$\hat{K}_{ij}^{12} = \sum_{k=1}^{N} \left(\frac{\partial K_{ik}^{11}}{\partial w_j} u_k + \frac{\partial K_{ik}^{12}}{\partial w_j} w_k + \frac{\partial K_{ik}^{13}}{\partial w_j} s_k \right) = \sum_k \frac{\partial K_{ik}^{12}}{\partial w_j} w_k$$
$$= \frac{\partial}{\partial w_j} \left\{ \sum_k \sum_m H_{im} \left[\frac{A_{xx} \tilde{W}_x}{2} w_k \frac{d\psi_k}{dx} \right] (x_{[m]}) \right\}$$
$$= \sum_m H_{im} \left[\frac{A_{xx} \tilde{W}_x}{2} \frac{d\psi_j}{dx} \right] (x_{[m]}) \tag{3.113}$$

Term 2: $\alpha = 2, \beta = 2$

$$
\begin{aligned}
\hat{K}_{ij}^{22} &= \sum_{k=1}^{N} \left(\frac{\partial K_{ik}^{21}}{\partial w_j} u_k + \frac{\partial K_{ik}^{22}}{\partial w_j} w_k + \frac{\partial K_{ik}^{23}}{\partial w_j} s_k \right) \\
&= \frac{\partial}{\partial w_j} \left\{ \sum_k \sum_m H_{im} \left[A_{xx} \tilde{W}_x u_k \frac{\mathrm{d}\psi_k}{\mathrm{d}x} \right] (x_{[m]}) \right\} \\
&+ \frac{\partial}{\partial w_j} \left\{ \sum_k \sum_m H_{im} \left[\left(S_{xx} + \frac{A_{xx} \tilde{W}_x^2}{2} \right) w_k \frac{\mathrm{d}\psi_k}{\mathrm{d}x} \right] (x_{[m]}) \right\} \\
&+ \frac{\partial}{\partial w_j} \left\{ \sum_k \sum_m H_{im} \left[S_{xx} s_k \psi_k + B_{xx} \tilde{W}_x s_k \frac{\mathrm{d}\psi_k}{\mathrm{d}x} \right] (x_{[m]}) \right\} \\
&= \sum_m H_{im} \left[A_{xx} \frac{\mathrm{d}\psi_j}{\mathrm{d}x} \left(\tilde{U}_x + \tilde{W}_x^2 \right) + B_{xx} \frac{\mathrm{d}\psi_j}{\mathrm{d}x} \tilde{S}_x \right] (x_{[m]})
\end{aligned}
\tag{3.114}
$$

Term 3: $\alpha = 3, \beta = 2$

$$
\begin{aligned}
\hat{K}_{ij}^{32} &= \sum_{k=1}^{N} \left(\frac{\partial K_{ik}^{31}}{\partial w_j} u_k + \frac{\partial K_{ik}^{32}}{\partial w_j} w_k + \frac{\partial K_{ik}^{33}}{\partial w_j} s_k \right) = \sum_k \frac{\partial K_{ik}^{32}}{\partial w_j} w_k \\
&= \sum_k \frac{\partial}{\partial w_j} \left\{ \sum_m H_{im} \left[\frac{B_{xx} \tilde{W}_x}{2} w_k \frac{\mathrm{d}\psi_k}{\mathrm{d}x} \right] (x_{[m]}) - \int_{\Omega_i^{\mathrm{CD}}} S_{xx} w_k \frac{\mathrm{d}\psi_k}{\mathrm{d}x} \,\mathrm{d}x \right\} \\
&= \sum_m H_{im} \left[\frac{B_{xx} \tilde{W}_x}{2} \frac{\mathrm{d}\psi_j}{\mathrm{d}x} \right] (x_{[m]})
\end{aligned}
\tag{3.115}
$$

3.4.1.6 Numerical Example

Consider a functionally graded beam of length $L = 100$ in, height $H = 1$ in, and width $B = 1$ in, whose elastic modulus is graded according to the power law in Eq. (3.100), with $E_1 = 30 \times 10^6$ psi, $E_2 = 10 \times 10^6$ psi, and $n = 0, 1$, and 10 (three cases). Take the Poisson's ratio to be $\nu = 0.3$. The beam is pinned at both ends (i.e., $u = w = 0$) and is subjected to a uniformly distributed transverse load of intensity $q_0 = 10$ lb/in. Analyze the beam using the DMCDM and comment on the effect of the element order on the numerical results. Use incremental loading with a load step of $\Delta q_0 = 0.25$ lb/in. For each load step, apply Newton's method with a convergence tolerance of $\varepsilon_{\text{tol}} = 10^{-3}$.

Solution. The problem is solved using the dual-mesh control-domain beam analysis program. Exploiting symmetry, we model only one half of the beam and impose the following boundary conditions:

$$\begin{aligned} u = w = 0 \quad & \text{at } x = 0 \quad \text{(pinned)} \\ u = \phi_x = 0 \quad & \text{at } x = L/2 \quad \text{(symmetry)} \end{aligned} \tag{3.116}$$

The input deck for the nonlinear analysis, using Newton's method with a mesh of five 4th-order elements, is presented below:

```
# Beam stiffness values (L/h = 100, n = 10)
AXX: 1.181818e+07
BXX: 7.575758e+05
DXX: 1.159674e+06
SXX: 3.787879e+06

# Mesh parameters
NEM: 5 # No. of elements in the mesh
P:    4 # Element order
X0:   0 # Global origin
L:   50 # Length of the computational domain

# Boundary conditions
EBCS: # Essential boundary conditions
 NODES: F(NNM)[1,1,NNM,NNM] # List of nodes
 DOFS: [1,2,1,3]            # List of DOFs
 VALS: [0,0,0,0]            # List of values

# Distributed loads
FX: 0.0000 # Axial
QZ: 1.0000 # Transverse

# Integration points
NGP: 2

# Nonlinear analysis settings
NONLIN:   2        # Newton's method
ITERMAX: 25        # Max. iterations before termination
BETA:     0.0      # Relaxation parameter
EPSILON: 1.0e-03   # Convergence tolerance
NLS:      40       # No. of load increments
LOAD_INCREMENTS: [0.25, 0.25, 0.25 ...(40 times)]
```

Though the GitHub repository where we have hosted the code contains comprehensive documentation and guidance on using the input file, it is worthwhile to state a few remarks here for the reader's convenience. First, anything typed after a # symbol in the input file is treated as a comment and is ignored by the input file parser. Second, the variable `NONLIN` specifies the type of analysis: a value of 0 corresponds to linear analysis, 1 invokes fixed-point iteration, and 2 invokes Newton's iteration. We recommend using Newton's method for nonlinear analyses since its rate of convergence is significantly better than that of fixed-point iteration.

Lastly, the essential boundary conditions are specified by listing the node–DOF–value triplets in the input field called `EBCS`. For the user's convenience, and to avoid manually identifying the last node, the `NODES` attribute of the `EBCS` field can be defined symbolically as a function of `NNM`, which denotes the number of nodes in the mesh (an internally computed quantity). To do so, use the syntax `F(NNM)` followed by the array; the program will internally replace the placeholder `NNM` with its actual numeric value once the mesh is generated.

Table 3.1 shows the linear analysis results (`NONLIN` = 0) for six different meshes, along with the exact solution from [15], for various values of the power-law index n. It is evident that analyses using linear elements severely underpredict the beam's deformation. This behavior is due to shear locking, a phenomenon that becomes more pronounced when the beam is slender (i.e., has a high L/h ratio, as in the present case where $L/h = 100$). The reader is referred to [12] for a detailed explanation of locking phenomena and their remedies.

One approach to mitigating locking is to increase the interpolation order p. From Table 3.1, we observe that an unreasonably large number of linear elements is required to obtain a reliable result. Quadratic and cubic elements perform considerably better, while the fourth-order element exactly recovers the analytical solution. This occurs because the exact solution for the transverse displacement is itself a fourth-degree polynomial in x [15].

It can be shown that, just as in the finite element method (FEM), the DMCDM also passes the patch test: if the exact solution is a polynomial of degree p, then a single pth-order DMCDM element yields the exact solution up to machine precision, provided the integration is performed exactly.

The results of the nonlinear analysis for different values of n are presented in Table 3.2 for a few selected load steps. The load–deflection curve is shown in Fig. 3.22. First, we observe the effect of including the von Kármán nonlinearity: the beam stiffens as the load increases. Second, note that $n = 0$ corresponds to a homogeneous beam made entirely of material 1. As n increases, the material distribution shifts toward material 2 (which has a smaller elastic modulus), and consequently, the beam exhibits larger deflections for the same applied load.

Table 3.1: Center deflection $w(L/2)$ (in inches) for different meshes and power-law indices obtained from a linear analysis with $q_0 = 1$ lb/in^2.

Mesh	$n = 0$	$n = 1$	$n = 10$
$p = 1$, NEM = 10	0.03998	0.06004	0.100189
$p = 1$, NEM = 100	0.46506	0.70880	1.027145
$p = 1$, NEM = 500	0.51847	0.79177	1.128297
$p = 2$, NEM = 5	0.74566	1.14144	1.614363
$p = 3$, NEM = 5	0.52381	0.80000	1.138915
$p = 4$, NEM = 1	0.52096	0.79565	1.132946
Exact	0.52096	0.79565	1.132946

Table 3.2: Center deflection $w(L/2)$ (in inches) versus load intensity q_0 for different power-law indices, obtained from a nonlinear analysis using Newton's method and a mesh of five 4th-order elements.

Load q_0	$n = 0$	$n = 1$	$n = 10$
0.5	0.22561	0.37347	0.45385
1.0	0.36856	0.55989	0.65934
2.5	0.60933	0.83847	0.97648
5.0	0.83138	1.08679	1.26540
7.5	0.98071	1.25313	1.46067
9.0	1.05331	1.33405	1.55599
9.8	1.08635	1.37089	1.59944
10.0	1.09695	1.38272	1.61340

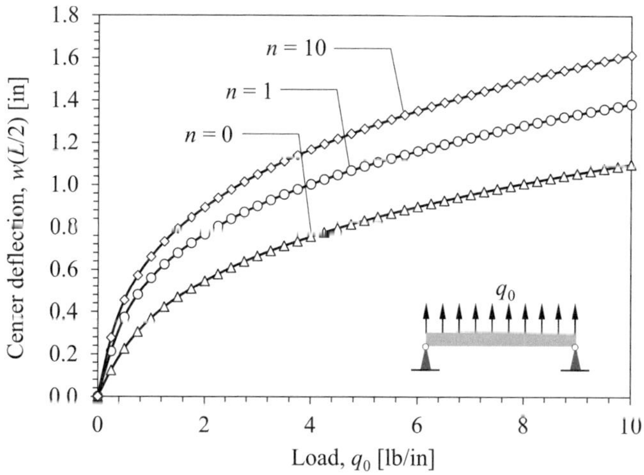

Fig. 3.22: Plots of center deflection $w(L/2)$ versus load q_0 for the functionally graded pinned beam for different values of the power-law index n.

3.4.2 Bending of Orthotropic Plates

In this section, we apply the DMCDM to perform nonlinear bending analysis of orthotropic, linearly elastic plates. The reader is referred to [10] for a comprehensive description of the various plate theories; here, we simply state the governing equations for the first-order shear deformation theory and demonstrate the use of the DMCDM by deriving the discretized dual-mesh control-domain equations and the tangent matrix coefficients, followed by a numerical example.

3.4.2.1 Assumptions

To analyze the bending of plates, we use the first-order shear deformation theory (FSDT), which extends the Timoshenko beam theory (TBT) to plate structures. In the FSDT, the displacement ansatz is given by

$$\begin{aligned} u_1 &= u(x, y) + z\phi_x(x, y) \\ u_2 &= v(x, y) + z\phi_y(x, y) \\ u_3 &= w(x, y) \end{aligned} \tag{3.117}$$

where (u, v, w) are the midplane displacements in the (x, y, z) coordinate directions, respectively, and ϕ_x and ϕ_y are the rotations of the transverse normal about the y- and x-axes. The functions $(u, v, w, \phi_x, \phi_y)$ constitute the unknown generalized displacements to be determined numerically.

We assume that the beam undergoes small strains and moderate rotations and therefore employ the von Kármán nonlinear strain–displacement relations. The reader is referred to [10, 12] for more detailed coverage of this topic.

3.4.2.2 Strong-Form of Governing Equations

The differential equations governing the deformation of plates, based on the first-order shear deformation theory and the von Kármán nonlinear strain–displacement relations, are given by [12]

$$\begin{aligned} \frac{\partial N_{xx}}{\partial x} + \frac{\partial N_{xy}}{\partial y} + f_x &= 0 \\ \frac{\partial N_{xy}}{\partial x} + \frac{\partial N_{yy}}{\partial y} + f_y &= 0 \\ \frac{\partial Q_x}{\partial x} + \frac{\partial Q_y}{\partial y} + \mathcal{N} + q &= 0 \\ \frac{\partial M_{xx}}{\partial x} + \frac{\partial M_{xy}}{\partial y} - Q_x &= 0 \\ \frac{\partial M_{xy}}{\partial x} + \frac{\partial M_{yy}}{\partial y} - Q_y &= 0 \end{aligned} \tag{3.118}$$

where

$$\mathcal{N} = \frac{\partial}{\partial x}\left(N_{xx}\frac{\partial w}{\partial x} + N_{xy}\frac{\partial w}{\partial y}\right) + \frac{\partial}{\partial y}\left(N_{xy}\frac{\partial w}{\partial x} + N_{yy}\frac{\partial w}{\partial y}\right) \tag{3.119}$$

The stress resultants appearing in Eqs. (3.118) are expressed in terms of the generalized displacements $(u, v, w, \phi_x, \phi_y)$ through the following constitutive relations:

$$
\begin{aligned}
N_{xx} &= A_{11}\left[\frac{\partial u}{\partial x}+\frac{1}{2}\left(\frac{\partial w}{\partial x}\right)^2\right]+A_{12}\left[\frac{\partial v}{\partial y}+\frac{1}{2}\left(\frac{\partial w}{\partial y}\right)^2\right]\\
N_{yy} &= A_{12}\left[\frac{\partial u}{\partial x}+\frac{1}{2}\left(\frac{\partial w}{\partial x}\right)^2\right]+A_{22}\left[\frac{\partial v}{\partial y}+\frac{1}{2}\left(\frac{\partial w}{\partial y}\right)^2\right]\\
N_{xy} &= A_{66}\left(\frac{\partial u}{\partial y}+\frac{\partial v}{\partial x}+\frac{\partial w}{\partial x}\frac{\partial w}{\partial y}\right),\quad M_{xx}=D_{11}\frac{\partial \phi_x}{\partial x}+D_{12}\frac{\partial \phi_y}{\partial y}\\
M_{yy} &= D_{12}\frac{\partial \phi_x}{\partial x}+D_{22}\frac{\partial \phi_y}{\partial y},\quad M_{xy}=D_{66}\left(\frac{\partial \phi_x}{\partial y}+\frac{\partial \phi_y}{\partial x}\right)\\
Q_x &= A_{55}\left(\phi_x+\frac{\partial w}{\partial x}\right),\quad Q_y=A_{44}\left(\phi_y+\frac{\partial w}{\partial y}\right)
\end{aligned}
\tag{3.120}
$$

where the plate stiffness coefficients (A_{ij}, D_{ij}) are given in Appendix 3 for an orthotropic, linearly elastic plate.

3.4.2.3 Integral Form of Governing Equations

Integrating Eqs. (3.118) over an arbitrary control domain Ω^{CD} and then applying the Gauss theorem yields

$$
\begin{aligned}
&\oint_{\Gamma^{\mathrm{CD}}}\begin{Bmatrix}N_{xx}\\ N_{xy}\end{Bmatrix}^{\mathrm{T}}\cdot\hat{\boldsymbol{n}}\,\mathrm{d}\Gamma+\iint_{\Omega^{\mathrm{CD}}} f_x\,\mathrm{d}\Omega=0\\
&\oint_{\Gamma^{\mathrm{CD}}}\begin{Bmatrix}N_{xy}\\ N_{yy}\end{Bmatrix}^{\mathrm{T}}\cdot\hat{\boldsymbol{n}}\,\mathrm{d}\Gamma+\iint_{\Omega^{\mathrm{CD}}} f_y\,\mathrm{d}\Omega=0\\
&\oint_{\Gamma^{\mathrm{CD}}}\begin{Bmatrix}Q_x+N_{xx}w_{,x}+N_{xy}w_{,y}\\ Q_y+N_{xy}w_{,x}+N_{yy}w_{,y}\end{Bmatrix}^{\mathrm{T}}\cdot\hat{\boldsymbol{n}}\,\mathrm{d}\Gamma+\iint_{\Omega^{\mathrm{CD}}} q\,\mathrm{d}\Omega=0\\
&\oint_{\Gamma^{\mathrm{CD}}}\begin{Bmatrix}M_{xx}\\ M_{xy}\end{Bmatrix}^{\mathrm{T}}\cdot\hat{\boldsymbol{n}}\,\mathrm{d}\Gamma-\iint_{\Omega^{\mathrm{CD}}} Q_x\,\mathrm{d}\Omega=0\\
&\oint_{\Gamma^{\mathrm{CD}}}\begin{Bmatrix}M_{xy}\\ M_{yy}\end{Bmatrix}^{\mathrm{T}}\cdot\hat{\boldsymbol{n}}\,\mathrm{d}\Gamma-\iint_{\Omega^{\mathrm{CD}}} Q_y\,\mathrm{d}\Omega=0
\end{aligned}
\tag{3.121}
$$

where $w_{,x}$ and $w_{,y}$ respectively denote $\partial w/\partial x$ and $\partial w/\partial y$. Substituting the constitutive equations (3.120) into Eqs. (3.121) and rearranging the expressions so that each group of terms corresponds to one of the five dependents $(u, v, w, \phi_x, \phi_y)$ yields the following integral equilibrium equations:

Equation 1:

$$\oint_{\Gamma^{CD}} \frac{\partial u}{\partial \boldsymbol{x}} \cdot \begin{bmatrix} A_{11} & 0 \\ 0 & A_{66} \end{bmatrix}^{\mathrm{T}} \cdot \hat{\boldsymbol{n}}\, \mathrm{d}\Gamma + \oint_{\Gamma^{CD}} \frac{\partial v}{\partial \boldsymbol{x}} \cdot \begin{bmatrix} 0 & A_{12} \\ A_{66} & 0 \end{bmatrix}^{\mathrm{T}} \cdot \hat{\boldsymbol{n}}\, \mathrm{d}\Gamma$$

$$+ \oint_{\Gamma^{CD}} \frac{1}{2} \frac{\partial w}{\partial \boldsymbol{x}} \cdot \begin{bmatrix} A_{11}\tilde{W}_x & A_{12}\tilde{W}_y \\ A_{66}\tilde{W}_y & A_{66}\tilde{W}_x \end{bmatrix}^{\mathrm{T}} \cdot \hat{\boldsymbol{n}}\, \mathrm{d}\Gamma + \iint_{\Omega^{CD}} f_x \, \mathrm{d}\Omega = 0 \tag{3.122}$$

Equation 2:

$$\oint_{\Gamma^{CD}} \frac{\partial u}{\partial \boldsymbol{x}} \cdot \begin{bmatrix} 0 & A_{66} \\ A_{12} & 0 \end{bmatrix}^{\mathrm{T}} \cdot \hat{\boldsymbol{n}}\, \mathrm{d}\Gamma + \oint_{\Gamma^{CD}} \frac{\partial v}{\partial \boldsymbol{x}} \cdot \begin{bmatrix} A_{66} & 0 \\ 0 & A_{22} \end{bmatrix}^{\mathrm{T}} \cdot \hat{\boldsymbol{n}}\, \mathrm{d}\Gamma$$

$$+ \oint_{\Gamma^{CD}} \frac{1}{2} \frac{\partial w}{\partial \boldsymbol{x}} \cdot \begin{bmatrix} A_{66}\tilde{W}_y & A_{66}\tilde{W}_x \\ A_{12}\tilde{W}_x & A_{22}\tilde{W}_y \end{bmatrix}^{\mathrm{T}} \cdot \hat{\boldsymbol{n}}\, \mathrm{d}\Gamma + \iint_{\Omega^{CD}} f_y \, \mathrm{d}\Omega = 0 \tag{3.123}$$

Equation 3:

$$\oint_{\Gamma^{CD}} \frac{\partial u}{\partial \boldsymbol{x}} \cdot \begin{bmatrix} A_{11}\tilde{W}_x & A_{66}\tilde{W}_y \\ A_{12}\tilde{W}_y & A_{66}\tilde{W}_x \end{bmatrix}^{\mathrm{T}} \cdot \hat{\boldsymbol{n}}\, \mathrm{d}\Gamma + \oint_{\Gamma^{CD}} \frac{\partial v}{\partial \boldsymbol{x}} \cdot \begin{bmatrix} A_{66}\tilde{W}_y & A_{12}\tilde{W}_x \\ A_{66}\tilde{W}_x & A_{22}\tilde{W}_y \end{bmatrix}^{\mathrm{T}} \cdot \hat{\boldsymbol{n}}\, \mathrm{d}\Gamma$$

$$+ \oint_{\Gamma^{CD}} \frac{\partial w}{\partial \boldsymbol{x}} \cdot \begin{bmatrix} A_{55} + \frac{1}{2}A_{11}\tilde{W}_x^2 + \hat{\mathcal{A}}\,\tilde{W}_y^2 & 0 \\ 0 & A_{44} + \frac{1}{2}A_{22}\tilde{W}_y^2 + \hat{\mathcal{A}}\,\tilde{W}_x^2 \end{bmatrix}^{\mathrm{T}} \cdot \hat{\boldsymbol{n}}\, \mathrm{d}\Gamma$$

$$+ \oint_{\Gamma^{CD}} \phi_x \begin{Bmatrix} A_{55} \\ 0 \end{Bmatrix}^{\mathrm{T}} \cdot \hat{\boldsymbol{n}}\, \mathrm{d}\Gamma + \oint_{\Gamma^{CD}} \phi_y \begin{Bmatrix} 0 \\ A_{44} \end{Bmatrix}^{\mathrm{T}} \cdot \hat{\boldsymbol{n}}\, \mathrm{d}\Gamma + \iint_{\Omega^{CD}} q \, \mathrm{d}\Omega = 0 \tag{3.124}$$

where we have defined

$$\hat{\mathcal{A}} = \frac{1}{2}A_{12} + A_{66}$$

Equation 4:

$$-\iint_{\Omega^{CD}} A_{55}\frac{\partial w}{\partial x}\,d\Omega + \oint_{\Gamma^{CD}} \frac{\partial \phi_x}{\partial \boldsymbol{x}} \cdot \begin{bmatrix} D_{11} & 0 \\ 0 & D_{66} \end{bmatrix}^{T} \cdot \hat{\boldsymbol{n}}\,d\Gamma - \iint_{\Omega^{CD}} A_{55}\,\phi_x\,d\Omega +$$
$$\oint_{\Gamma^{CD}} \frac{\partial \phi_y}{\partial \boldsymbol{x}} \cdot \begin{bmatrix} 0 & D_{12} \\ D_{66} & 0 \end{bmatrix}^{T} \cdot \hat{\boldsymbol{n}}\,d\Gamma = 0 \tag{3.125}$$

Equation 5:

$$-\iint_{\Omega^{CD}} A_{44}\frac{\partial w}{\partial y}\,d\Omega + \oint_{\Gamma^{CD}} \frac{\partial \phi_x}{\partial \boldsymbol{x}} \cdot \begin{bmatrix} 0 & D_{66} \\ D_{12} & 0 \end{bmatrix}^{T} \cdot \hat{\boldsymbol{n}}\,d\Gamma$$
$$+\oint_{\Gamma^{CD}} \frac{\partial \phi_y}{\partial \boldsymbol{x}} \cdot \begin{bmatrix} D_{66} & 0 \\ 0 & D_{22} \end{bmatrix}^{T} \cdot \hat{\boldsymbol{n}}\,d\Gamma - \iint_{\Omega^{CD}} A_{44}\,\phi_y\,d\Omega = 0 \tag{3.126}$$

In Eqs. (3.122)–(3.126),

$$\frac{\partial}{\partial \boldsymbol{x}}\left(u, v, w, \phi_x, \phi_y\right) \equiv \begin{bmatrix} \frac{\partial}{\partial x} & \frac{\partial}{\partial y} \end{bmatrix}\left(u, v, w, \phi_x, \phi_y\right) \tag{3.127}$$

Also note that Eqs. (3.122)–(3.126) are linearized, and the following notations are used: $\tilde{W}_x$ denotes $\partial w/\partial x$, and $\tilde{W}_y$ denotes $\partial w/\partial y$, both evaluated using the solution from the previous iteration and therefore treated as known.

3.4.2.4 Dual Mesh Control Domain Model

To discretize the integral equations (3.122)–(3.126), we replace the dependent variables $(u, v, w, \phi_x, \phi_y)$ with their finite element approximations (we assume that all five variables are interpolated using the same set of interpolation functions):

$$u(x) \approx \sum_{j=1}^{N} u_j\psi_j(x), \quad v(x) \approx \sum_{j=1}^{N} v_j\psi_j(x), \quad w(x) \approx \sum_{j=1}^{N} w_j\psi_j(x),$$
$$\phi_x(x) \approx \sum_{j=1}^{N} s_j^1\psi_j(x), \quad \phi_y(x) \approx \sum_{j=1}^{N} s_j^2\psi_j(x) \qquad x \in \Omega^{(e)} \tag{3.128}$$

where $(u_j, v_j, w_j, s_j^1, s_j^2)$, $j = 1, \ldots, N$ are the unknown nodal degrees of freedom of the element. Substitution of the finite element interpolations given in Eq. (3.128) into Eqs. (3.122)–(3.126) results in a discrete system of $5N$ algebraic equations, which can be cast in matrix form following the template shown in Eq. (3.96) (here

$\mathcal{P} = 5$):

$$\begin{bmatrix} \boldsymbol{K}^{11} & \boldsymbol{K}^{12} & \boldsymbol{K}^{13} & \boldsymbol{K}^{14} & \boldsymbol{K}^{15} \\ \boldsymbol{K}^{21} & \boldsymbol{K}^{22} & \boldsymbol{K}^{23} & \boldsymbol{K}^{24} & \boldsymbol{K}^{25} \\ \boldsymbol{K}^{31} & \boldsymbol{K}^{32} & \boldsymbol{K}^{33} & \boldsymbol{K}^{34} & \boldsymbol{K}^{35} \\ \boldsymbol{K}^{41} & \boldsymbol{K}^{42} & \boldsymbol{K}^{43} & \boldsymbol{K}^{44} & \boldsymbol{K}^{45} \\ \boldsymbol{K}^{51} & \boldsymbol{K}^{52} & \boldsymbol{K}^{53} & \boldsymbol{K}^{54} & \boldsymbol{K}^{55} \end{bmatrix} \begin{Bmatrix} \boldsymbol{u} \\ \boldsymbol{v} \\ \boldsymbol{w} \\ \boldsymbol{s}^1 \\ \boldsymbol{s}^2 \end{Bmatrix} = \begin{Bmatrix} \boldsymbol{F}^1 \\ \boldsymbol{F}^2 \\ \boldsymbol{F}^3 \\ \boldsymbol{F}^4 \\ \boldsymbol{F}^5 \end{Bmatrix} \tag{3.129}$$

Each stiffness submatrix $\boldsymbol{K}^{\alpha\beta}$ in the above system is of order $N \times N$, while the force subvector $\boldsymbol{F}^{\alpha}$ is of dimension N. The components of the element stiffness submatrices are given by

$$K_{ij}^{11} = \int_{\Gamma_i^{\mathrm{CD}}} \frac{\partial \psi_j}{\partial \boldsymbol{x}} \cdot \begin{bmatrix} A_{11} & 0 \\ 0 & A_{66} \end{bmatrix}^{\mathrm{T}} \cdot \hat{\boldsymbol{n}}\, \mathrm{d}\Gamma\,, \quad K_{ij}^{12} = \int_{\Gamma_i^{\mathrm{CD}}} \frac{\partial \psi_j}{\partial \boldsymbol{x}} \cdot \begin{bmatrix} 0 & A_{12} \\ A_{66} & 0 \end{bmatrix}^{\mathrm{T}} \cdot \hat{\boldsymbol{n}}\, \mathrm{d}\Gamma$$

$$K_{ij}^{13} = \frac{1}{2} \int_{\Gamma_i^{\mathrm{CD}}} \frac{\partial \psi_j}{\partial \boldsymbol{x}} \cdot \begin{bmatrix} A_{11} \tilde{W}_x & A_{12} \tilde{W}_y \\ A_{66} \tilde{W}_y & A_{66} \tilde{W}_x \end{bmatrix}^{\mathrm{T}} \cdot \hat{\boldsymbol{n}}\, \mathrm{d}\Gamma$$

$$K_{ij}^{21} = \int_{\Gamma_i^{\mathrm{CD}}} \frac{\partial \psi_j}{\partial \boldsymbol{x}} \cdot \begin{bmatrix} 0 & A_{66} \\ A_{12} & 0 \end{bmatrix}^{\mathrm{T}} \cdot \hat{\boldsymbol{n}}\, \mathrm{d}\Gamma\,, \quad K_{ij}^{22} = \int_{\Gamma_i^{\mathrm{CD}}} \frac{\partial \psi_j}{\partial \boldsymbol{x}} \cdot \begin{bmatrix} A_{66} & 0 \\ 0 & A_{22} \end{bmatrix}^{\mathrm{T}} \cdot \hat{\boldsymbol{n}}\, \mathrm{d}\Gamma$$

$$K_{ij}^{23} = \frac{1}{2} \int_{\Gamma_i^{\mathrm{CD}}} \frac{\partial \psi_j}{\partial \boldsymbol{x}} \cdot \begin{bmatrix} A_{66} \tilde{W}_y & A_{66} \tilde{W}_x \\ A_{12} \tilde{W}_x & A_{22} \tilde{W}_y \end{bmatrix}^{\mathrm{T}} \cdot \hat{\boldsymbol{n}}\, \mathrm{d}\Gamma$$

$$K_{ij}^{31} = \int_{\Gamma_i^{\mathrm{CD}}} \frac{\partial \psi_j}{\partial \boldsymbol{x}} \cdot \begin{bmatrix} A_{11} \tilde{W}_x & A_{66} \tilde{W}_y \\ A_{12} \tilde{W}_y & A_{66} \tilde{W}_x \end{bmatrix}^{\mathrm{T}} \cdot \hat{\boldsymbol{n}}\, \mathrm{d}\Gamma$$

$$K_{ij}^{32} = \int_{\Gamma_i^{\mathrm{CD}}} \frac{\partial \psi_j}{\partial \boldsymbol{x}} \cdot \begin{bmatrix} A_{66} \tilde{W}_y & A_{12} \tilde{W}_x \\ A_{66} \tilde{W}_x & A_{22} \tilde{W}_y \end{bmatrix}^{\mathrm{T}} \cdot \hat{\boldsymbol{n}}\, \mathrm{d}\Gamma$$

$$K_{ij}^{33} = \int_{\Gamma_i^{\mathrm{CD}}} \frac{\partial \psi_j}{\partial \boldsymbol{x}} \cdot \begin{bmatrix} A_{55} + \frac{1}{2} A_{11} \tilde{W}_x^2 + \hat{\mathcal{A}}\, \tilde{W}_y^2 & 0 \\ 0 & A_{44} + \frac{1}{2} A_{22} \tilde{W}_y^2 + \hat{\mathcal{A}}\, \tilde{W}_x^2 \end{bmatrix}^{\mathrm{T}} \cdot \hat{\boldsymbol{n}}\, \mathrm{d}\Gamma$$

$$K_{ij}^{34} = \int_{\Gamma_i^{\mathrm{CD}}} \psi_j \begin{Bmatrix} A_{55} \\ 0 \end{Bmatrix}^{\mathrm{T}} \cdot \hat{\boldsymbol{n}}\, \mathrm{d}\Gamma\,, \qquad K_{ij}^{35} = \int_{\Gamma_i^{\mathrm{CD}}} \psi_j \begin{Bmatrix} 0 \\ A_{44} \end{Bmatrix}^{\mathrm{T}} \cdot \hat{\boldsymbol{n}}\, \mathrm{d}\Gamma$$

$$K_{ij}^{43} = -\iint\limits_{\Omega_i^{\mathrm{CD}}} A_{55} \frac{\partial \psi_j}{\partial x}\, \mathrm{d}\Omega$$

$$K_{ij}^{44} = \int\limits_{\Gamma_i^{\mathrm{CD}}} \frac{\partial \psi_j}{\partial \boldsymbol{x}} \cdot \begin{bmatrix} D_{11} & 0 \\ 0 & D_{66} \end{bmatrix}^{\mathrm{T}} \cdot \hat{\boldsymbol{n}}\, \mathrm{d}\Gamma - \iint\limits_{\Omega_i^{\mathrm{CD}}} A_{55}\, \psi_j\, \mathrm{d}\Omega$$

$$K_{ij}^{45} = \int\limits_{\Gamma_i^{\mathrm{CD}}} \frac{\partial \psi_j}{\partial \boldsymbol{x}} \cdot \begin{bmatrix} 0 & D_{12} \\ D_{66} & 0 \end{bmatrix}^{\mathrm{T}} \cdot \hat{\boldsymbol{n}}\, \mathrm{d}\Gamma$$

$$K_{ij}^{53} = -\iint\limits_{\Omega_i^{\mathrm{CD}}} A_{44} \frac{\partial \psi_j}{\partial y}\, \mathrm{d}\Omega\,, \quad K_{ij}^{54} = \int\limits_{\Gamma_i^{\mathrm{CD}}} \frac{\partial \psi_j}{\partial \boldsymbol{x}} \cdot \begin{bmatrix} 0 & D_{66} \\ D_{12} & 0 \end{bmatrix}^{\mathrm{T}} \cdot \hat{\boldsymbol{n}}\, \mathrm{d}\Gamma$$

$$K_{ij}^{55} = \int\limits_{\Gamma_i^{\mathrm{CD}}} \frac{\partial \psi_j}{\partial \boldsymbol{x}} \cdot \begin{bmatrix} D_{66} & 0 \\ 0 & D_{22} \end{bmatrix}^{\mathrm{T}} \cdot \hat{\boldsymbol{n}}\, \mathrm{d}\Gamma - \iint\limits_{\Omega_i^{\mathrm{CD}}} A_{44}\, \psi_j\, \mathrm{d}\Omega \tag{3.130}$$

The force vector components are given by

$$F_i^1 = -\iint\limits_{\Omega_i^{\mathrm{CD}}} f_x\, \mathrm{d}\Omega\,, \quad F_i^2 = -\iint\limits_{\Omega_i^{\mathrm{CD}}} f_y\, \mathrm{d}\Omega\,, \quad F_i^3 = -\iint\limits_{\Omega_i^{\mathrm{CD}}} q\, \mathrm{d}\Omega\,, \tag{3.131}$$

$$F_i^4 = F_i^5 = 0$$

3.4.2.5 Tangent Matrix Coefficients

Recall Eq. (3.97) to compute the tangent matrix coefficients for a nonlinear discrete model with multiple degrees of freedom per node (repeated below for ease of reference):

$$T_{ij}^{\alpha\beta} = K_{ij}^{\alpha\beta} + \hat{K}_{ij}^{\alpha\beta} = K_{ij}^{\alpha\beta} + \sum_{\gamma=1}^{\mathcal{P}} \sum_{m=1}^{N} \frac{\partial K_{im}^{\alpha\gamma}}{\partial \Delta_j^{\beta}} \Delta_m^{\gamma} \quad \begin{cases} i, j = 1, \ldots, N \\ \alpha, \beta = 1, \ldots, \mathcal{P} \end{cases}$$

where $\Delta_j^1 = u_j$, $\Delta_j^2 = v_j$, $\Delta_j^3 = w_j$, $\Delta_j^4 = s_j^1$, and $\Delta_j^5 = s_j^2$. Since the nonlinear terms depend only on w_j, it follows that $\boldsymbol{T}^{\alpha 1} = \boldsymbol{K}^{\alpha 1}$, $\boldsymbol{T}^{\alpha 2} = \boldsymbol{K}^{\alpha 2}$, $\boldsymbol{T}^{\alpha 4} = \boldsymbol{K}^{\alpha 4}$, and $\boldsymbol{T}^{\alpha 5} = \boldsymbol{K}^{\alpha 5}$ for all values of α. The tangent coefficients $\boldsymbol{T}^{\alpha 3}$, which do receive contributions from the second term in Eq. (3.97), are calculated as explained next.

It is helpful to pre-compute the following in advance

$$\frac{\partial \tilde{W}_x}{\partial w_j} = \frac{\partial}{\partial w_j}\left(\frac{\partial w}{\partial x}\right) = \frac{\partial}{\partial w_j}\left(\sum_{k=1}^{p+1} w_k \frac{\partial \psi_k}{\partial x}\right) = \sum_{k=1}^{p+1} \frac{\partial w_k}{\partial w_j}\frac{\partial \psi_k}{\partial x} = \sum_{k=1}^{p+1} \delta_{kj}\frac{\partial \psi_k}{\partial x} = \frac{\partial \psi_j}{\partial x}$$

Similarly,

$$\frac{\partial \tilde{W}_y}{\partial w_j} = \frac{\partial \psi_j}{\partial y}$$

Term 1: $\alpha = 1, \beta = 3$

$$\begin{aligned}
\hat{K}_{ij}^{13} &= \sum_m \left(\frac{\partial K_{im}^{11}}{\partial w_j} u_m + \frac{\partial K_{im}^{12}}{\partial w_j} v_m + \frac{\partial K_{im}^{13}}{\partial w_j} w_m + \frac{\partial K_{im}^{14}}{\partial w_j} s_m^1 + \frac{\partial K_{im}^{15}}{\partial w_j} s_m^2\right) \\
&= \sum_m \frac{\partial K_{im}^{13}}{\partial w_j} w_m = \sum_m \frac{\partial}{\partial w_j}\left\{\int_{\Gamma_i^{\mathrm{CD}}} \frac{w_m}{2}\frac{\partial \psi_m}{\partial \boldsymbol{x}} \cdot \begin{bmatrix} A_{11}\tilde{W}_x & A_{12}\tilde{W}_y \\ A_{66}\tilde{W}_y & A_{66}\tilde{W}_x \end{bmatrix}^{\mathrm{T}} \cdot \hat{\boldsymbol{n}}\, \mathrm{d}\Gamma\right\} \\
&= \int_{\Gamma_i^{\mathrm{CD}}} \frac{1}{2}\begin{Bmatrix} \tilde{W}_x \\ \tilde{W}_y \end{Bmatrix} \cdot \begin{bmatrix} A_{11}\,\psi_{j,x} & A_{12}\,\psi_{j,y} \\ A_{66}\,\psi_{j,y} & A_{66}\,\psi_{j,x} \end{bmatrix}^{\mathrm{T}} \cdot \hat{\boldsymbol{n}}\, \mathrm{d}\Gamma
\end{aligned} \tag{3.132}$$

Term 2: $\alpha = 2, \beta = 3$

$$\begin{aligned}
\hat{K}_{ij}^{23} &= \sum_m \left(\frac{\partial K_{im}^{21}}{\partial w_j} u_m + \frac{\partial K_{im}^{22}}{\partial w_j} v_m + \frac{\partial K_{im}^{23}}{\partial w_j} w_m + \frac{\partial K_{im}^{24}}{\partial w_j} s_m^1 + \frac{\partial K_{im}^{25}}{\partial w_j} s_m^2\right) \\
&= \sum_m \frac{\partial K_{im}^{23}}{\partial w_j} w_m = \sum_m \frac{\partial}{\partial w_j}\left\{\int_{\Gamma_i^{\mathrm{CD}}} \frac{w_m}{2}\frac{\partial \psi_m}{\partial \boldsymbol{x}} \cdot \begin{bmatrix} A_{66}\tilde{W}_y & A_{66}\tilde{W}_x \\ A_{12}\tilde{W}_x & A_{22}\tilde{W}_y \end{bmatrix}^{\mathrm{T}} \cdot \hat{\boldsymbol{n}}\, \mathrm{d}\Gamma\right\} \\
&= \int_{\Gamma_i^{\mathrm{CD}}} \frac{1}{2}\begin{Bmatrix} \tilde{W}_x \\ \tilde{W}_y \end{Bmatrix} \cdot \begin{bmatrix} A_{66}\,\psi_{j,y} & A_{66}\,\psi_{j,x} \\ A_{12}\,\psi_{j,x} & A_{22}\,\psi_{j,y} \end{bmatrix}^{\mathrm{T}} \cdot \hat{\boldsymbol{n}}\, \mathrm{d}\Gamma
\end{aligned} \tag{3.133}$$

Term 3: $\alpha = 3, \beta = 3$

$$\begin{aligned}
\hat{K}_{ij}^{33} &= \sum_m \left(\frac{\partial K_{im}^{31}}{\partial w_j} u_m + \frac{\partial K_{im}^{32}}{\partial w_j} v_m + \frac{\partial K_{im}^{33}}{\partial w_j} w_m + \frac{\partial K_{im}^{34}}{\partial w_j} s_m^1 + \frac{\partial K_{im}^{35}}{\partial w_j} s_m^2\right) \\
&= \sum_m \left(\frac{\partial K_{im}^{31}}{\partial w_j} u_m + \frac{\partial K_{im}^{32}}{\partial w_j} v_m + \frac{\partial K_{im}^{33}}{\partial w_j} w_m\right)
\end{aligned}$$

$$
\begin{aligned}
&= \sum_m \int_{\Gamma_i^{\rm CD}} \left\{ u_m \frac{\partial \psi_m}{\partial \boldsymbol{x}} \cdot \frac{\partial}{\partial w_j} \begin{bmatrix} A_{11}\tilde{W}_x & A_{66}\tilde{W}_y \\ A_{12}\tilde{W}_y & A_{66}\tilde{W}_x \end{bmatrix}^{\rm T} \right. \\
&\qquad + v_m \frac{\partial \psi_m}{\partial \boldsymbol{x}} \cdot \frac{\partial}{\partial w_j} \begin{bmatrix} A_{66}\tilde{W}_y & A_{12}\tilde{W}_x \\ A_{66}\tilde{W}_x & A_{22}\tilde{W}_y \end{bmatrix}^{\rm T} \\
&+ \left. w_m \frac{\partial \psi_m}{\partial \boldsymbol{x}} \cdot \frac{\partial}{\partial w_j} \begin{bmatrix} \frac{1}{2}A_{11}\tilde{W}_x^2 + \hat{\mathcal{A}}\,\tilde{W}_y^2 \ldots & 0 \\ 0 & \frac{1}{2}A_{22}\tilde{W}_y^2 + \hat{\mathcal{A}}\,\tilde{W}_x^2 \ldots \end{bmatrix}^{\rm T} \right\} \cdot \hat{\boldsymbol{n}}\, d\Gamma \\
&= \int_{\Gamma_i^{\rm CD}} \left\{ \begin{Bmatrix} \tilde{U}_x \\ \tilde{U}_y \end{Bmatrix} \cdot \begin{bmatrix} A_{11}\psi_{j,x} & A_{66}\psi_{j,y} \\ A_{12}\psi_{j,y} & A_{66}\psi_{j,x} \end{bmatrix}^{\rm T} + \begin{Bmatrix} \tilde{V}_x \\ \tilde{V}_y \end{Bmatrix} \cdot \begin{bmatrix} A_{66}\psi_{j,y} & A_{12}\psi_{j,x} \\ A_{66}\psi_{j,x} & A_{22}\psi_{j,y} \end{bmatrix}^{\rm T} \right. \\
&+ \left. \begin{Bmatrix} \tilde{W}_x \\ \tilde{W}_y \end{Bmatrix} \cdot \begin{bmatrix} A_{11}\tilde{W}_x\psi_{j,x} + 2\hat{\mathcal{A}}\,\tilde{W}_y\psi_{j,y} & 0 \\ 0 & A_{22}\tilde{W}_y\psi_{j,y} + 2\hat{\mathcal{A}}\,\tilde{W}_x\psi_{j,x} \end{bmatrix}^{\rm T} \right\} \cdot \hat{\boldsymbol{n}}\, d\Gamma
\end{aligned} \tag{3.134}
$$

3.4.2.6 Numerical Example

Consider the nonlinear bending analysis of a clamped, isotropic plate ($E = 10^6$ and $\nu = 0.3$) of radius $a = 100$ in and thickness $H = 10$ in, subjected to a uniformly distributed transverse load of intensity $q_0 = 1.2 \times 10^4$ lb/in^2. Analyze the problem using only one quadrant of the plate as the computational domain (exploiting symmetry). Plot the center deflection as a function of the load. Solve using incremental loading with a load step of $\Delta q_0 = 250$ lb/in^2. For each load step, use Newton's method with a convergence tolerance of $\varepsilon_{\rm tol} = 10^{-3}$.

Solution. Figure 3.23(a) shows the computational domain and boundary conditions for this problem. Symmetry boundary conditions are applied along the x- and y-axes, while the circular edge is clamped. In Fig. 3.23(b), we show a sample mesh of three 6th-order elements used to analyze the quarter-plate model. This mesh was generated using gmsh [2], a powerful open-source 3D finite element mesh generator. The mesh files (containing the nodal coordinates, element connectivity, and boundary condition data) are provided in the GitHub repository referenced in Subsect. 3.3.5. The problem is solved using the dual-mesh control-domain plate analysis program, and the input deck for a sample nonlinear run is presented below:

```
# Mesh file path
MESH_FILE: Problem_02\Mesh_P6

# Plate stiffness values (a/H=100)
A11: 1.098901e+07
A22: 1.098901e+07
A12: 3.296703e+06
```

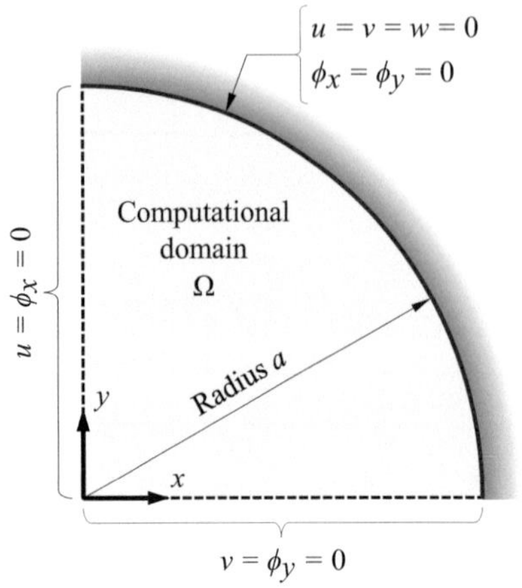

(a) Quarter-plate domain with boundary conditions

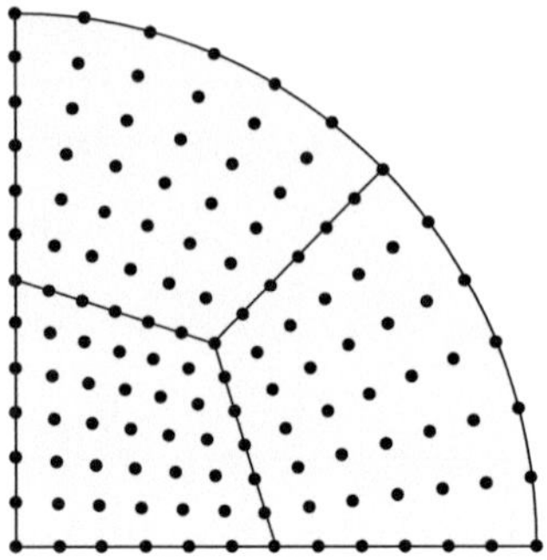

(b) Mesh of three 6th-order elements (49 nodes each)

Fig. 3.23: Computational domain and mesh for the circular clamped plate analysis.

```
A44: 3.205128e+06
A55: 3.205128e+06
A66: 3.846154e+06
D11: 9.157509e+07
D22: 9.157509e+07
D12: 2.747253e+07
D66: 3.205128e+07

# Distributed load
FX: 0.0000
FY: 0.0000
QZ: 1.0000

# Integration points
NGP_PATH:  2 # for line integrals
NGP_PATCH: 2 # for area integrals

# Nonlinear analysis options
NONLIN:  2   # Newton's method
ITERMAX: 25
BETA:    0.0
EPSILON: 1.0e-03
NLS:     48
LOAD_INCREMENTS: [250,250,250,250,250,250,250,250,
                  250,250,250,250 ...(48 times)]
```

Table 3.3 shows the results of the analysis for four different meshes. Each mesh has the same number and placement of elements as the one shown in Fig. 3.23, but the element order is increased (this type of refinement is referred to as p-refinement

Table 3.3: Center deflection $w(0, 0)$ (in inches) obtained from a linear analysis with $q_0 = 1$ lb/in^2.

a/H	Mesh A [1]	Mesh B [2]	Mesh C [3]	Exact Solution
10	0.018021	0.017842	0.017843	0.017843
100	10.990924	17.123955	17.070338	17.070300

[1] Mesh of three $p = 2$ elements
[2] Mesh of three $p = 4$ elements
[3] Mesh of three $p = 6$ elements

in finite element terminology). The center displacement $w(0, 0)$ is plotted against a dimensionless load parameter $P = q_0 a^4/Eh^4$ in Fig. 3.24, and the axial stiffening effect is clearly visible. A few points on this curve are tabulated in Table 3.4 for the reader to verify. Higher order elements accurately capture curved geometries, such as the one in the present problem, unlike bilinear elements, which are quadrilaterals with straight edges. As a result, bilinear meshes approximate curved boundaries only in a piecewise linear manner, whereas higher order elements provide a much more faithful representation of the actual domain.

3.5 Closing Remarks

In this section, we have outlined the theory and implementation details of the dual mesh control domain method (DMCDM) for one- and two-dimensional problems. The developments show that DMCDM admits a modular structure, and many of the standard procedures used in the finite element method such as assembly, enforce-

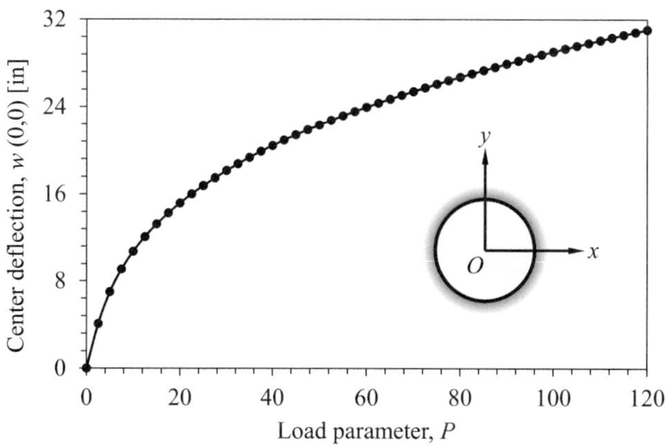

Fig. 3.24: Plot of center deflection $w(0, 0)$ versus load parameter P for the clamped circular plate.

Table 3.4: Center deflection $w(0,0)$ (in inches) for various levels of the load parameter P for the plate with $a/H = 10$, using Mesh 3 (three 6th-order elements).

Load parameter, P	Center deflection (in inches)
5	6.9857
25	16.7691
50	22.3239
70	25.4202
95	28.5033
115	30.5798
120	31.0590

ment of boundary conditions, and nonlinear solution strategies can be used without significant modification. In practice, an existing finite element code can be adapted to DMCDM by replacing only the routine that evaluates the element stiffness matrix and force vector.

We also illustrated how the method extends naturally to higher order elements through vectorization and the rail system approach. Higher order elements help mitigate locking and allow curved geometries to be represented more accurately. Although our examples focused on functionally graded beams and plates, the method has already been applied to linear elasticity [5], finite deformation with material nonlinearity [1], buckling and free vibration of plates [6], and bending and free vibration of doubly curved shells [7]. The approach is suitable for time dependent problems as well, and readers are referred to [3] for further details.

We hope the accompanying code and examples will encourage readers to explore the method further and adapt it to their own applications.

Acknowledgements The authors dedicate this book chapter to the fond memory of Professor Cristóvão Mota Soares and his contributions to the field of composite materials and structures. The first and second authors gratefully acknowledge the support of the O'Donnell Foundation Chair IV, while the last author acknowledges the support of the J. N. Reddy Chair in Applied Mechanics.

Appendix A

In the following tables, we list the nonzero entries of the correspondence (or path-to-node mapping) matrices. Since all nonzero values are either +1 or −1, we simply specify the corresponding (row, column) pairs for each case. This listing is provided only for illustration and completeness. We do not recommend that readers manually copy these entries into their own code. Instead, the programs provided at the repository referenced in Subsect. 3.3.5 include a preprocessor function that computes these matrices for any arbitrary order p and passes them internally to the element subroutine without requiring any user intervention.

1. Bilinear element ($p = 1, N_p = 2$) see Tables 3.5 and 3.6

Table 3.5: Nonzero entries of $\boldsymbol{H}_{\mathrm{H}}$ for $p = 1$ element.

Entries with value +1	(3, 1) (4, 2)
Entries with value −1	(1, 1) (2, 2)

Table 3.6: Nonzero entries of $\boldsymbol{H}_{\mathrm{V}}$ for $p = 1$ element.

Entries with value +1	(1, 1) (3, 2)
Entries with value −1	(2, 1) (4, 2)

2. Biquadratic element ($p = 2, N_p = 6$) see Tables 3.7 and 3.8

Table 3.7: Nonzero entries of $\boldsymbol{H}_{\mathrm{H}}$ for $p = 2$ element.

Entries with value +1	(4, 1) (5, 2) (6, 3) (7, 4) (8, 5) (9, 6)
Entries with value −1	(1, 1) (2, 2) (3, 3) (4, 4) (5, 5) (6, 6)

Table 3.8: Nonzero entries of $\boldsymbol{H}_{\mathrm{V}}$ for $p = 2$ element.

Entries with value +1	(1, 1) (4, 2) (7, 3) (2, 4) (5, 5) (8, 6)
Entries with value −1	(2, 1) (5, 2) (8, 3) (3, 4) (6, 5) (9, 6)

3. Bicubic element ($p = 3, N_p = 12$) see Tables 3.9 and 3.10

Appendix B

The beam stiffnesses A_{xx}, B_{xx}, D_{xx}, and S_{xx} for the functionally graded beam of rectangular cross-section described in Subsect. 3.4.1 are given by [12]

Table 3.9: Nonzero entries of $\boldsymbol{H}_{\mathrm{H}}$ for $p = 3$ element.

Entries with value +1	(5, 1)	(6, 2)	(7, 3)	(8, 4)
	(9, 5)	(10, 6)	(11, 7)	(12, 8)
	(13, 9)	(14, 10)	(15, 11)	(16, 12)
Entries with value −1	(1, 1)	(2, 2)	(3, 3)	(4, 4)
	(5, 5)	(6, 6)	(7, 7)	(8, 8)
	(9, 9)	(10, 10)	(11, 11)	(12, 12)

Table 3.10: Nonzero entries of $\boldsymbol{H}_{\mathrm{V}}$ for $p = 3$ element.

Entries with value +1	(1, 1)	(5, 2)	(9, 3)	(13, 4)
	(2, 5)	(6, 6)	(10, 7)	(14, 8)
	(3, 9)	(7, 10)	(11, 11)	(15, 12)
Entries with value −1	(2, 1)	(6, 2)	(10, 3)	(14, 4)
	(3, 5)	(7, 6)	(11, 7)	(15, 8)
	(4, 9)	(8, 10)	(12, 11)	(16, 12)

$$
\begin{aligned}
A_{xx} &= E_2 A_0 \frac{M+n}{1+n}, \quad B_{xx} = E_2 B_0 \frac{n(M-1)}{2(1+n)(2+n)} \\
D_{xx} &= E_2 I_0 \frac{\left(6+3n+3n^2\right) M + \left(8n+3n^2+n^3\right)}{6+11n+6n^2+n^3}, \quad S_{xx} = \frac{K_s E_2 A_0}{2\,(1+\nu)} \frac{M+n}{1+n}
\end{aligned}
\tag{3.135}
$$

where $A_0 = bH$ is the area of cross section, $B_0 = bH^2$, $I_0 = bH^3/12$ is the second moment of area, K_s is the shear correction factor, and M is the modulus ratio $M = E_1/E_2$. For beams of rectangular cross section, the shear correction factor is equal to $K_s = 5/6$.

Appendix C

The stiffness coefficients (A_{ij}, Q_{ij}) where $i, j = 1, 2, 6$ of an orthotropic linearly elastic plate of thickness H are given by

$$
\begin{aligned}
A_{ij} &= Q_{ij}\, H, \quad D_{ij} = Q_{ij} \frac{H^3}{12}, \quad A_{44} = K_s G_{23} H, \quad A_{55} = K_s G_{13} H \\
Q_{11} &= \frac{E_1}{1-\nu_{12}\nu_{21}}, \quad Q_{22} = Q_{11} \frac{E_2}{E_1}, \quad Q_{12} = \nu_{12}\, Q_{22}, \quad Q_{66} = G_{12}
\end{aligned}
\tag{3.136}
$$

where (E_1, E_2) are the principal moduli, G_{12} is the in-plane shear modulus and (G_{13}, G_{23}) are the transverse shear moduli.

References

[1] Areias P, Srinivasa A, Moleiro F, Reddy JN (2025) Finite strain analysis with the dual mesh control domain method. International Journal for Numerical Methods in Engineering 126(1):e7654, DOI 10.1002/nme.7654
[2] Geuzaine C, Remacle JF (2009) Gmsh: A 3-D finite element mesh generator with built-in pre-and post-processing facilities. International Jjournal for Numerical Methods in Engineering 79(11):1309–1331, DOI 10.1002/nme.2579
[3] Heblekar TY, Reddy JN (2025) An improved dual mesh control domain formulation for the unsteady flow of viscous incompressible fluids. Physics of Fluids 37(3), DOI 10.1063/5.0259696
[4] Heblekar TY, Reddy JN, Srinivasa AR (2024) Analysis of nonlinear problems using the Dual Mesh Control Domain Method with arbitrary meshes. Computer Methods in Applied Mechanics and Engineering 427:117044, DOI 10.1016/j.cma.2024.117044
[5] Jiao Z, Heblekar T, Wang G, Xu R, Chen W, Reddy JN (2023) Analysis of plane elasticity problems using the dual mesh control domain method. Computer Methods in Applied Mechanics and Engineering 416:116342, DOI 10.1016/j.cma.2023.116342
[6] Jiao Z, Heblekar T, Wang G, Xu R, Reddy JN (2024) Static, free vibration and buckling analysis of functionally graded plates using the dual meshcontrol domain method. Composite Structures 305:107575, DOI 10.1016/j.compstruc.2024.107575
[7] Jiao Z, Zhao S, Wang G, Xu R, Yang J (2025) Bending and vibration analyses of functionally graded auxetic doubly curved shells via dual mesh control domain model. International Journal of Solids and Structures p 113526, DOI 10.1016/j.ijsolstr.2025.113526
[8] Kamermans M Legendre–Gauss quadrature. `https://pomax.github.io/bezierinfo/legendre-gauss.html`
[9] Nampally P, Ruocco E, Reddy JN (2021) Bending analysis of functionally graded rectangular plates using the dual mesh control domain method. International Journal for Computational Methods in Engineering Science and Mechanics 22(5):425–437, DOI 10.1080/15502287.2021.1890279
[10] Reddy JN (2006) Theory and Analysis of Elastic Plates and Shells. CRC press
[11] Reddy JN (2013) An Introduction to Continuum Mechanics. Cambridge university press
[12] Reddy JN (2014) An Introduction to Nonlinear Finite Element Analysis: with Applications to Heat Transfer, Fluid Mechanics, and Solid Mechanics, 2nd edn. OUP Oxford
[13] Reddy JN (2019) A dual mesh finite domain method for the numerical solution of differential equations. International Journal for Computational Methods in Engineering Science and Mechanics 20(3):212–228, DOI 10.1080/15502287.2019.1610987
[14] Reddy JN (2019) An Introduction to the Finite Element Method, 4th edn. McGraw-Hill New York

[15] Reddy JN (2022) Theories and Analyses of Beams and Axisymmetric Circular Plates. CRC Press
[16] Reddy JN (2024) Computational Methods in Engineering: Finite Difference, Finite Volume, Finite Element, and Dual Mesh Control Domain Methods. CRC Press
[17] Reddy JN (2024) Computational Methods in Engineering: Finite Difference, Finite Volume, Finite Element, and Dual Mesh Control Domain Methods. CRC Press
[18] Reddy JN, Nampally P (2020) A dual mesh finite domain method for the analysis of functionally graded beams. Composite structures 251:112648, DOI 10.1016/j.compstruct.2020.112648
[19] Reddy JN, Nampally P, Srinivasa AR (2020) Nonlinear analysis of functionally graded beams using the dual mesh finite domain method and the finite element method. International Journal of Non-Linear Mechanics 127:103575, DOI 10.1016/j.ijnonlinmec.2020.103575
[20] Reddy JN, Nampally P, Phan N (2021) Dual mesh control domain analysis of functionally graded circular plates accounting for moderate rotations. Composite Structures 257:113153, DOI 10.1016/j.compstruct.2020.113153

Chapter 4
Mechanism-based Analysis and Modelling of Delamination Damage Growth in Composite Laminates

Shuxin Li, Qingfeng Duan, Dongfeng Cao, and Haixiao Hu

Abstract Delamination is one of the most common and dangerous failure modes of composite laminates widely used for engineering components. Two distinguishing delamination mechanisms associated with intrinsic damage and extrinsic fiber bridging within the two delamination zones ahead and behind delamination crack tip respectively for both unidirectional (UD) and multidirectional composite laminates are identified. A new mechanism-based cohesive zone model (MB-CZM) with two constitutive laws related to the two distinctive delamination mechanisms is proposed and validated. The application limitations of the current empirical analysis and modelling methods are clarified and a mechanism-based modification of the fracture toughness calculation method is developed and verified with the MB-CZM and numerical simulation of the experimental results. The advantage of mechanisms-based analysis and modelling methodology over the current empirical or semi-empirical analysis and modeling approaches are highlighted and further applications are discussed. The good agreement of the analysis and simulation results with the experimental results provided the verification and demonstration of the capabilities of the proposed mechanism-based analysis and modelling methods. Therefore, the proposed methods can be extended for investigations of delamination initiation and propagation of composite laminates under complex loading conditions such as cyclic and fatigue loading.

Key words: Delamination, Cohesive zone model, Fracture toughness, Simulations, Composite laminates

---

Shuxin Li (Corresponding Author) · Qingfeng Duan · Dongfeng Cao
State Key Laboratory of Materials Synthesis and Processing, Wuhan University of Technology, Wuhan 430070, P. R. China, e-mail: lishuxin@whut.edu.cn, absinthe1223@163.com, cao_dongf@whut.edu.cn

Haixiao Hu
Hubei Key Laboratory of Theory and Application of Advanced Materials Mechanics, School of Science, Wuhan University of Technology, Wuhan 430070, P. R. China, e-mail: hhx1987@whut.edu.cn

H. Altenbach et al. (eds.), *Composite Structures and Technology*,
Advanced Structured Materials 255,
https://doi.org/10.1007/978-3-032-24656-1_4

4.1 Introduction

Carbon fiber reinforced polymer (CFRP) composite laminates are widely used in engineering applications such as in aerospace, transportation, wind power, ocean engineering and infrastructure fields [9, 12]. This is attributed to the CFRP advantages such as high strength/stiffness-to-weight ratio and extensive design optimization spaces with different stacking sequence and fiber orientations [15, 41]. However, the damage and failure mechanisms of composite laminates under complex engineering operation conditions are significantly different compared with metallic structures due to the highly intrinsic anisotropic and inhomogeneous feature of composite laminates. Delamination is one of the most common and dangerous failure modes due to lack of reinforcement in the thickness direction of the ply-by-ply formulation of composite laminates used for engineering components [42]. The significant losses of stiffness and strength induced by delamination growth could have dramatically detrimental effect on the structural integrity and operational safety level. Consequently, delamination initiation and propagation under complex engineering operation conditions gains significant attention during the design and analysis of composite structures, especially for low velocity impact (LVI) with compression after impact (CAI) as one of the most critical design criterion and certification requirements [55].

Even with extensive studies in both academic fundamental science research and industrial application development [23, 32, 43, 44, 58] in the past decades, it is still not able to predict delamination damage initiation and propagation in composite laminates under engineering applications to the similar level as damage tolerance analysis of metallic structure based on fracture mechanics. In line with the FAA and EASA airworthiness regulations [29], the conservative damage 'no-growth' criterion is currently used in the design and certification of commercial aircraft such as Boeing 787 and Airbus A350WXB with over 50% of composite structures. Experimental results demonstrated that propagation of delamination damage under fatigue loading can occur far below the static fracture loads [4, 6]. Consequently, engineering composite structures designed with the 'no-growth' criterion often carry additional weight to mitigate the risk of structures failure from manufacture and/or service-induced delamination damages such as introduced by LVI. To fully take advantage of the lightweight characteristics of composite materials and to reduce the cost burden induced by the current design practices, it is vital to gain further comprehensive understanding of delamination initiation and propagation mechanisms. Then mechanism-based delamination analysis methods can be developed to ensure the safety operation of composite structures during service without excessive testing to demonstrate damage 'no-growth'.

In engineering applications such as in aerospace industry, structural design requires testing for the generation of material data and design allowable such as strength and fracture toughness for sizing the structure and for validation of designed composite components to meet safety regulations. For composite laminates, fracture toughness is one of the most critical mechanical properties for engineering applications and tests are performed to quantify the resistance of composite laminate to failure by delamination. Recent years, applications of the current ASTM D5528 method [2] have been extended in two aspects. One is extended for delamination propagation and another one is for multidirectional (delamination between plies of dissimilar orientation) composite laminates. However, there is no comprehensive investigation on the applicability of the current ASTM method for the extended cases.

Compared with the experimental approach, numerical simulations using finite element modelling are less expensive and the validated numerical modelling can be used for wide range of composite laminates to provide comprehensive understanding of the delamination damage mechanisms. The FAA and EASA certification regulations for the commercial aircraft composite structures such as B787 are based on analysis supported by testing [1]. Consequently, significant efforts have been devoted to the development of modelling and numerical simulations of the delamination failure behavior of laminate composites. Thus, development of so-called 'double build blocks' approach has been gained great interests in the last decades [7, 29], where 'pyramid' modelling paired with the 'pyramid' testing is developed to reduce the testing requirements. The 'analysis supported by testing' approach defined in FAA and EASA airworthiness regulation for the design certification of commercial aircraft composite structures highlights the importance of modelling development. It is because modelling methods are not only less expensive compared with testing, but also more importantly could be used for enabling prediction of the behavior of an arbitrary lay-up stacking sequences of composite laminates based on the intrinsic deformation and damage mechanism and a limited set of basic material input data. For example, the classical composite laminate modelling approach can predict the stiffness of any lay-up stacking sequences based on the fiber 0° and transverse 90° direction material data.

In terms of modelling approaches, the cohesive zone model (CZM) [48] has been widely used for the delamination initiation and propagation of the composite laminates under various loading conditions [13, 14, 27, 28]. It is well known that successful modelling depends on both selection of appropriate constitutive equations and determination of the material properties and parameters in the modelling. For the CZM approach, the key factors are the selection of the traction-separation laws and determination of the strength and fracture toughness of the interlaminar interface represented by the cohesive elements. In this aspect, there are still many issues even for the fundamental Mode I delamination characterization with the double cantilever beam (DCB) specimen.

It was recognized from the very beginning in the development of ASTM D5528 that it is not possible to obtain a single-valued measurement of Mode I fracture toughness for composite laminate with DCB test. The experimental observation of the delamination resistance curve (R-curve) phenomenon from DCB test is attributed to the additional fiber bridging effect [25, 28, 49]. This fiber bridging mechanism results from growing the delamination between two 0° unidirectional plies and is considered in ASTM D5528 to be an artifact of the DCB test on unidirectional (UD) materials [2]. The key factor influencing the extension of the current ASTM method to delamination propagation and multidirectional composite laminates is also the fiber bridging phenomenon. Consequently, extensive investigations, both in experimental and modelling approaches, have been conducted on the characterization and qualification of the fiber bridging phenomenon observed in the DCB testing [19, 25, 33].

The experimental delamination R-curve is a direct reflection of the influence of fiber bridging as additional toughening mechanism. Furthermore, the experimental results also indicated that the shape of the R-curve is related to the geometry of the test piece, especially the thickness and the layup of the test piece [19, 36]. The use of R-curve can only describe the increase of delamination fracture toughness of composite materials from a macroscopic perspective and cannot effectively characterize the mechanical properties of composite laminate from the aspect of fiber bridging mechanism. Therefore, more and more researchers have begun to study the bridging stress distribution in the delamination expansion area to describe the variation of the interlaminar fracture characteristics of composite materials with the delamination propagation [44, 50]. Based on the experimental results, various fiber bridging stress models are proposed to describe the stress distribution of fiber bridging and its effect on Mode I delamination fracture toughness [43, 44, 47, 50–52, 58]. The results show that the non-linear exponential bridging stress distribution model has the best calculation accuracy and convergence among these three bridging models [47, 50–52].

The traditional cohesive zone model (CZM) with bilinear traction-separation law was developed and widely used in the numerical simulations for delamination initiation and growth under the conditions of no fiber bridging [55]. Consequently, various modifications for the traction-separation law are proposed to consider the additional fiber bridging effect in the Mode I delamination of DCB test, including three-linear [14, 38, 50, 54], four-linear [58] and multi-linear [44] laws. A significant and important common feature of the previous proposed CZMs had been based on different level of linear approximation of the fiber bridging characteristics. However, extensive experimental results and analysis investigations clearly indicates that the real fiber bridging stress distribution exhibits strongly nonlinear feature [20, 22, 30, 38, 44, 47, 50, 51]. Therefore, there are nature limitations in the previous CZMs to fully represent the nonlinear characteristics of fiber bridging observed in the Mode I delamination with DCB specimen.

One of the fundamental material property and parameter inputs of these CZMs is the fracture toughness, e.g., the R-curves, calculated by the current ASTM method. It is well known that successful numerical simulations depend on both selection of appropriate constitutive equations in modelling and the material properties and parameters associated with the modelling. Since inputs of the R-curves are not real experimental data but calculated from the load-displacement experimental data by a given method, such as ASTM D5528 standard [2], the applicability of the calculation method will determine the reliability and representative of the calculated R-curves. Therefore, any errors introduced by the fracture toughness calculation method will influence the accuracy of the CZM modelling ability to predict the experimental results such as the load-displacement curves, which is a common observation in the literatures [14, 16, 43, 44, 58]. Consequently, it is vital to remove any limitations in the fracture toughness calculation method to enable successful modelling and simulations.

The objectives of this study are identification of the intrinsic and extrinsic mechanisms of delamination, development of mechanism-based modelling and mechanism-based fracture toughness calculation method. Firstly, two distinguishing delamination mechanisms associated with intrinsic damage and extrinsic fiber bridging within the two delamination zones ahead and behind delamination crack tip respectively, for both the UD and multidirectional composite laminates are identified. Consequently, a new mechanism-based cohesive zone model (MB-CZM) with two traction-separation relations to individually represent the two distinctive delamination mechanisms is proposed and validated. Then, a mechanism-based modification of the fracture toughness calculation method is developed and verified with the MB-CZM and numerical simulation of the experimental results. Finally, the advantage of mechanisms-based analysis and modelling methodology over the previous empirical or semi-empirical analysis and modeling approaches are highlighted and further applications are discussed.

4.2 Intrinsic and Extrinsic Mechanisms of Delamination Growth

There have been nearly 30 years extensive investigations on Mode I delamination of continuous fiber-reinforced composite materials using DCB specimen with ASTM D5528 standard. It is valuable to identify the discrepancies between the previous empirical studies and the mechanisms of delamination.

4.2.1 Delamination R-Curve Feature and Toughening Mechanisms

The ASTM D5528 test method which was originally approved in 1994 was initially published for determination of the opening Mode I interlaminar fracture toughness, G_{IC}, of continuous fiber-reinforced composite materials using the double cantilever beam (DCB) specimen (Fig. 4.1). Fracture toughness is classified as intrinsic material properties and a critical parameter in the theoretical analysis and numerical modelling (e.g., CMD and CZM) for engineering applications.

From the very beginning in the development of ASTM D5528, it was recognized that it is not possible to obtain a single-valued measure of Mode I fracture toughness K_{IC} for composite laminate with DCB specimen testing. The experimental observation of the delamination resistance curve (R-curve) phenomenon, as shown in Fig. 4.1b, is attributed to the additional fiber bridging effect. Consequently, the generic significance of G_{IC} propagation values calculated beyond the end of the implanted insert is questionable and an initiation value of G_{IC} measured from the implanted insert is preferred in ASTM D5528 as the material toughness of the composite materials. Therefore, the fiber bridging phenomenon is not new but rec-

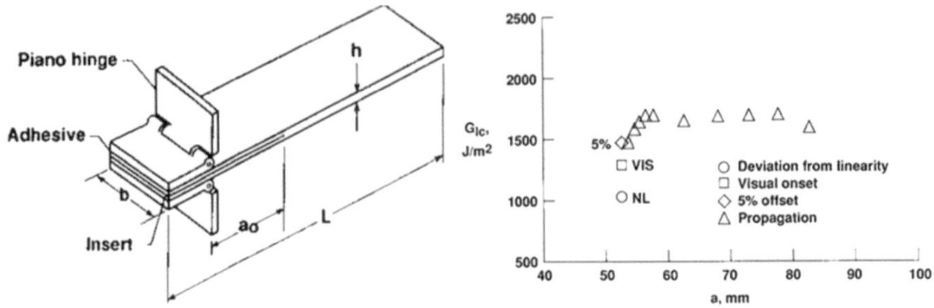

(a) Double Cantilever Beam Specimen.

(b) Delamination Resistance Curve (R-curve).

Fig. 4.1: Mode I fracture toughness with ASTM D5528 test standard [2].

ognized over 30 years ago. The recent extensive studies on the fiber bridging of DCB test were partially driven by the development of CZM and its increasing publicity for modeling the interface behavior of materials and structures such as composite laminates and components.

However, it is also recognized during the development of ASTM D5528 standard that even the initiation value (G_{IC}) preferred as the material toughness of the composite material in the standard is difficult to determine accurately. Three definitions for an initiation value of G_{IC} have been evaluated in the round-robin testing during the development of ASTM D5528 [2]. It can be seen from Fig. 4.1b that there are significant differences among the G_{IC} values determined by the three definitions.

In addition to the UD specimen defined in ASTM D5528, extending investigations for DCB specimen with different stacking sequence layups in the past 30 years confirmed that it is difficult to accurately determine of initial value G_{INI} based on the delamination resistance R-curve [41, 45]. This is because the R-curve is only the phenomenological representation of the material toughness influenced by the fiber bridging. The fracture toughness, G_{IC}, required in the CZM as the intrinsic material properties should be determined based on the microscopic delamination mechanisms ahead of crack tip excluding the fiber bridging contributions.

The typical R-curve characteristics and the associated fiber bridging features observed from the Mode I delamination tests with DCB specimen are given in Fig. 4.2. The global nonlinear strain energy release rate after onset of delamination initiation is attributed to the gradually development of fiber bridging behind the delamination crack tip. The local delamination mechanisms underneath the global R-curve phenomenon had been identified from the microscopic analysis such as fractographic examination of the fracture surfaces of the failed specimen [17]. There are two distinctive toughening mechanisms within two distinguishing delamination zones. The first toughening mechanism is directly related to the fundamental structure and bonding of the base material, such as matrix bonding and matrix/fiber bonding between the composite plies. The second toughening mechanism is related to the fiber bridging which just holds the two opening fracture surfaces together after matrix/fiber debonding and the delamination crack formed. The first delamination

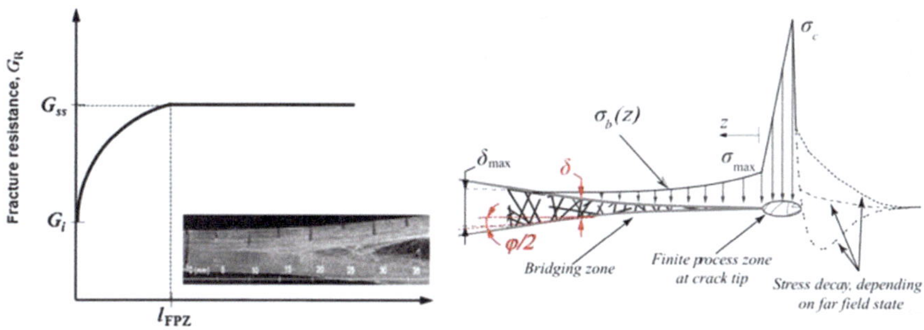

(**a**) Typical R-curve phenomenon. (**b**) Toughening mechanisms in crack-tip.

Fig. 4.2: (a) Global R-curve phenomenon [43] associated with (b) two toughening mechanisms within two delamination zones.

zone is the crack front damage or finite process zone (FPZ) associated with both first and second delamination mechanisms. Whilst the second delamination zone is the fiber bridging zone (FBZ) linked to the second delamination mechanism only.

From material science point of view, the first delamination mechanism can be classified as intrinsic mechanism whilst the second delamination mechanisms as extrinsic mechanism. Consequently, the fracture toughness, G_{IC}, determined from the ASTM D5528 standard test and used in the CZM should be only related to the intrinsic mechanism in the first delamination zone ahead of crack tip. However, most of the previous studies [14, 23, 44, 58] only qualitatively recognize the two mechanisms and two delamination zones but not quantitatively investigate the two mechanisms and two zones ın details.

4.2.2 Damage Mechanisms in Delamination Zone Ahead Crack Tip

Based on the inherent heterogeneity material (matrix and fiber) characteristics and the fractographic examination of the failed testing specimen [17, 32], the deformation and damage at the delamination crack tip of the composite laminate can be related to two different damage mechanisms and three toughening mechanisms, as shown in Fig. 4.3. The first damage mechanism is the interlaminar matrix cracking between the interface piles, which is related to the fracture toughness of the cured polymer matrix (e.g., epoxy). The second damage mechanism is the fiber/matrix interfacial separations, both continue fiber/matrix and staggered fiber/matrix interface debonding, between the interlaminar piles, which is related to the fracture toughness of fiber/matrix interface. The staggered fiber/matrix interface debonding introduce initial fiber bridging ahead of crack tip and thereafter results in the third toughening mechanism ahead the crack tip [21, 25, 34, 37, 39, 49, 59].

Consequently, the fracture toughness associated with these two mechanisms is related to the inherent material properties of a given composite material system

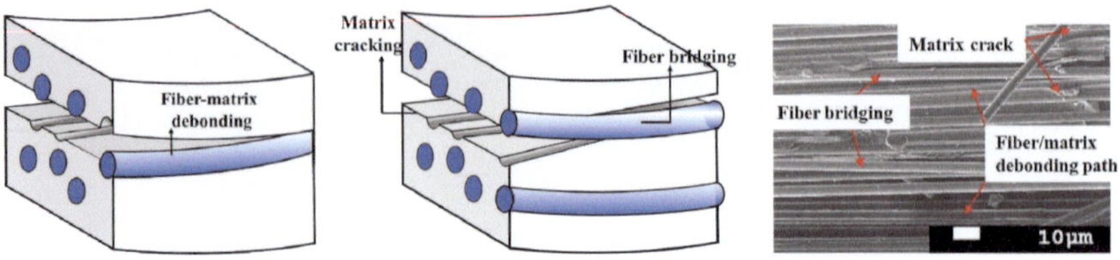

(a) Crack front damage zone mechanism. **(b)** SEM image of fracture surface [32].

Fig. 4.3: Schematic of damage mechanisms in the delamination damage zone.

(fiber and matrix). The original objective of ASTM D5528 was to determine the strain energy release rate, G_{IC}, associated with these two damage mechanisms observed in most applications of composite structures. Consequently, the initiation value of strain energy release rate, G_{INI}, obtained from the DCB testing includes three parts associated with the different damage and toughening mechanisms

$$G_{INI} = G_{I-MC} + G_{I-FD} + G_{I-FB}, \tag{4.1}$$

where G_{I-MC} represent strain-energy release rate associated with the matrix cracking damage mechanism, and G_{I-FD} represents the strain energy release rate related to the matrix/fiber interfacial separation damage mechanism. G_{I-FB} represent strain-energy release rate associated fiber bridging ahead of the crack tip. The combination of G_{I-MC} and G_{I-FD} should be regarded as the strain energy release rate, G_{IC}, for a composite laminate with a given fiber and matrix material system. The value of G_{IC} as an intrinsic material property should be independent to the specimen geometry such as thickness.

4.2.3 Fiber Bridging Mechanisms in Delamination Zone Behind Crack Tip

It is commonly recognized from numerous previous studies [16, 20, 22, 30, 37, 38, 40, 44, 50, 51, 54] that the fiber bridging stress distribution in Mode I delamination with DCB tests exhibit strong nonlinear characteristics. Consequently, the fiber bridging stress σ_{br} can be accurately described as a nonlinear function of the initial crack tip opening displacement (ICTOD) represented by δ determined from the DCB tests in the fiber bridging zone, as shown in Fig. 4.4.

Due to the complex nonlinear stress distribution induced by the fiber bridging, the delamination energy release rate $G_{br}(\delta)$ as a function of delamination crack opening in the fiber bridging zone (ICTOD) δ can be expressed as:

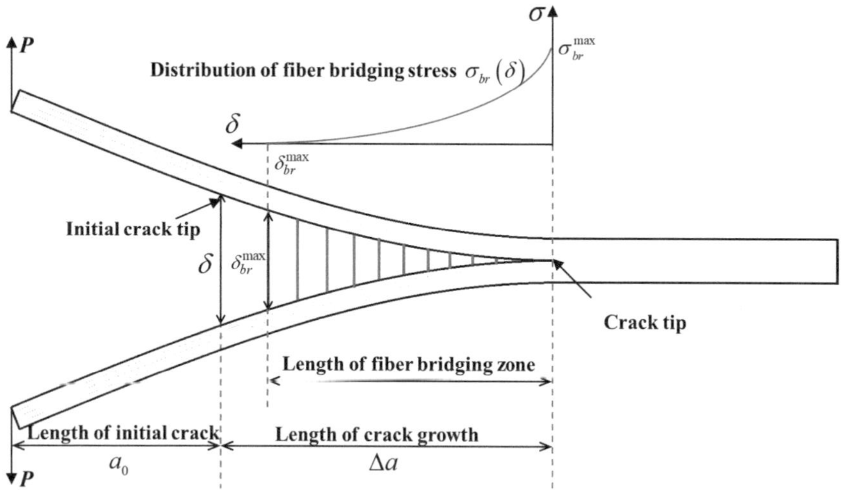

Fig. 4.4: Schematic of distribution of fiber bridging stress with crack tip opening displacement.

$$G_{br}(\delta) = \int_{0}^{\delta_{br}^{max}} \sigma_{br}(\delta)\, d\delta \tag{4.2}$$

The bridging traction stress in the bridging zone of the DCB specimen is assumed to be only the function of the normal separation of the delamination crack tip in the fiber bridging zone. By differentiation of Eq. (4.2) with respect to the CTOD (δ), the bridging traction stress is obtained as:

$$\sigma_{br}(\delta) = \frac{dG_{br}(\delta)}{d\delta} \tag{4.3}$$

Numerous previous experimental studies and analytical investigations [20, 22, 30, 37, 38, 44, 45, 47, 50, 51, 54] on the fiber bridging stress distribution have confirmed the nonlinear characteristics of the traction -separation relationship for Mode I delamination in the DCB tests. Consequently, various bridging stress models used to describe the fiber bridging effect [23, 43, 44, 47, 50, 51, 58]. Investigation results show that the non-linear exponential bridging stress distribution model has the advantages of being more accurate among different bridging models [44, 47, 50, 51, 58]. As an example, a combination of a strong and weak exponential function has been widely to describe the strain energy release rate $G_{br}(\delta)$ as [16, 44]:

$$G_{br}(\delta) = G_a(1 - e^{-\frac{\delta}{\delta_a}}) + G_b(1 - e^{-\frac{\delta}{\delta_b}}) \tag{4.4}$$

Consequently, the fiber bridging traction stress $\sigma_{br}(\delta)$ in the fiber bridging zone behind of the crack tip of the mode I delamination in DCB test can also be express as a combination of two exponential functions as:

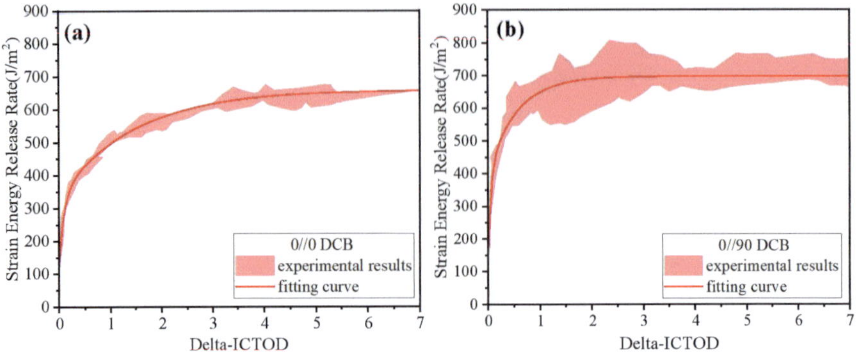

Fig. 4.5: Corresponding R-curves for different ply-angle; (a) 0//0, (b) 0//90.

$$\sigma_{br}(\delta) = \frac{G_a}{\delta_a} e^{-\frac{\delta}{\delta_a}} + \frac{G_b}{\delta_a} e^{-\frac{\delta}{\delta_b}} = \sigma_{br\_a}(\delta) + \sigma_{br\_b}(\delta) \tag{4.5}$$

Figure 4.5, as examples, presents the correlations with the experimental results for the two 0//0 and 0//90 basic composite laminate ply-angle layups [16]. It is evident that good correlations can be achieved with the proposed exponential function and similar correlation results also attainted for other two 0//30 and 0//45 composite laminate layups [16].

4.2.4 Total Fracture Toughness and Empirical Calculations

The R-curves obtained from the experiments represent the variation of total strain energy release rate with delamination propagation. Consequently, the experimental results can be correlated with the total fracture toughness in terms of strain energy release rate, G_I, and expressed as:

$$G_I = G_{INI} + G_{br}(\delta) \tag{4.6}$$

Based on fracture mechanics concept, the fracture toughness (G) is defined as the rate of change of the system's potential energy with respect to the crack surface area. ASTM D5528 [2] integrates this concept with the characteristics of DCB specimens to analyze crack propagation, assuming that the crack in the DCB specimen propagates uniformly under constant displacement or load. When delamination undergoes a small extension (Δa), the fracture toughness of the specimen can be determined according to linear elastic fracture mechanics by the following expression

$$G_I = \frac{1}{b}\frac{dU}{da}, \tag{4.7}$$

where b is the width of the DCB specimen, a is the delamination length.

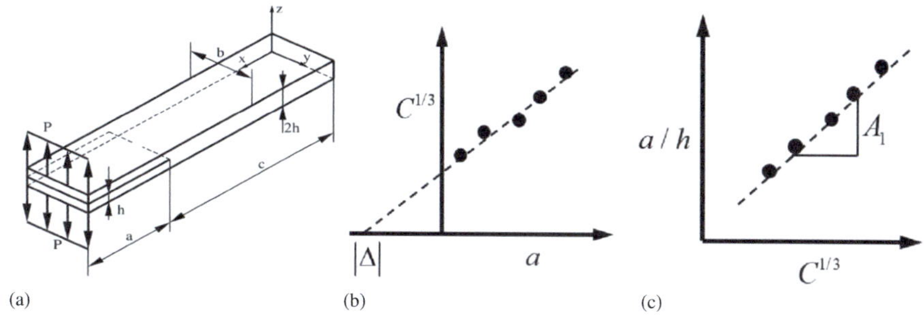

Fig. 4.6: (a) Schematic diagram of DCB specimen geometry [53]; (b) Modified Beam Theory for Δ; (c) Modified Compliance Calibration for A_1.

One of the most used current Mode I delamination fracture toughness calculation methods in ASTM D5528 standard is derived from Eq. (4.7) based on the modified beam theory (MBT) and is defined as [32]

$$G_{IC} = \frac{3P\delta}{2b(a + |\Delta|)}, \tag{4.8}$$

where P is the reaction load, δ is the applied displacement, b is the width of the specimen, and a represents the crack length from the loading point to the delamination front. An empirical correction factor ($|\Delta|$) is recommended by ASTM D5528 to adjust the crack length a to account for the beam end rotation. This correction factor is determined based on the assumption of a linear relationship between the cube root of the beam compliance $C^{1/3}$ and the crack length a as illustrated in Fig. 4.6b. The ASTM-D5528 standard also proposes the Modified Compliance Calibration (MCC) method [32] which also includes an empirical slope parameter A_1 to correct the end rotation as shown in Fig. 4.6c

$$G_{IC} = \frac{3P^2C^{2/3}}{2A_1bh} \tag{4.9}$$

Since the fracture toughness in terms of R-curve are indirectly calculated from the experimental data with a given method, any errors introduced by the given method will influence the accuracy calculation results. Further clarifications will be provided in the following sections.

4.2.5 Delamination Mechanisms in Multidirectional Composite Laminates

Even though ASTM D5528 was mainly developed for unidirectional (UD) composite laminates, it has been used extensively for multidirectional composite laminates with different interlaminar fibre ply angles lay-up [41, 44, 60, 62]. There are various previous investigations on the effects of ply angle on fracture toughness, but conflicting results and discrepancies are reported in the literature [11, 16, 22, 44, 56, 58, 60–62]. Consequently, more insight into the physics and micro-mechanics of delamination features related to composite laminates with the different lay-ups composite is required.

A comprehensive study was carried out recently focusing on the effect of interlaminar fiber ply angle on Mode I initial delamination fracture behavior [10]. A series of composite laminate DCB tests (0//0, 0//30, 0//45, 0//60, 0//90) were conducted in accordance with ASTM D5528 to systematically investigate the delamination damage mechanisms. The experimental results indicate that there are significant differences in delamination propagation behavior among specimens with different interlaminar fiber ply angles [10], which are in line with the previous studies [41, 44, 60, 62]. Fig. 4.7 illustrates the experimental results of the R-curve characteristics of different interlaminar fibre ply angle composite laminate DCB tests (0//0, 0//30, 0//45, 0//60, 0//90). The results were normalised with the G_{INI} of the basic UD 0//0 composite laminate originally used in ASTM 5528. There is a strong influence of interlaminar fiber ply angle on both delamination initiation and propagation behavior.

Figure 4.8 illustrate the global delamination propagation path in the composite laminate DCB specimens with different interlaminar fiber ply angles. Specifically, the 0//0 specimen exhibited smooth crack propagation characteristics. For the 0//θ specimens with different interlaminar fiber ply angle (θ = 30, 45, 60, and 90), the extent of the fiber bridging region increases progressively with the increasing of interlaminar fiber ply angle. The delamination crack tends to branch or propagate along the fiber direction of the adjacent layer, forming a "sawtooth" or "wavy" path in the repetition in the 0//45 specimens. Meantime, for the 0//60 and 0//90 specimens, the crack tends to deflect into the adjacent layer and generate a tortuous propagation path (such as a "Zig-zag" path) that requires overcoming additional fiber breakage and matrix tearing. The experimental results indicate that the interlaminar fiber ply angle

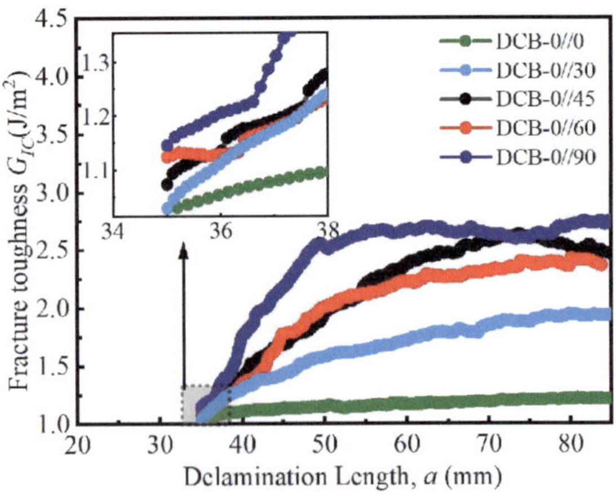

Fig. 4.7 Normalized fracture toughness R-curve of DCB specimens with different interlaminar fiber ply angles: (a) 0//0; (b) 0//30; (c) 0//45; (d) 0//60; (e) 0//90.

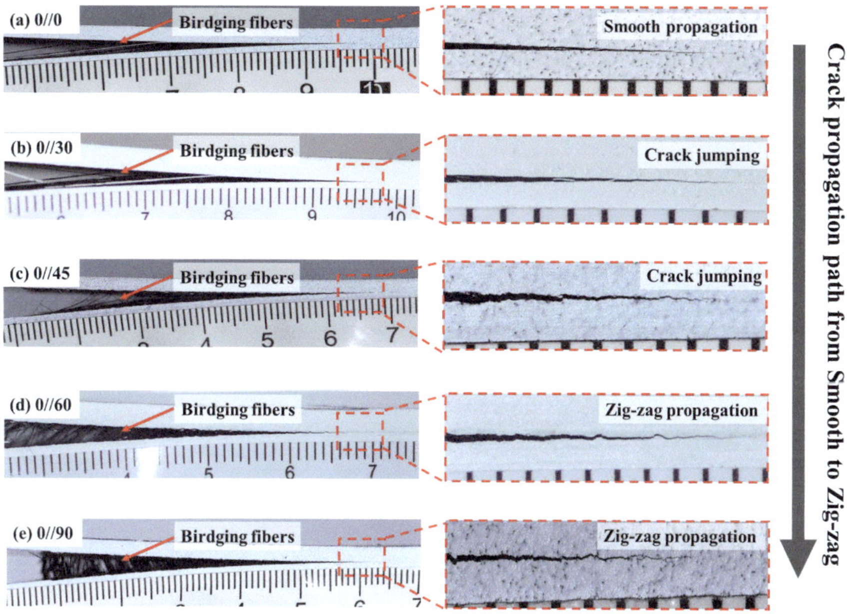

Fig. 4.8: Delamination propagation path in DCB specimens with different interlaminar fiber ply angles: (a) 0//0; (b) 0//30; (c) 0//45; (d) 0//60; (e) 0//90.

has a significant influence on the morphology of delamination propagation paths, and the transition in crack path morphology notably affects the crack propagation area.

The microscopic fracture surface analyses further reveal that the primary damage modes during delamination crack propagation ahead of delamination crack tip are matrix cracking and fiber/matrix debonding [10]. The microscopic images reveal fractured matrix regions alongside exposed fibers resulting from fiber/matrix separation. Compared with the 0//0 specimens, the fibers in the 0//θ specimens impose a directional constraint that forces the crack tip to undergo continuous jumps in the thickness direction, thereby forming a deeply concave surface and increasing the crack propagation path area. Three-dimensional fracture surface features are observed in [0//θ] specimens, resulting from interlaminar crack propagation deviation and migration. Furthermore, the crack propagation path area increases with θ, confirming a positive correlation between interlaminar initial fracture toughness and fiber ply angle.

Interpretation of the failure mechanisms based on the macroscopic delamination propagation paths and the microscopic SEM analysis is used to build up the three-dimensional feature as shown in Fig. 4.9 for 0//0, 0//45 and 0//90 angle ply laminate configurations. Compared with the 0//0 specimen with flat and straight delamination propagation path and microscopic cross-sectional morphology, the interlaminar crack propagation deviation and migration with the tortuous paths in the $0//\theta$ specimens results in significant larger fracture area for a given crack extension length.

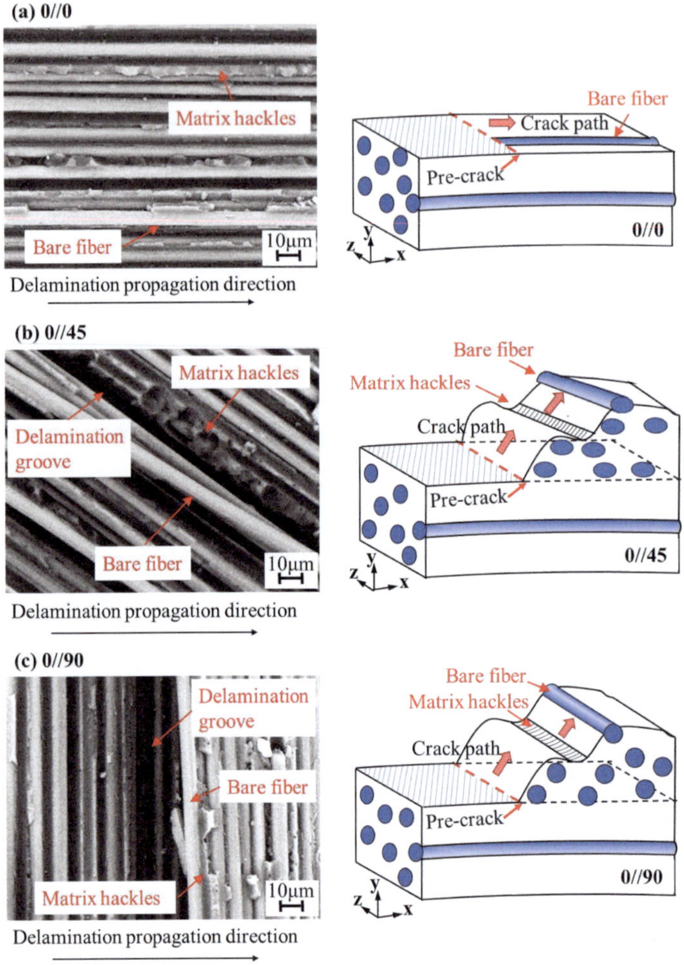

Fig. 4.9: Fracture surface morphology images and interpretation of delamination crack propagation mechanism model: (a) 0//0; (b) 0//45; (c) 0//90.

As defined in Eq. (4.2), the Mode I delamination fracture toughness (G_I) is defined as the incremental change in the total elastic energy (U) of the structure with respect to the increase in crack length. In the case of DCB specimens, this is typically expressed using an incremental change in crack length (da) because the specimen width (b) is constant. However, this approach assumes that the crack propagates along a linear path, thereby neglecting the complexities introduced by variations in interlaminar fiber orientation. Consequently, a mapping relationship among interlaminar fiber ply angle, crack propagation path area, and fracture toughness was established [10]. Based on this mapping relationship, an analytical method for Mode I initiation fracture toughness for 0//θ composite laminate DCB experiments was established. The new method can correlate and predict the fracture toughness of any 0//θ composite laminate with different fiber ply angles based on the experimental data of the basic 0//0 and 0//90 experimental results [10].

4.3 A New Mechanism-Based Cohesive Zone Model Approach

The limitations of the current widely used semi-empirical modelling approaches with superposed CZM is presented firstly. Then, the Mechanism-based cohesive zone model (MB-CZM) based on the decomposition of the two delamination zones and the two toughening mechanisms identified in the previous sections is developed.

4.3.1 Semi-Empirical Modelling Approach – Superposed CZM

Classical CZM widely used for modelling the delamination initiation and propagation of composite laminates under different loading conditions normally adopt the bilinear traction-separation law as shown in Fig. 4.10a [48]. However, modification of the classical CZM is required for the delamination of composite laminates using the DCB specimen due to the extrinsic fiber bridging toughening mechanisms discussed in the above section. Fig. 4.10b-d present the reproductions of the commonly modified CZM including tri-linear CZM [54], four-linear CZM [58], ML-CZM (multi-linear cohesive zone model) [43] respectively.

All the previously modified CZM can be classified as superposed CZM approaches since the FPZ ahead of crack tip and the fiber bridging zone (FBZ) behind crack tip are combined as an artificial crack in the modelling. Consequently, a number of limitations can be identified related to the delamination mechanisms discussed in the previous sections. Firstly, numerous previous experimental studies and analytical investigations on the fiber bridging stress distribution have confirmed the nonlinear characteristics of the traction -separation relationship for Mode I delamination in the DCB tests [20, 22, 30, 36]. However, the nonlinear characteristics of fiber bridging cannot be fully represented with linear approximation such as tri-linear, four-linear, and multi-linear in most of the previous CZMs [43, 54, 58]. Different level of errors will be naturally generated by the CZM with the linear approximation. For robust simulation of the delamination growth, the CZM approach should necessarily have the nonlinear capability to implement any arbitrary shape of nonlinear behavior of fiber bridging.

Secondly, in all these so called superposed CZM approaches [23, 38, 43, 44, 51, 54, 58], the two delamination zones are combined as one pseudo fracture process zone (FPZ) and treated as an artificial crack tip in fracture toughness analysis which is not in line with the traditional fracture mechanics principle. It is desired to distinguish the two different delamination mechanisms and to determine the fracture toughness associated with the intrinsic mechanisms only.

4.3.2 A New Mechanism-Based Cohesive Zone Model for Delamination

The sketch of the Mechanism-based cohesive zone model (MB-CZM) [16] with decoupled cohesive elements is shown in Fig. 4.11. Instead of using a single cohesive element to describe the combination of the two different toughening mechanisms in the previous CZM methods [23, 38, 43, 44, 51, 54, 58], the proposed method employs two separate cohesive elements to describe the two different toughening mechanisms respectively.

The 1st cohesive element E1, with the classical bilinear cohesive traction-separation relation, is used to describe the toughening mechanism associated with the traditional delamination damage mechanism. The 2nd cohesive element E2, with linear-nonlinear traction-separation relation, is used to describe the toughening mechanisms associated with additional fiber bridging. Overcoming the limitation of the linear approximation widely used in the semi-empirical tri-linear or four-linear and ML-CZM methods [16, 20, 22, 30, 38, 44, 45, 47, 50, 51, 54], the decoupled

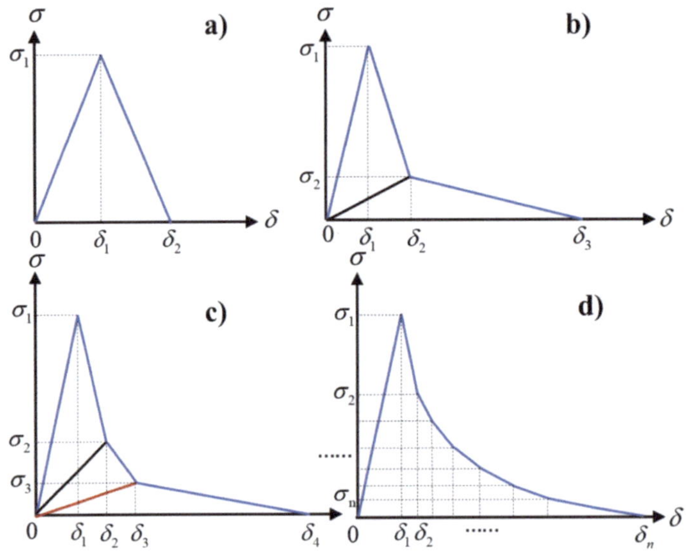

Fig. 4.10: Various modified CZM for Mode I delamination (a) bilinear without fiber bridging [48], (b) tri-linear with fiber bridging [54], (c) four-linear with fiber bridging [58], (d) multi-linear with fiber bridging [43].

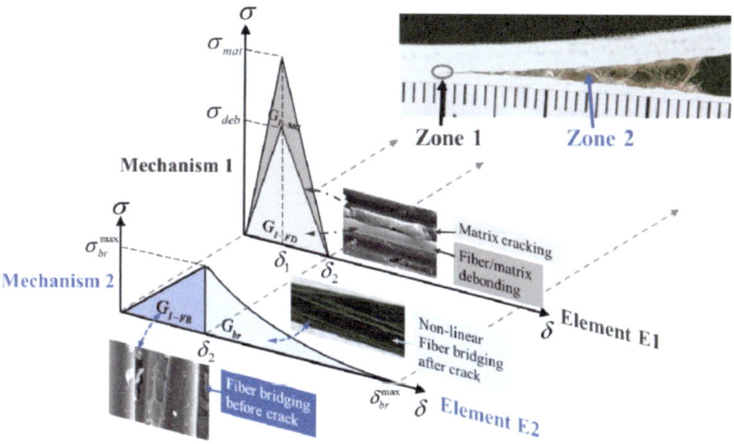

Fig. 4.11: Sketch of the proposed MB-CZM based on two toughening mechanisms and two delamination zones and represented by two cohesive elements [16].

linear-nonlinear traction-separation relation associated with the 2nd cohesive element can accurately describe the experimental observed complex nonlinear fiber bridging characteristics.

Based on the different damage and toughening mechanisms within the two delamination zones, the strain energy release rate of delamination initiation and propagation during the DCB testing can be categorized into intrinsic and extrinsic mechanism respectively. The intrinsic mechanism associated with material properties of composite laminate is represented by the first cohesive element E1. Whilst the extrinsic mechanism associated with fiber bridging is represented by the second cohesive element E2. Consequently, the global strain energy release rate represented by the phenomenological R-curve obtained from the experiments can be de-coupled into the following three stages associated with the energies of the two cohesive elements and the two delamination zones.

Stage 1 delamination initiation with damage evolution in zone 1 only

$$G_I = G_{I-MC} + G_{I-FD} + G_{I-FB} = G_{E1} + G_{E2} \tag{4.10}$$

At the point of delamination initiation and formation of a new fracture surface, the fracture toughness can be divided into two parts according to the difference of damage and toughening mechanism

$$\begin{cases} G_{E1} = G_{I-MC} + G_{I-FD} = G_{IC} \\ G_{E2} = G_{I-FB} = G_{br}(tip) \\ G_{INI} = G_{IC} + G_{br}(tip) \end{cases} \tag{4.11}$$

Stage 2 delamination propagation with damage evolution in zone 1 and fiber bridging evolution in zone 2

$$G_I(\delta) = G_{IC} + G_{I-FB} + G_{br}(\delta) = G_{E1} + G_{E2} \tag{4.12}$$

At the point with delamination crack propagation

$$\begin{cases} G_{E1} = G_{I-MC} + G_{I-FD} = G_{IC} \\ G_{E2} = G_{BR} = G_{I-FB} + G_{br} = G_{br}(tip) + \int_{\delta_2}^{\delta_{br}^{\max}} \sigma_{br}(\delta)\, d\delta \end{cases} \tag{4.13}$$

In the MB-CZM method, the nonlinear evolution of bridging stress $\sigma_{br}(\delta)$ is induced between the opening displacement of element E1 failure point and the opening displacement of the maximum ICTOD of the fiber bridging zone.

Stage 3 delamination propagation with stable damage in zone 1 and stable fiber bridging in zone 2

$$G_I(\delta) = G_{ss} = G_{IC} + G_{br} = G_{E1} + G_{E2} \tag{4.14}$$

with

$$\begin{cases} G_{E1} = G_{I-MC} + G_{I-FD} = G_{IC} \\ G_{E2} = G_{BR} = G_{I-FB} + G_{br}(\delta) = G_{br}(tip) + \int_{\delta_2}^{\delta} \sigma_{br}(\delta) d\delta \end{cases} \tag{4.15}$$

For cohesive element E1, the classical bilinear traction separation law widely used for describe the interlaminar deformation and damage evolution is adopted in this work

$$\begin{cases} \sigma_{E1}(\delta) = \dfrac{\sigma_{\max}^c}{\delta_1}\delta & 0 \le \delta \le \delta_1 \\ \sigma_{E1}(\delta) = \dfrac{\sigma_{\max}^c}{\delta_1 - \delta_2}\delta - \dfrac{\sigma_{\max}^c}{\delta_1 - \delta_2}\delta_2 & \delta_1 \le \delta \le \delta_2 \end{cases} \tag{4.16}$$

with

$$\begin{cases} \delta_1 = \dfrac{\sigma_{\max}^c}{K_{E1}} \\ \delta_2 = \dfrac{2G_{IC}}{\sigma_{\max}^c} \end{cases} \tag{4.17}$$

The damage variables of the cohesive elements E1 can thus be expressed as:

$$D_{E1} = \begin{cases} 0 & 0 \le \delta \le \delta_1 \\ \dfrac{\delta_2(\delta - \delta_1)}{\delta(\delta_2 - \delta_1)} & \delta_1 \le \delta \le \delta_2 \end{cases} \tag{4.18}$$

Where $\sigma_{\max}^{c}$ and δ_1 are the maximum interface strength (transverse strength S_{33} for basic UD laminate) and corresponding opening displacement of the damage initiation, and δ_2 is the opening displacement when the cohesive element is completed failed with damage growth. K_{E1} is the initial interfacial stiffness of cohesive element E1 in the classical bilinear CZM for the quasi-brittle composite materials.

For cohesive element E2, the fiber bridging strain energy release rate is fitted to the energy distribution curve (R-curve) with open displacement using an exponential function [44] to obtain

$$G_{br}(\delta) = G_a(1 - e^{-\frac{\delta-\delta_2}{\delta_a}}) + G_b(1 - e^{-\frac{\delta-\delta_2}{\delta_b}}). \tag{4.19}$$

G_a, δ_a, G_b and δ_b are the fitting parameters, derived from the J integral, from which the bridge stress represented by the cohesive element E2 can be obtained as

$$\begin{cases} \sigma_{E2}(\delta) = \dfrac{\sigma_{br}^{\max}}{\delta_2}\delta & 0 \le \delta \le \delta_2 \\ \sigma_{E2}(\delta) = \dfrac{G_a}{\delta_a} e^{-\frac{\delta-\delta_2}{\delta_a}} + \dfrac{G_b}{\delta_a} e^{-\frac{\delta-\delta_2}{\delta_b}} & \delta_2 \le \delta \le \delta_{br}^{\max} \end{cases} \tag{4.20}$$

with

$$\begin{aligned} K_{E2} &= \frac{\sigma_{br}^{\max}}{\delta_2} \\ \sigma_{br}^{\max} &= \frac{G_a}{\delta_a} e^{-\frac{0}{\delta_a}} + \frac{G_b}{\delta_a} e^{-\frac{0}{\delta_b}} = \frac{G_a}{\delta_a} + \frac{G_b}{\delta_a}. \end{aligned} \tag{4.21}$$

$\sigma_{br}^{\max}$ is the maximum fiber bridging interface strength and K_{E2} is the initial pseudo-stiffness of cohesive element E2 in the nonlinear CZM for the fiber bridging. The damage variables of the cohesive elements E2 representing the degradation of the fiber bridging traction can thus be expressed as

$$D_{E2} = \begin{cases} 0 & 0 \le \delta \le \delta_2 \\ 1 - \dfrac{\sigma_{E2}(\delta)}{K_{E2}\delta} & \delta_2 \le \delta \le \delta_{br}^{\max} \end{cases} \tag{4.22}$$

All physics-related parameters can be confirmed according to the following flowchart Fig. 4.12, as detailed in [16].

The proposed MB-CZM uses two cohesive elements with two traction-separation relations to individually represent the two delamination mechanisms. It therefore can provide a better description of the quasi-brittle linear elastic behavior of composite material and the nonlinear characteristics of fiber bridging respectively, rather than only the empirical combined effects in the previous superposed CZM approaches [43, 54, 58].

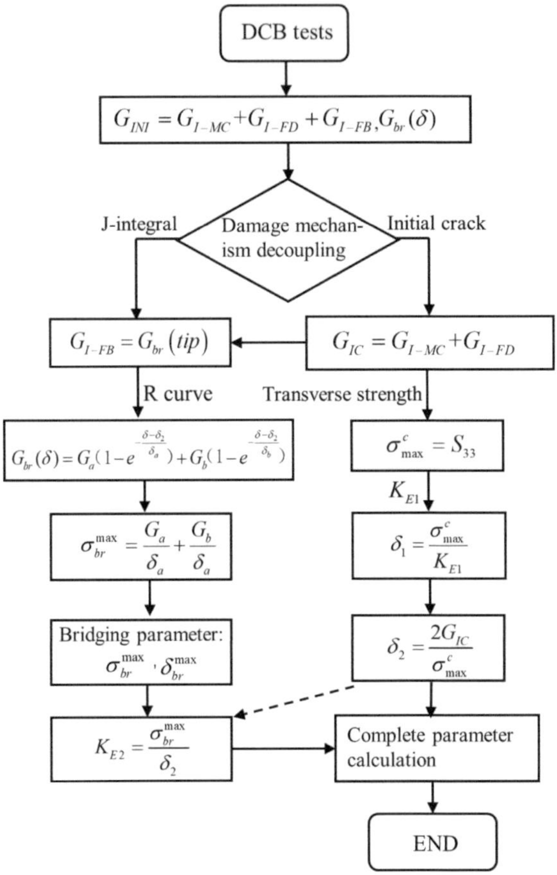

Fig. 4.12 Flowchart for calculating parameters in the new MB-CZM

4.3.3 Implementation and MB-CZM and Validation

The implementation of the MB-CZM can be readily carried out with ABAQUS commercial software combined with the UMAT subroutines developed for the MB-CZM [16]. The first set of cohesive elements (COH3D8) representing the intrinsic material toughness mechanism is firstly inserted into the interface of the upper and lower composite laminate layers of the DCB specimen along the delamination propagation path. Then the second set of cohesive elements (COH3D8) representing the fiber bridging mechanism is inserted into the same interface and connected to the same nodes with the first set of cohesive elements. The details of the finite element model are shown in Fig. 4.13.

The bilinear traction-separation law associated with the first set of cohesive elements and the linear-exponential traction-separation law associated with the second set of cohesive elements developed in this work are coded separately via UMAT subroutine and implemented with ABAQUS commercial software. Two sets of ma-

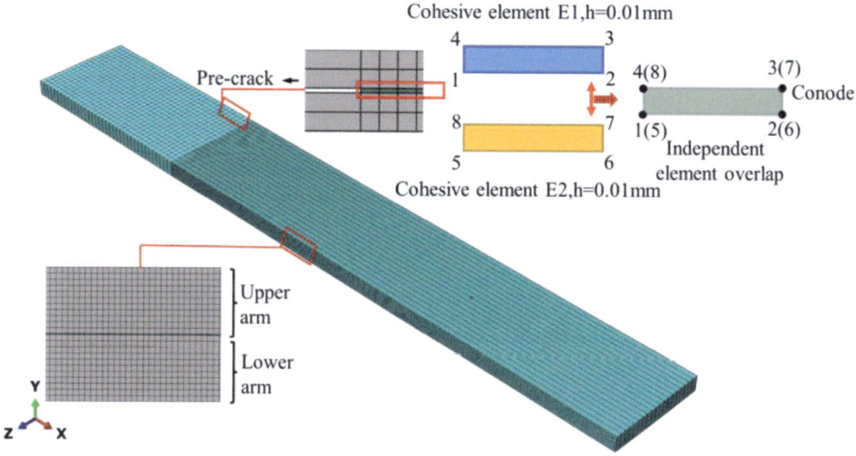

Fig. 4.13: Finite element model of the DCB specimen for MB-CZM.

terial property and modelling parameters are determined for the bilinear traction-separation law and the linear exponential traction-separation law in the next section. Numerical simulations are carried out and compared with the experimental results to validate the proposed MB-CZM approach. Mode I delamination characteristics of four groups of DCB specimen with different stacking sequence layups in the order of $[0_{11}//\theta//0_{12}]$ with $\theta = 0°, 30°, 45°$ and $90°$ respectively had been experimentally investigated by Shokrieh et al [45]. The experimental results were widely used to verify the various semi-empirical CZM approaches [44, 57] with directly fitted parameters.

The load-displacement curves predicted by the FE analysis for the DCB specimen with the four different interface fiber angles including 0//0, 0//30, 0// 45 and 0//90 are shown in Fig. 4.14. The numerical simulation results are compared with the corresponding experimental results for each DCB test for validation of the proposed MB-CZM and the numerical modelling procedures. The damage onset within the crack tip is denoted by the beginning of the degradation of the global stiffness which results in nonlinearity in the load-displacement curve. The additional load bearing capacity could be attributed to the toughening mechanism of fiber bridging behind the crack tip generated by delamination propagation within the DCB specimen.

The comparisons of the global load-displacement curves demonstrates that the numerical simulation results obtained with the MB-CZM work are in reasonably good agreement with the corresponding experimental data for all the DCB specimens with four different interface fiber angles. The comparisons in Fig. 4.14 illustrates that the new MB-CZM can capture the Mode I delamination growth global deformation behavior in multidirectional composite laminates with the effect of large-scale fiber bridging. Further validations of the MB-CZM and the numerical modelling procedures proposed in this study is presented in Fig. 4.15 with the comparison of the global delamination resistance curves (R-curve) obtained from the FE analysis

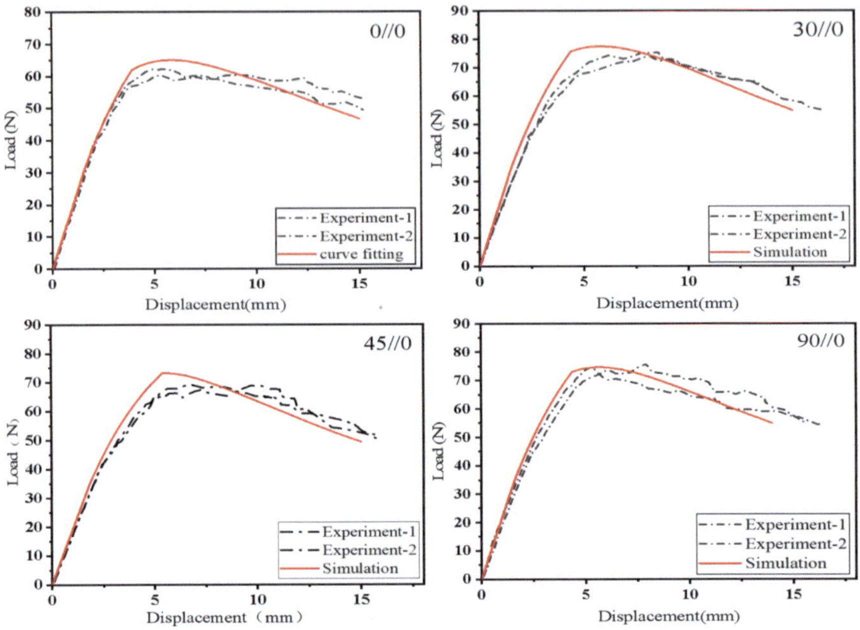

Fig. 4.14: Comparison between experimental and simulation results with MB-CZM.

and the experimental results. All the simulation results are in good agreement with the experimental data.

The load-displacement curves are commonly used for the verification of the previous CZM approaches [23, 43, 44, 50, 51, 54, 58] and there are always different levels of error. There are limited comparisons of the R-curves between model calculation and experimental results in the literature. This is because even the parameters are identified by fitting the R-curves with nonlinear feature, the linear simplification (e.g., tri-linear, four-linear) in the previous CZMs [23, 43, 44, 54, 58] cannot describe the nonlinear feature and then fully reproduce the R-curves. On the other hand, the proposed MB-CZM can fully reproduce the R-curves as shown in Fig. 4.15. This is one of the advantages of the proposed MB-CZM compared with the previous superposed CZMs.

4.4 Mechanism-Based Fracture Toughness Calculation Method

The fact is that the only real experimental data obtained from the DCB testing are the load-displacement curves and the corresponding delamination length. The values of the fracture toughness in the form of R-curve are not direct experimental data but calculated with a given method. Consequently, any errors introduced by the fracture

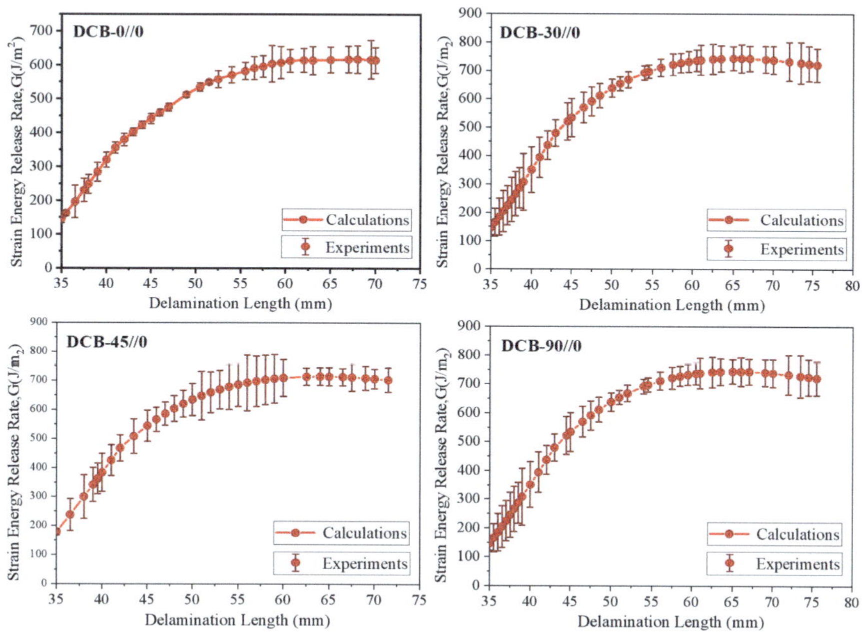

Fig. 4.15: Comparison of the experimental and numerical simulation R-curves.

toughness calculation method will influence the accuracy of the CZM modelling ability.

4.4.1 Current Method Application and Limitations

The most used current Mode I delamination fracture toughness calculation methods in ASTM D5528 standard [2] is based on the modified beam theory (MBT). The well-known rigidly fixed Cantilever Beam Theory (CBT) model together with the elastic foundation beam model, is shown in Fig. 4.16.

The derivation of the ASTM method was based on the Euler-Bernoulli beam theory. In this case, the bending moment of the cantilever beam can be expressed as:

$$M_I(x) = Px \tag{4.23}$$

From the basic theory of cantilever beams, the compliance of the specimen with delamination crack length a in the DCB tests can be easily derived as:

$$C = \frac{\delta}{P} = \frac{2w}{P} = \frac{8a^3}{bEh^3} \tag{4.24}$$

The above equation clearly indicates the linear relationship between the cube root of compliance $C^{1/3}$ and the delamination crack length a, which is the theoretical background for the ASTM D5528 method as shown in Fig. 4.6 (b) in Sect. 4.4 [2].

The fracture toughness of the Mode I delamination for the DCB testing is calculated based on the principle of linear elastic fracture mechanics and is given by

$$G_{IC}^{I} = \frac{P^2}{2b}\left(\frac{dC}{da}\right) \tag{4.25}$$

However, it was recognized during the development of current ASTM D5528 method that the fracture toughness calculation Eqs. (4.7) and (4.8) with the assumption of built-in cantilever beam to represent the DCB specimen will generate unacceptable errors. This is because the end at the delamination crack tip in the DCB specimen is not fully fixed, additional rotation will be introduced at the crack tip in the DCB testing. Therefore, various modified beam theories using such as the elastic foundation (Fig. 4.16b) are proposed to consider the root rotation in the crack-tip regions [36]. For simplicity, an empirical engineering correction approaches based on round-robin excises was adopted in the ASTM D5528 method to considering the influence of end rotation in the DCB specimen [32]. An empirical additional crack length, Δ, is suggested to be added to the value of crack length a to introduce the modified crack length as:

$$a_M = a + |\Delta| \tag{4.26}$$

The empirical correction factor (Δ) recommended by ASTM D5528 is determined based on the assumption of a linear relationship between the cube root of the beam compliance $C^{1/3}$ and the crack length a as illustrated in Fig. 4.18b. However, the extrinsic mechanisms related to the nonlinear fiber bridging distribution in the delamination zone behind the crack tip could strongly influence the linear relationship assumed in ASTM D5528. Experimental studies had been carried out to verify the applicability of the assumption.

As example, the experimental beam compliance plotted against the experimental measured delamination crack length for the two types of specimens with 0//0 and 0//90 interface respectively are presented in Fig. 4.17 [17]. It is observed that there is almost a linear relationship between the cube root of the beam compliance and

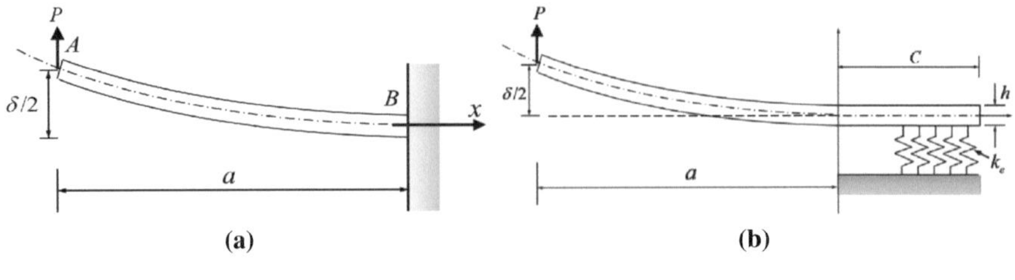

Fig. 4.16: DCB specimen models. (a) Rigidly cantilever beam model; (b) Elastic foundation beam model..

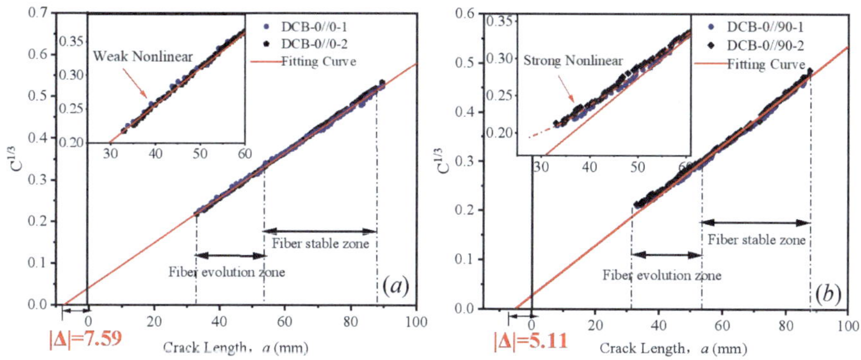

(a) Tests with 0//0 specimen. **(b)** Tests with 0//90 specimen.

Fig. 4.17: Current method compliance $C^{1/3}$ vs crack length.

the crack length for the 0//0 specimen as illustrated in Fig. 4.17a. But a strong nonlinear relationship is observed for the 0//90 specimen, as shown in Fig. 4.17b. This nonlinear behavior is primarily evident in the fiber bridging evolution zone, whereas a linear relationship still exists in the fiber bridging stable zone.

In line with ASTM D5528 method, the mathematically fitting curves without considering the nonlinear feature are also presented in Fig. 4.17. Different correction factors are then obtained for the 0//0 and 0//90 specimen respectively. It is clear that the correction factor for the 0//90 specimen is smaller than that for the 0//0 specimen. This is attributed to the strong fiber bridging in the 0//90 specimen introducing higher restriction on the end rotation of the DCB specimen.

Clearly, if the ASTM D5528 method is used just by direct mathematical fitting of the experimental data, the correction factor will strongly depend on the data set (number and distribution) associated with the delamination crack length. For example, the calculated experimental compliance results from the fiber bridging evolution stage and the stable fiber bridging stage are correlated with the corresponding delamination length for the tests with 0//0 and 0//90 specimen in Figs. 4.18a and b, respectively. The mathematically fitted correction factor for the fiber bridging evolution stage is generally bigger than that for the stable fiber bridging stage. This is attributed to the additional constraint introduced by the fiber bridging behind the delamination crack tip which provides restriction for the root rotation of the DCB beam.

Obviously, the fracture toughness (G_{IC}) calculated using the ASTM current method with various correction factors $|\Delta|$, $|\Delta_E|$ and $|\Delta_S|$ will be significantly different. For example, the fracture toughness (G_{IC}) calculated with $|\Delta_S|$ is about 100% higher than that calculated with $|\Delta_E|$. The significant differences among the values of the fracture toughness calculated with the different correction factors $|\Delta|$, $|\Delta_E|$ and $|\Delta_S|$ for the 0//90 is not acceptable and reveals the limitation of the current method. Consequently, modification of the ASTM method is necessary to remove

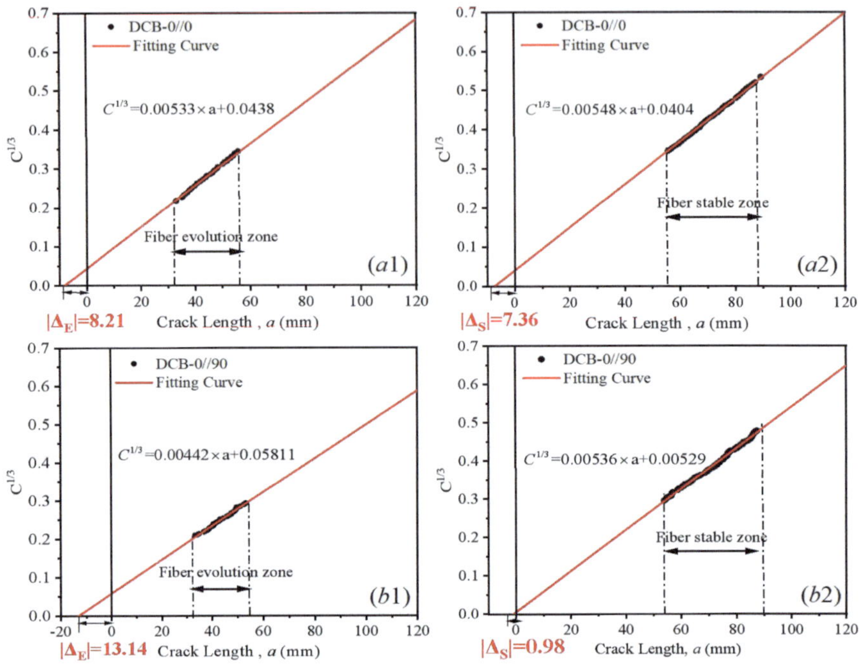

Fig. 4.18: Compliance $C^{1/3}$ vs delamination length a. (a1) 0//0 fiber bridging evolution, (a2) 0//0 stable fiber bridging; (b1) 0//90 fiber bridging evolution, (b2) 0//90 stable fiber bridging.

the limitation and enable its universal applicability for fracture toughness calculation with different level of fiber bridging.

4.4.2 Modification of Current Method with Delamination Mechanisms

The principle of the ASTM D5528 standard is to combine the theoretical solution based on the rigid fixed beam model with an empirical correction factor to obtain the fracture toughness calculation methods. A similar principle for the development of the current ASTM D5528 method is adopted in the method modification process. Based on the delamination mechanisms identified in Subsect. 4.2.2, the crack initiation and propagation process in the DCB specimen can be divided into three stages. Stage I is the initiation stage with damage developed ahead of the artificial crack tip and there is no fiber bridging behind the crack tip. Stage II is the fiber bridging evolution stage where the progressive fiber bridging traction is developed. Stage III is the final stage when fiber bridging reached to stable condition. The compliance of stage I and III are the starting and finishing stage of stage II respectively. Consequently, the derivation of the compliance of the DCB beam considering fiber bridging is carried out for stage II first in this section.

Figure 4.19 presents the cantilever beam model used in the theoretical derivation. In addition to the applied load P at the free end of the beam, the progressive development of the fiber bridging and the increasing range of traction force acting (from Fig. 4.19a to Fig. 4.19b) behind the delamination crack tip is also illustrated. At this stage, the length of the cantilever beams a range up to $a_0 \leq a \leq a_0 + a_{br}$, where a_{br} is the length of the stable full bridging zone as shown in Fig. 4.19b and a_0 is the initial crack length.

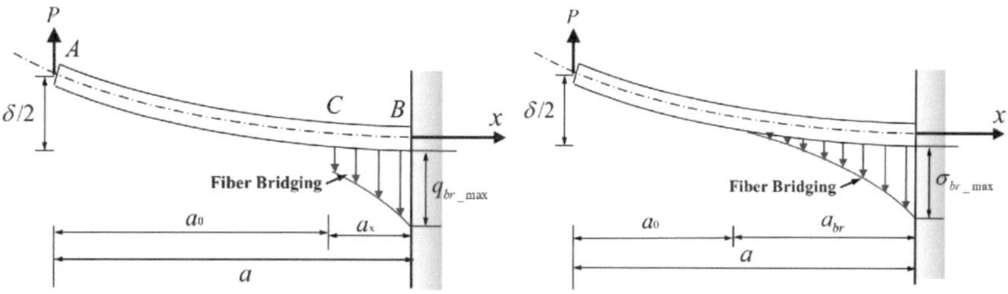

(**a**) Fiber bridging evolution. (**b**) Stable fiber bridging.

Fig. 4.19: The fixed end cantilever beam subjected to applied load and distributed fiber bridging.

As example, a non-linear exponential fiber bridging stress distribution function along the length of the fiber bridging zone is defined as

$$\sigma_{br}(x) = -q_a e^{\frac{x-a}{a_a}} \qquad a_0 \leq x \leq a, \tag{4.27}$$

where q_a, a_a are the fiber bridging stress distribution parameters used to characterize the non-linear distribution of fiber bridging.

Therefore, the combined bending moment at the delamination crack tip induced by both external applied load and the additional fiber bridging traction force can be obtained. From the classic beam theory [13, 26], the fundamental equation governing the beam deflection (w) and rotation angle (θ) with the overall applied bending moment (M) can be expressed as

$$EI\frac{\partial w^2}{\partial x^2} = EI\frac{\partial \theta}{\partial x} = -M(x) \tag{4.28}$$

By integration of above equation with the overall applied bending moment given in Eq. (4.27), and applying the boundary conditions$\theta_{\mathrm{II}}(a) = 0$ and $w_{\mathrm{II}}(a) = 0$ for the fixed end cantilever beam, the deflection of the cantilever beam can be obtained. Then, the beam compliance in stage II with delamination crack length within the fiber bridging evolution zone ($a_0 < a < a_0 + a_{br}$) can be expressed as

$$
\begin{aligned}
C_{II} &= \frac{\delta}{P} = \frac{2w_{II}|_{x=0}(x)}{P} \\
&= \frac{8a^3}{Ebh^3} - \frac{2}{P}\left\{ \frac{q_a a_a}{2} e^{\frac{a_0-a}{a_a}} \left(\frac{a_0^3}{3} - a^2 a_0 + \frac{2a^3}{3} \right) \right. \\
&\quad - q_a a_a^4 e^{\frac{a_0-a}{a_a}} + q_a a_a^4 + q_a a_a^3 (a_0 - a) \\
&\quad \left. + \left(-q_a a_a e^{\frac{a_0-a}{a_a}} a_0 + q_a a_a^2 e^{\frac{a_0-a}{a_a}} \right) \left(\frac{a_0^2}{2} - a a_0 + \frac{a^2}{2} \right) \right\}
\end{aligned}
\tag{4.29}
$$

For delamination initiation stage I where $a = a_0$, the above beam compliance is reduced to the compliance used in the current ASTM method [32] as

$$
C_{II} = \frac{\delta}{P} = \frac{2w}{P} = \frac{8a^3}{bEh^3} = C_I, \qquad a = a_0 \tag{4.30}
$$

The cube root of compliance $C_I^{1/3}$ is linearly related to the crack length a at the delamination initiation stage I, as expected from the principle of the current ASTM D5528 method [2]. When fiber bridging reaches to the stable condition in stage III, the distribution of the additional fiber bridging force behind the delamination crack tip does not change anymore. The compliance of the DCB beam at stage III can be derived with the same procedure as in stage II and expressed as

$$
C_{III} = \frac{\delta}{P} = \frac{2w_2|_{x=0}(x)}{P} = \frac{8a^3}{Ebh^3} - \frac{2}{P}\left(q_a a_a^3(-a_{br}) + q_a a_a^4 \right) \tag{4.31}
$$

The compliance in stage III includes two parts. The first part is generated by the external applied load P and is the same as that in stage I and II. The second part is introduced by the stable distributed fiber bridging traction force behind the delamination crack tip. The compliances at each of the stages derived in above sections can be correlated with the experimental data to determine the fiber bridging stress distribution parameters, λ_a, λ_b and C_{br}. As an example, the correlation results are presented in Fig. 4.20 for the 0//90 composite laminate specimen since more significant fiber bridging effects have been observed in the experiments as shown in Fig. 4.22b.

Based on the experimental correlation results, the empirical correction factors for the three stages from delamination initiation to propagation with progressive fiber bridging evolution and with stable fiber bridging condition can be determined as [17]

$$
\Delta_i = \begin{cases} |\Delta_{\mathrm{I}}| = g_{\mathrm{I}}(a) = 6.61, & i = \mathrm{I} \\ |\Delta_{\mathrm{II}}| = g_{\mathrm{II}}(a) = 81.87\exp(-0.08017a) + 0.31, & i = \mathrm{II} \\ |\Delta_{\mathrm{III}}| = g_{\mathrm{III}}(a) = 0.98, & i = \mathrm{III} \end{cases} \tag{4.32}
$$

Therefore, instead a single constant correction factor used in the current ASTM method, the influence of the fiber bridging behind the delamination crack tip which introduce restriction to the root rotation of the DCB specimen is considered in this work with multiple correction factors corresponding to different levels of fiber bridging. Following the same principle deployed in the current ASTM D5528 method, the modified three-stage fracture toughness can be determined as

$$G_{IC}^{I} = \frac{P^2}{2b}\left(\frac{24}{Ebh^3}(a + |\Delta_I|)^2\right), \qquad a = a_0, \tag{33-a}$$

$$\begin{aligned} G_{IC}^{E} = {} & \frac{P^2}{2b}\left(\frac{24}{Ebh^3}a_M^2\right) - \frac{2P}{b}\left\{\frac{q_a a_a}{2} e^{\frac{a_0 - a_M}{a_a}}\right. \\ & \times \left(-2a_M a_0 + 2a_M^2 - \frac{a_0^3}{3a_a} + \frac{a_M^2 a_0}{a_a} - \frac{2a_M^3}{3a_a}\right) \\ & + q_a a_a^3\left(e^{\frac{a_0 - a_M}{a_a}} - 1\right) + \left(\frac{a_0^2}{2} - a_M a_0 + \frac{a_M^2}{2}\right. \\ & \left.\left. + a_0 a_a - a_M a_a\right)\left(\frac{q_a a_a a_0}{a_a} - q_a a_a\right) e^{\frac{a_0 - a_M}{a_a}}\right\} \end{aligned} \qquad a_0 < a \leq a_0 + a_{br}, \tag{33-b}$$

$$G_{IC}^{S} = \frac{P^2}{2b}\left(\frac{24}{Ebh^3}(a + |\Delta_{III}|)^2\right), \qquad a_0 + a_{br} \leq a \tag{33-c}$$

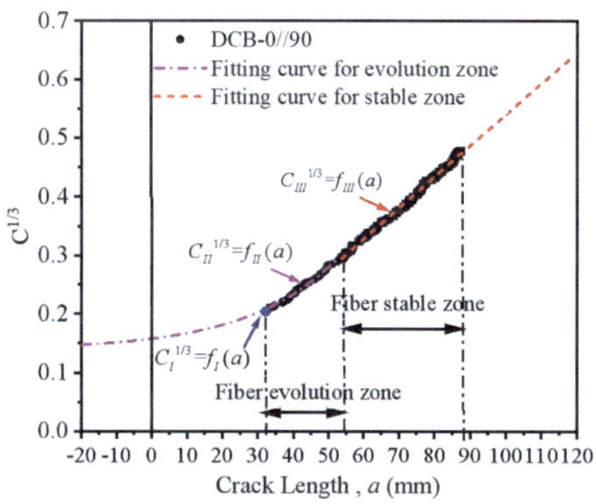

(a) Nonlinear compliance.

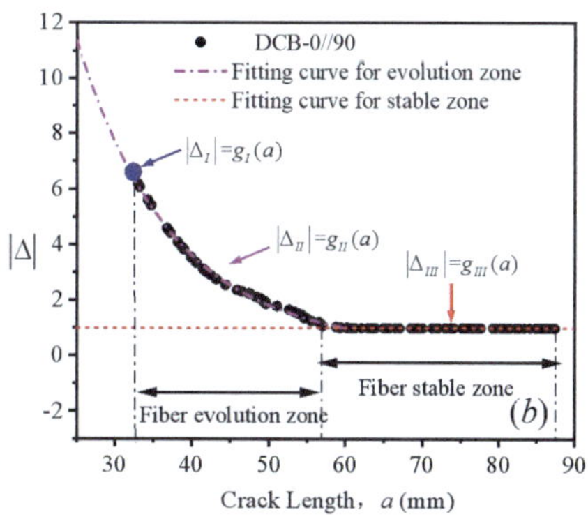

(b) Nonlinear correction factor.

Fig. 4.20: Correlation of nonlinear responses.

4.4.3 Verification of Mechanisms-Based Methodology

Since the values of the fracture toughness in the form of R-curve are not direct experimental data but calculated with a given method. Consequently, any errors introduced by the fracture toughness calculation method will influence the accuracy of the CZM modelling ability to predict the experimental results in terms of the experimental load-displacement curves. Therefore, the fracture toughness calculation method can be indirectly verified by the mechanisms based CZM modelling.

Firstly, the simulation results of the fracture toughness in terms of strain energy release rate G_{IC} are presented in Fig. 4.21 and compared with the calculation results based on the experimental data and parametrical fitting results. Like all the previous work [14, 23, 43, 44, 58], regardless of what calculation method used for the fracture toughness, the simulation results of the R-curves are always in good agreement with calculated results which sometimes are referred as 'experimental' results. This is expected since the basic parameters in the MB-CZM are obtained by data fitting with the calculated R-curve. Good agreement between modelling simulation and calculated R-curve can only verify the correctness of the modelling itself, it is not sufficient to verify the representative of modelling results to the experimental behavior.

The real experimental data obtained from the DCB testing are the load-displacement curves associated with delamination initiation and propagation. The load-displacement curves predicted by the FE analysis for the DCB specimen with both interface fiber angles of 0//0 and 0//90 are shown in Fig. 4.22 with the fracture toughness calculated with the ASTM current method and the modified method proposed in this work. The predicted load-displacement curves are then compared with the experimental data directly obtained from each of the DCB tests.

Figure 4.22a shows that the load-displacement curves predicted by the modelling simulation of the DCB tests with 0//0 specimen using the fracture toughness calculated by both methods are very similar. This is because there is less fiber bridging developed during the delamination initiation and propagation. Consequently, the influence of the fiber bridging on the compliance of the DCB beam and the root rotation

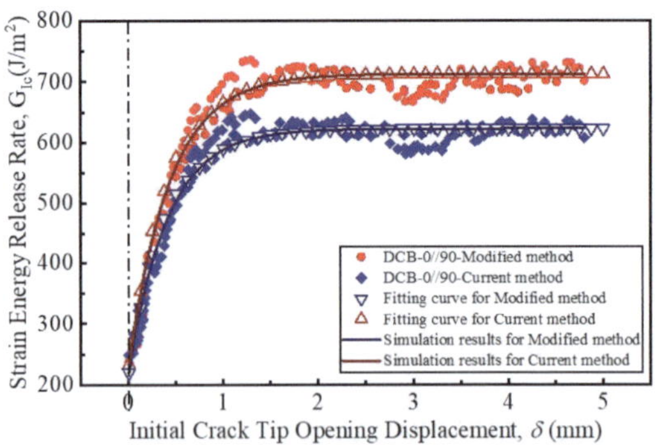

Fig. 4.21 Comparison between simulation R-curve with experimental and fitting curves.

correction factor is insignificant, and the facture toughness calculated with the ASTM current method and the modified method are similar. Fig. 4.22b indicates that there is a significant difference in the load-displacement curves of the DCB tests with 0//90 specimen predicted by the modelling simulations using fracture toughness calculated by the ASTM current method and the modified method. The modelling simulation predicted load-displacement results based on the fracture toughness calculated by the modified method developed in this work are much closer to the experimental data compared to the results based on the ASTM current method. The good agreement of the modelling simulation prediction with the experimental results provides verification of the proposed modified method, which is more universal, compared with the ASTM current method.

4.5 Advantage of Mechanisms Based Analysis Modelling and Further Applications

The common feature of the previous semi-empirical superposed CZM approaches [23, 43, 44, 50, 51, 58] is to combine the two delamination zones ahead and behind crack tip as one pseudo fracture process zone and treated as an artificial crack tip. Since the two toughening mechanisms associated with the intrinsic material facture toughness and the extrinsic fiber bridging effect are superposed together, it is difficult to identify the local deformation and fracture mechanism. The MB-CZM employs two distinct cohesive elements and two individual traction-separation laws to describe the two different toughening mechanisms respectively. Consequently, the numerical simulation results can be used to understand the local deformation and fracture mechanisms and the Mode I delamination initiation and propagation process during the DCB testing [16].

Taking the DCB specimen with basic 0//0 interface fiber angle as an example [16], the local deformation and fracture mechanisms associated with the quasi-brittle

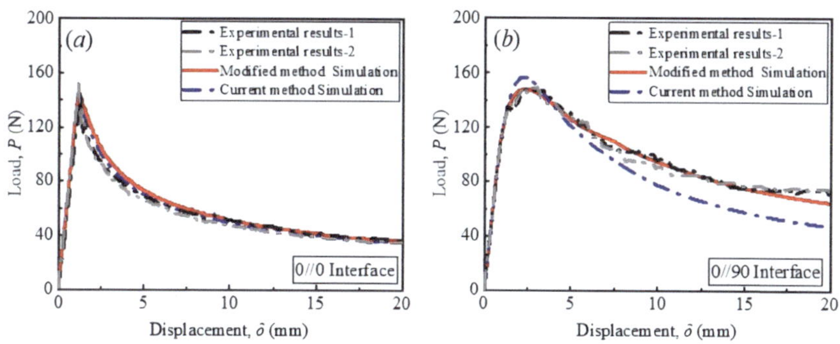

(**a**) current fracture toughness calculation method. (**b**) modified fracture toughness calculation method.

Fig. 4.22: Comparison of experimental results with modelling predictions.

linear feature of composite material and the non-linear feature of fiber bridging are examined with the numerical simulation results. Figure 4.23a provides the damage distribution ahead of crack tip represented by element E1, and the fiber bridging status (D = 1.0 indicating broken fiber bridging) behind crack tip represented by E2 at the steady state of delamination propagation (green circular in Fig. 4.23b. It

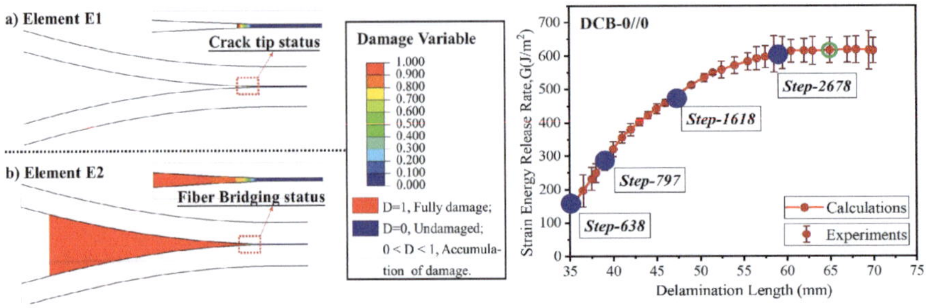

(a) Damage state at steady propagation stage. (b) Different stage during propagation.

Fig. 4.23: Damage states ahead and behind crack tip at delamination steady propagation stage.

can be seen that the quasi-brittle linear fracture zone associated with the damage mechanisms of matrix cracking and fiber/matrix debonding ahead of the crack tip is relatively small. Whilst the non-linear fiber bridging zone is significantly large with new fiber bridging near the crack front (D = 0) and old broken fiber bridging (D = 1.0) in the vicinity of the end of the fiber bridging zone (FBZ). Consequently, a constant steady strain energy release rate is achieved, as shown in Fig. 4.23b.

The progressive development of the quasi-brittle linear fracture zone ahead of the crack and the non-linear fiber bridging zone behind the crack tip is investigated with the numerical simulation results for the four stages on the R-curve as shown in Fig. 4.23b The stage A, B, C and D is related to the onset of delamination initiation (step-638), delamination length of 2 mm (step-797) and 10mm (step-1618) and 22 mm (step-2678) respectively.

The stress distribution ahead of crack tip represented by element E1 is plotted in Fig. 4.24. It is observed that the stress distribution is divided by two different zones which are related to different mechanisms. The first zone is the stress decay zone (SDZ) where the elastic stress is gradually decreased to zero in the far field. The second stress distribution zone is the finite process zone (FPZ) within which damage initiation and propagation results in delamination growth. The numerical simulation results indicates that the FPZ is independent to delamination length whilst the SDZ is gradually reduced with increasing of the delamination length. The stress decay zone (SDZ) is the results of stress concentration induced by the crack tip. The typical stress distribution is well recognized in linear elastic fracture mechanics (LEFM). The numerical simulation results indicated that the stress distribution and the decay within SDZ is influenced by the delamination length. This is attributed to the additional toughening mechanism introduced by the fiber bridging. The size of SDZ is reduced in proportional with the length of fiber bridging and the traction force induced by the fiber bridging.

The finite process zone (FPZ) starts when the traction stress equals to the material transverse tensile strength (S_{33} = 45 MPa) and ends when traction stress equals to zero at the crack tip. Material damage associated with matrix cracking and fiber matrix debonding initiates at the boundary of FPZ and SDZ and then progressively grows within the FPZ. Delamination initiation and propagation occurs when the strain energy release rate in the FPZ reaches to the interlaminar fracture toughness G_{IC}. Consequently, the size of the FPZ which is directly related to interlaminar fracture toughness G_{IC} should be independent to the delamination length. The results demonstrate the capability of the proposed MB-CZM to realistically describe the local deformation and damage mechanism in the FPZ ahead of crack tip. This is because the energy released in the FPZ is directly related to generation of new fracture surface which is the fracture toughness of the composite material. However, the change of the elastic stress distribution in the SDZ with delamination growth

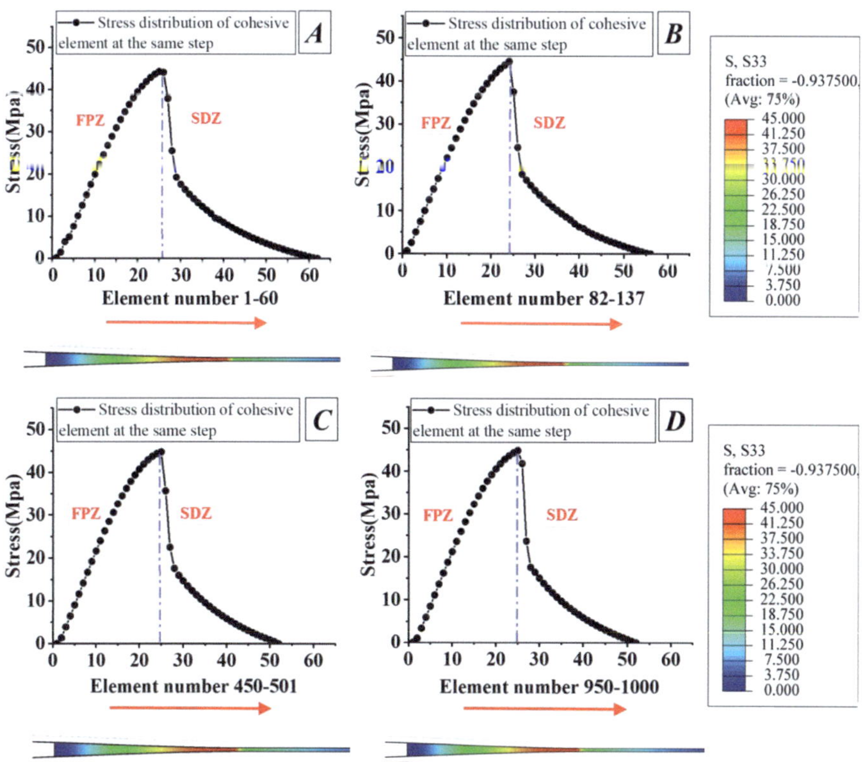

Fig. 4.24: Local stress distribution ahead of crack tip at different stages of delamination propagation.

is attributed to the additional development of fiber bridging. The bigger the fiber bridging zone size, the smaller the SDZ size.

The stress distribution of fiber bridging ahead and behind crack tip represented by element E2 at different delamination lengths at stage A, B, C and D as shown in Fig. 4.23b is also plotted in Fig. 4.25. It is clear that the fiber bridging zone is developed with delamination growth. At the onset of delamination initiation (A) the fiber bridging is restricted within the finite process zone (FPZ) ahead of crack tip with a linear stress distribution. Then a small zone of nonlinear fiber bridging

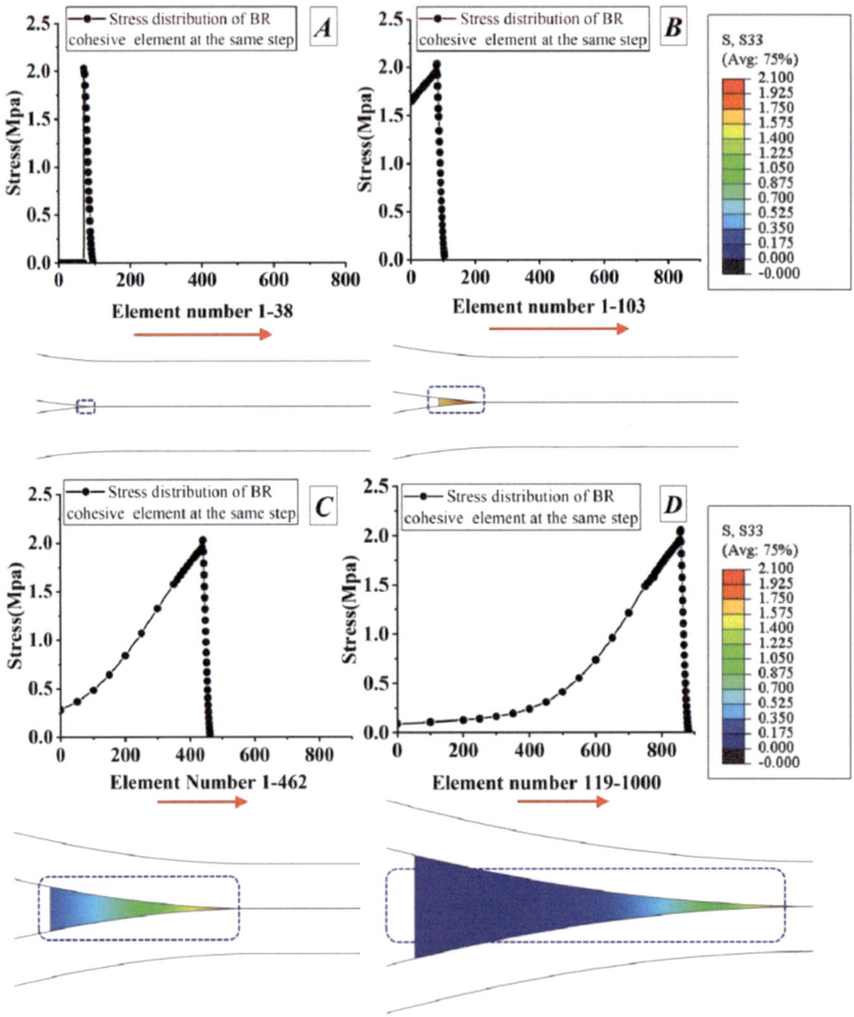

Fig. 4.25: Local stress distribution (fiber bridging stress) behind crack tip at different stages of delamination propagation.

stress distribution is developed in the delamination zone behind the crack tip at stage B. The nonlinear fiber bridging stress distribution zone is progressively increased at stage C and D respectively. The progressive development of the nonlinear fiber bridging stress distribution zone is attributed to the gradually nonlinear rise of the strain energy release rate in the R-curve shown in Fig. 4.23b. The experimental R-curves from extensive investigations [16, 20, 22, 30, 38, 44, 45, 47, 50, 51, 54] all exhibit the nonlinear characteristics. Therefore, the numerical simulation results clearly demonstrate the advantage of the proposed MB-CZM with separate nonlinear function for the fiber bridging stress distribution over the previous linear approximation commonly used tri-linear CZM [54], four-linear CZM [58] and multi-linear CZM [43].

The numerical simulation results of the overall fiber bridging stress evolution and distribution as a function of the opening displacement for the four different ply-angles are plotted in Fig. 4.26. All the fiber bridging stress distribution exhibits strong non-linear characteristics which are in agreement with the previous experimental studies and analytical investigations [16, 20, 22, 23, 30, 38, 43–45, 47, 51, 54, 58]. Furthermore, it is well recognized that fiber bridging density is progressively reduced away from the delamination crack tip. Consequently, the concept of short fiber dense area and long fiber dense area as shown in the picture inserted in Fig. 4.26 have been proposed in many previous investigations [58]. Various simplifications based on this concept have been used to approximately represent the experimental observed non-linear characteristics of fiber bridging with tri-linear [54], four-linear [58] and multi-linear [43] functions in the previous CZMs. The inserted picture in Fig. 4.26a illustrated the simplification with tri-linear function (red lines) for the approximation of the non-linear fiber bridging stress distribution [58]. It is evident that the linear simplification is not possible to fully describe a nonlinear characteristics of fiber bridging. The bi-exponential functions in Eq. (4.5) used in the MB-CZM proposed in this work can realistically represent the non-linear characteristics of the short fiber dense area and long fiber dense area of the whole fiber bridging. The numerical simulation results illustrated in Fig. 4.26 (b) demonstrates the full capabilities of the MB-CZM proposed in this work to fully represent non-linear characteristics fiber bridging observed in the experiments [3, 5, 8, 18, 24, 31, 35, 46].

It is demonstrated that mechanism-based analysis and modelling by identification of the two distinguishing delamination mechanisms within the two delamination zones associated with Mode I fracture toughness testing with DCB specimen can overcome the limitations of the current widely used semi-empirical superposed CZM. The MB-CZM not only can predict the global load-displacement deformation and the delamination propagation R-curve characteristics, but also can reveal the local progressive material damage evolution and stress distribution both ahead of the delamination crack tip and the fiber bridging stress distribution behind the delamination crack tip. Consequently, the mechanism-based analysis modelling can be potentially extended to further applications for engineering composite structures under complex loading conditions such as cyclic and fatigue loading.

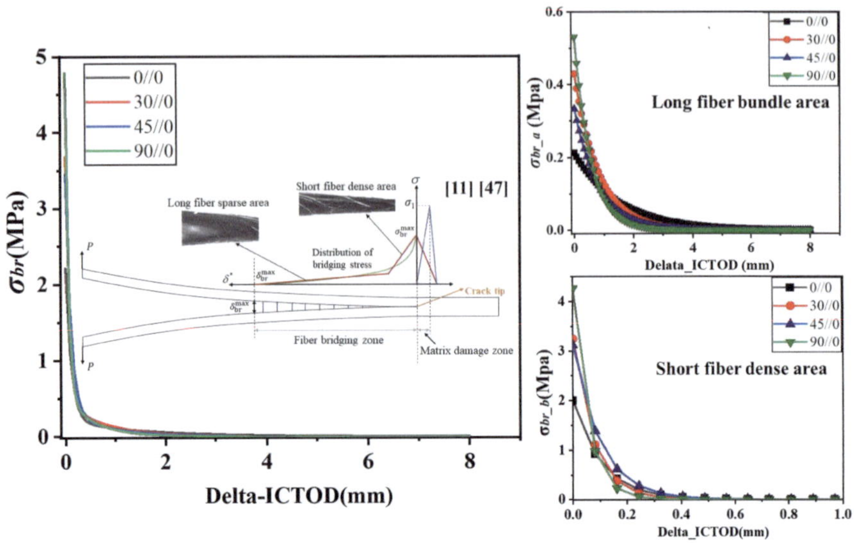

(a) Overall fiber bridging stress distribution.

(b) Short and long fiber dense distribution.

Fig. 4.26: Fiber bridging stress distribution of different layup ply-angles [31, 46, 58].

4.6 Conclusions

Delamination is one of the most common and dangerous failure modes of composite laminates used for engineering components. Consequently, understanding delamination mechanisms and development of mechanism-based analysis and modelling methods are vital for reliable assessment of delamination to support FAA and EASA airworthiness regulations. Two distinguishing delamination mechanisms within the two delamination zones associated with Mode I fracture toughness testing of composite laminates using the well-known ASTM standard double cantilever specimen (DCB) had been identified for both UD and multidirectional composite laminates. The delamination initiation and propagation energies associated with the two delamination mechanisms within the two delamination zones have been analysis and discussed. The limitations of the current empirical or semi-empirical analyzed and modelling approaches have been identified.

A new Mechanism-based cohesive zone model (MB-CZM) has been proposed and implemented in the finite element analysis via two UMAT subroutines and used in the numerical simulations. The proposed MB-CZM develops two traction-separation laws associated with two cohesive elements to individually represent the two distinctive delamination mechanisms. Consequently, the new MB-CZM can overcome the limitations of the current widely used semi-empirical superposed CZM to fully represent the quasi-brittle linear feature of composite material and the non-linear feature of fiber bridging. The good agreement of the simulation results with the experimental results provides the verification and demonstration of the capabilities of the proposed MB-CZM.

The application of the current ASTM D5528 method for calculating Mode I delamination fracture toughness has been assessed based on experimental and theoretical analysis. It has revealed that nonlinear characteristics fiber bridging evolution behind the delamination crack tip will violate the basic assumption of the current ASTM method. Consequently, modifications have been proposed to remove the limitations based on the delamination mechanisms following same principle of the current ASTM method. The modified method is verified with Mechanism-based CZM modelling and numerical simulations. The significant improvement of the prediction results by the modified method over the current method demonstrated its applicability.

The proposed MB-CZM employed two distinct cohesive elements and two individual traction-separation laws to describe the two distinguishing delamination mechanisms associated with intrinsic damage and extrinsic fiber bridging within the two delamination zones ahead and behind delamination crack tip respectively. Consequently, the proposed MB-CZM not only can predict the global load-displacement deformation and the delamination propagation R-curve characteristics, but also can reveal the local progressive material damage evolution and stress distribution both ahead of the delamination crack tip and the fiber bridging stress distribution behind the delamination crack tip. The applicability and capabilities of the proposed MB-CZM have been verified by the good agreement between the simulation and the experimental results. Therefore, the proposed MB-CZM can be extended for engineering composite structures under complex loading conditions such as cyclic and fatigue loading.

Acknowledgements The authors gratefully acknowledge financial support from National Key Research and Development Program of China (2024YFF0308400), Major Program (JD) of Hubei Province (2023BAA028), Foshan Xianhu Laboratory Project (XHD2024-12134142-02), National Natural Science Foundation of China (Grant No. 52273080,12302481, 12172265).

References

[1] AC 20-107B (2010) Composite aircraft structures. Federal Aviation Administration

[2] ASTM D5528-13 (2013) Standard Test Method for Mode I Interlaminar Fracture Toughness of Unidirectional Fiber-Reinforced Polymer Matrix Composites. ASTM International, p 1–12

[3] Airoldi A, Dávila CG (2012) Identification of material parameters for modelling delamination in the presence of fibre bridging. Composite Structures 94(11):3240-3249, DOI 10.1016/j.compstruct.2012.05.014

[4] Amiri A, Burkart V, Yu A, Webster D, Ulven C (2018) 13 - The potential of natural composite materials in structural design. In: Jawaid M, Thariq M (eds)

Sustainable Composites for Aerospace Applications: Woodhead Publishing Series in Composites Science and Engineering. Woodhead Publishing, p 269-291, DOI 10.1016/B978-0-08-102131-6.00013-X

[5] Blondeau C, Pappas GA, Botsis J (2021) Crack propagation in CFRP laminates under mode I monotonic and fatigue loads: A methodological study. Composite Structures 256:113002, DOI 10.1016/j.compstruct.2020.113002

[6] Bramsiepe K, Handojo V, Meddaikar Y, Schulze M, Klimmek T (2018) Loads and Structural Optimization Process for Composite Long Range Transport Aircraft Configuration, 2018 Multidisciplinary Analysis and Optimization Conference, AIAA 2018-3572, Session: Aeroelastic and Aero-Structures Optimization II, DOI 10.2514/6.2018-3572

[7] Brunner A, Alderliesten R, Pascoe JA (2022) In-Service Delaminations in FRP Structures under Operational Loading Conditions: Are Current Fracture Testing and Analysis on Coupons Sufficient for Capturing the Essential Effects for Reliable Predictions? Materials 16(1):248, DOI 10.3390/ma16010248

[8] Canal LP, Alfano M, Botsis J (2017) A multi-scale based cohesive zone model for the analysis of thickness scaling effect in fiber bridging. Composites Science and Technology 139:90-98, DOI 10.1016/j.compscitech.2016.11.027

[9] Cao D, Hu H, Duan Q, Song P, Li S (2019) Experimental and three-dimensional numerical investigation of matrix cracking and delamination interaction with edge effect of curved composite laminates. Composite Structures 225:111154, DOI 10.1016/j.compstruct.2019.111154

[10] Cao D, Shao L, Hu H, Duan Q, Xia J, Chen H, Cai W, Zheng K, Ji Y, Li S (2025) An analytical method for determining Mode I initiation fracture toughness in multidirectional composite laminates: Interlaminar fiber ply angle effects. Theoretical and Applied Fracture Mechanics 139:105079, DOI 10.1016/j.tafmec.2025.105079

[11] Chai H (1984) The characterization of Mode I delamination failure in non-woven, multidirectional laminates. Composites 15(4):277–290, DOI 10.1016/0010-4361(84)90708-0

[12] Christensen R (1979) Mechanics of Composite Materials. John Wiley & Sons, New York

[13] Cao D, Duan Q, Hu H, Zhong Y, Li S (2018) Computational investigation of both intra-laminar matrix cracking and inter-laminar delamination of curved composite components with cohesive elements. Composite Structures 192:300-309, DOI 10.1016/j.compstruct.2018.02.072

[14] Dávila CG, Rose CA, Camanho PP (2009) A procedure for superposing linear cohesive laws to represent multiple damage mechanisms in the fracture of composites. International Journal of Fracture. 158(2):211-223, DOI 10.1007/s10704-009-9366-z

[15] Du S, Guan Z (2008) Strategic considerations for development of advanced composite technology for large commercial aircraft in China (in Chin.). Acta Materiae Compositae Sinica 25(1):1-10

[16] Duan Q, Hu H, Cao D, Cai W, Li S (2024) A new Mechanism-based cohesive zone model for Mode I delamination coupled with fiber bridging of composite laminates. Composite Structures 332, DOI 10.1016/j.compstruct.2024.117931
[17] Duan Q, Hu H, Cao D, Cai W, Xia J, Li S (2024) Modified mode I fracture toughness calculation method for composite laminate with large scale fiber bridging, Composite Structures 349-350:118521, DOI 10.1016/j.compstruct.2024.118521
[18] Farmand-Ashtiani E, Alanis D, Cugnoni J, Botsis, J (2015) Delamination in cross-ply laminates: Identification of traction–separation relations and cohesive zone modeling. Composites Science and Technology 119:85-92, DOI 10.1016/j.compscitech.2015.09.025
[19] Farmand-Ashtiani E, Cugnoni J, Botsis J (2015) Specimen thickness dependence of large scale fiber bridging in Mode I interlaminar fracture of carbon epoxy composite. International Journal of Solids and Structures 55:58-65, DOI 10.1016/j.ijsolstr.2014.03.031
[20] Frossard G, Cugnoni J, Gmür T, Botsis J (2017) An efficient method for fiber bridging traction identification based on the R-curve: Formulation and experimental validation. Composite Structures. 175:135-144, DOI 10.1016/j.compstruct.2017.04.032
[21] Frossard G, Cugnoni J, Gmür T, Botsis J (2016) Mode I interlaminar fracture of carbon epoxy laminates: Effects of ply thickness. Composites Part A: Applied Science and Manufacturing 91:1-8, DOI 10.1016/j.compositesa.2016.09.009
[22] Gong Y, Chen X, Li W, Zhao L, Tao J, Zhang J, Hu N (2021) Delamination in carbon fiber epoxy DCB laminates with different stacking sequences: R-curve behavior and bridging traction-separation relation. Composite Structures 262:113605, DOI https://doi.org/10.1016/j.compstruct.2021.113605
[23] Gong Y, Hou Y, Zhao L, Li W, Zhang J, Hu N (2020) A modified Mode I cohesive zone model for the delamination growth in DCB laminates with the effect of fiber bridging. International Journal of Mechanical Sciences 176:105514, DOI 10.1016/j.ijmecsci.2020.105514
[24] Gong Y, Zhang B, Mukhopadhyay S, Hallett SR (2018) Experimental study on delamination migration in multidirectional laminates under mode II static and fatigue loading, with comparison to mode I. Composite Structures 201:683-698, DOI 10.1016/j.compstruct.2018.06.081
[25] Gutkin R, Laffan ML, Pinho ST, Robinson P, Curtis PT (2011) Modelling the R-curve effect and its specimen-dependence. International Journal of Solids and Structures 48(11-12):1767-1777, DOI 10.1016/j.ijsolstr.2011.02.025
[26] Hashemi S, Kinloch AJ, Williams JG (1989) Corrections needed in double-cantilever beam tests for assessing the interlaminar failure of fibre-composites. Journal of Materials Science Letters 8(2):125–129, DOI 10.1007/BF00730701
[27] Heidari-Rarani M, Ghasemi AR (2017) Appropriate shape of cohesive zone model for delamination propagation in ENF specimens with R-curve effects. Theoretical and Applied Fracture Mechanics 90:174-181, DOI 10.1016/j.tafmec.2017.04.009

[28] Hu H, Cao D, Cao Z, Li S (2021) Experimental and numerical investigations of wrinkle effect on failure behavior of curved composite laminates. Composite Structures 261:113541, DOI 10.1016/j.compstruct.2021.113541
[29] Jones R, Kinloch AJ, Michopoulos JG, Brunner AJ, Phan N (2017) . Composite Structures 180:316-333, DOI 10.1016/j.compstruct.2017.07.097
[30] Manshadi BD, Farmand-Ashtiani E, Botsis J, Vassilopoulos AP (2014) An iterative analytical/experiment al study of bridging in delamination of the double cantilever beam specimen. Composites Part A: Applied Science and Manufacturing 61:43-50, DOI 10.1016/j.compositesa.2014.02.001
[31] de Morais AB, Pereira AB (2007) Application of the effective crack method to Mode I and Mode II interlaminar fracture of carbon/epoxy unidirectional laminates. Composites Part A: Applied Science and Manufacturing 38(3):785-794, DOI 10.1016/j.compositesa.2006.09.001
[32] Khan R (2019) Fiber bridging in composite laminates: A literature review. Composite Structures. 229, DOI 10.1016/j.compstruct.2019.111418
[33] Liu C, Bai R, Lei Z, Di J, Dong D, Gao T, Jiang H, Yan C (2020) Study on mode-I fracture toughness of composite laminates with curved plies applied by automated fiber placement. Materials & Design 195, DOI 10.1016/j.matdes.2020.108963
[34] Khan R, Alderliesten R, Benedictus R (2014) Two-parameter model for delamination growth under Mode I fatigue loading (Part A: Experimental study). Composites Part A: Applied Science and Manufacturing. 65:192-200, DOI 10.1016/j.compositesa.2014.06.007
[35] Monticeli FM, Voorwald HJ, Cioffi MOH (2022) The influence of carbon-glass/epoxy hybrid composite under mode I fatigue loading: Hybrid fiber bridging zone model. Composite Structures 286:115274, DOI 10.1016/j.compstruct.2022.115274
[36] Pappas G, Botsis J (2016) Intralaminar fracture of unidirectional carbon/epoxy composite: experimental results and numerical analysis. International Journal of Solids and Structures 85-86:114-124, DOI 10.1016/j.ijsolstr.2016.02.007
[37] Pappas GA, Botsis J (2020) Towards a geometry independent traction-separation and angle relation due to large scale bridging in DCB configuration. Composites Science and Technology 197:108172, DOI 10.1016/j.compscitech.2020.108172
[38] Pappas G, Joncas S, Michaud V, Botsis J (2017) The influence of through-thickness reinforcement geometry and pattern on delamination of fiber-reinforced composites: Part II – Modeling. Composite Structures. 181:379-390, DOI 10.1016/j.compstruct.2017.08.096
[39] Peng L, Zhang J, Zhao L, Bao R, Yang H, Fei B (2011) Mode I delamination growth of multidirectional composite laminates under fatigue loading. Journal of Composite Materials 45(10):1077-90, DOI 10.1177/002199831038502
[40] Rice JR, Rosengren GF (1968) Plane strain deformation near a crack tip in a power-law hardening material. Journal of the Mechanics and Physics of Solids 16(1):1-12, DOI 10.1016/0022-5096(68)90013-6

[41] Sebaey TA, Blanco N, Costa J, Lopes CS (2012) Characterization of crack propagation in Mode I delamination of multidirectional CFRP laminates. Composites Science and Technology 72(11):1251-1256, DOI 10.1016/j.compscitech.2012.04.01

[42] Sela N, Ishai O (1989) Interlaminar fracture toughness and toughening of laminated composite materials: a review. Composites 20(5):423-435, DOI 10.1016/0010-4361(89)90211-5

[43] Shokrieh MM, Daneshjoo Z, Fakoor M (2016) A modified model for simulation of Mode I delamination growth in laminated composite materials. Theoretical and Applied Fracture Mechanics 82:107-16, DOI 10.1016/j.tafmec.2015.12.012

[44] Shokrieh MM, Salamat-talab M, Heidari-Rarani M (2017) Dependency of bridging traction of DCB composite specimen on interface fiber angle. Theoretical and Applied Fracture Mechanics 90:22-32 DOI 10.1016/j.tafmec.2017.02.009

[45] Shokrieh MM, Salamat-talab M, Heidari-Rarani M (2016) Effect of interface fiber angle on the R-curve behavior of E-glass/epoxy DCB specimens. Theoretical and Applied Fracture Mechanics 86:153-160, DOI 10.1016/j.tafmec.2016.06.006

[46] Sorensen L, Botsis J, Gmür T, Cugnoni J (2007) Delamination detection and characterisation of bridging tractions using long FBG optical sensors. Composites Part A: Applied Science and Manufacturing. 38(10):2087-2096, DOI 10.1016/j.compositesa.2007.07.009

[47] Sorensen L, Botsis J, Gmür T, Humbert L (2008) Bridging tractions in Mode I delamination: Measurements and simulations. Composites Science and Technology 68(12):2350-2358, DOI 10.1016/j.compscitech.2007.08.024

[48] Sørensen BF, Jacobsen TK (2003) Determination of cohesive laws by the J integral approach. Engineering Fracture Mechanics 70(14):1841 1858, DOI 10.1016/S0013-7944(03)00127-9

[49] Spearing SM, Evans AG (1992) The role of fiber bridging in the delamination resistance of fiber-reinforced composites. Acta Metallurgica et Materialia 40(9):2191-219, DOI 10.1016/0956-7151(92)90137-4

[50] Stutz S, Cugnoni J, Botsis J (2011) Crack - fiber sensor interaction and characterization of the bridging tractions in Mode I delamination. Engineering Fracture Mechanics 78(6):890-900, DOI 10.1016/j.engfracmech.2011.01.014

[51] Stutz S, Cugnoni J, Botsis J (2011) Studies of Mode I delamination in monotonic and fatigue loading using FBG wavelength multiplexing and numerical analysis. Composites Science and Technology. 71(4):443-449, DOI 10.1016/j.compscitech.2010.12.016

[52] Suo Z, Bao G, Fan B (1992) Delamination R-curve phenomena due to damage. J. Mech. Phys. Solids. 40(1):1-16, DOI 10.1016/0022-5096(92)90198-B

[53] Szekrényes A, Uj J (2005) Advanced beam model for fiber-bridging in unidirectional composite double-cantilever beam specimens. Engineering Fracture Mechanics 72(17):2686–2702, DOI 10.1016/j.engfracmech.2005.05.001

[54] Tamuzs V, Tarasovs S, Vilks U (2001) Progressive delamination and fiber bridging modeling in double cantilever beam composite specimens. Engineering Fracture Mechanics 68(5):513-525, DOI 10.1016/S0013-7944(00)00131-4
[55] Tan W, Falzon BG, Chiu LNS, Price M (2015) Predicting low velocity impact damage and compression-after-impact (CAI) behaviour of composite laminates. Composites Part A: Applied Science and Manufacturing 71:212-226, DOI 10.1016/j.compositesa.2015.01.025
[56] Wilkins DJ, Eisenmann JR, Camin RA, Margolis WS, Benson RA (1982) Characterizing delamination growth in graphite-epoxy. Damage in Composite Materials, ASTM Special Technical Publication 775, American Society for Testing and Materials, p 168–183
[57] Ye J, Gong Y, Tao J, Cao T, Zhao L, Zhang J, Hu N (2022) Efficiently determining the R-curve and bridging traction-separation relation of Mode I delamination in a simple way. Composite Structures 288:115388, DOI 10.1016/j.compstruct.2022.115388
[58] Yin S, Gong Y, Li W, Zhao L, Zhang J, Hu N (2020) A novel four-linear cohesive law for the delamination simulation in composite DCB laminates. Composites Part B: Engineering 180:107526, DOI 10.1016/j.compositesb.2019.107526
[59] Zhao L, Gong Y, Zhang J, Chen Y, Fei B (2014) Simulation of delamination growth in multidirectional laminates under Mode I and mixed Mode I/II loadings using cohesive elements. Composite Structures 116:509-522, DOI 10.1016/j.compstruct.2014.05.042
[60] Zhao L, Wang Y, Zhang J, Gong Y, Lu Z, Hu N, Xu J (2017) An interface-dependent model of plateau fracture toughness in multidirectional CFRP laminates under Mode I loading. Composites Part B: Engineering 131:196–208, DOI 10.1016/j.compositesb.2017.07.077
[61] Zhao M, Zhao Y, Wang A, Chang Z, Zhang J, Wang Z (2024) Investigation of the Mode I delamination behavior of double-double laminate carbon fiber reinforced composite. Composites Science and Technology 248:110463, DOI 10.1016/j.compscitech.2024.110463
[62] Zou L, Gong Y, Tian D, Zhao L, Zhang J, Hu N (2023) Effect of interface angle on Mode I delamination damage behavior of multidirectional fully isotropic laminates with the same global stiffness. Thin-Walled Structures 182:110211, DOI 10.1016/j.tws.2022.110211

Chapter 5
Auxetic Functionally Graded Beams Reinforced with Carbon Nanotubes: Prediction and Clustering of Characteristics and Mechanical Responses

M. Amélia R. Loja, Joaquim I. Barbosa, André R.D. Carvalho, and Inês J.C. Barbosa

Abstract The present work is focused on the analysis of the static and free vibration of through-thickness functionally graded composite beams, where the material graded mixture joins a metallic phase, carbon nanotubes and a ceramic phase. In the beam plane containing its length and width a re-entrant honeycomb auxetic configuration is considered. The first-order shear deformation theory is used to perform the analyses, considering case-specific shear correction factors. Numerical applications are considered with a parametric study perspective to characterise the influence of different parameters in the mentioned mechanical responses. The predictions obtained are then analysed using a non-supervised machine learning method, the K-means clustering method, to identify groupings based on shared responses or characteristics, and subsequently to support decision-making for possible solutions.

Key words: Functionally graded materials, Auxetic materials, Carbon nanotubes, First-order shear deformation theory, K-means clustering

5.1 Introduction

Functionally graded composite materials (FGMs) have gained attention in advanced engineering applications due to their ability to tailor their properties in a 3D space. This flexibility allows for customized mechanical, thermal, and dynamic properties according to specific application requirements. Among composite materials, the auxetics exhibit significant capabilities in terms of energy absorption, shear resistance,

M. Amélia R. Loja (corresponding author) · Joaquim I. Barbosa · André R.D. Carvalho · Inês J.C. Barbosa
IDMEC, Instituto Superior de Engenharia de Lisboa, Instituto Politécnico de Lisboa & CIMOSM, ISEL, IPL – Centro de Investigação em Modelação e Optimização de Sistemas Multifuncionais, Av. Conselheiro Emídio Navarro 1, 1959-007 Lisboa, Portugal, e-mail: amelia.loja@isel.pt, joaquim.barbosa@isel.pt, andre.carvalho@isel.pt, ines.barbosa@isel.pt

H. Altenbach et al. (eds.), *Composite Structures and Technology*,
Advanced Structured Materials 255,
https://doi.org/10.1007/978-3-032-24656-1_5

and fracture toughness, which makes them a relevant solution that should be taken into account, particularly when designing impact-resistant and lightweight solutions. The synergy of functionally graded materials and auxetic materials, along with the integration of carbon nanotubes (CNT), enables advanced materials architectures, where geometrical and material characteristics constitute effective design variables that should be considered beyond the traditional geometrical variables associated with the geometrical modelling of structures. In the review work by El-Galy et al.[16], the functionally graded materials were typified according to their composition, fabrication methods and envisaged industrial applications. In this latter context, the authors pointed out the current trends that include additive manufacturing to achieve a more precise mixture evolution of materials. Despite well-known advantages such as the mitigation of stress concentrations and the ability to deal with high thermal gradients, the production cost and interfacial adhesion are noted challenges. In another review proposed by Yadav et al. [61], current FGM applications were analysed as well as additive manufacturing-based fabrication techniques, including materials, processes and parameters. The authors also addressed the potential challenges and research gaps.

The study of functionally graded materials started some decades ago, in the eighties of the last century [22], and since then, an intense research work has been developed. Among the significant number of published works, one notable example is the pioneer work on the analysis of functionally graded plates developed by Reddy [40]. Later, Ferreira et al. [18], considering as a theoretical basis a third-order shear deformation theory, used collocation multiquadric radial basis functions to model and analyse the static behaviour of FGM plates. Chi and Chung [11] presented a theoretical study of FGM plates under to transverse loadings, using a higher-order shear deformation theory. The authors concluded that the material mixture evolution may significantly reduce the transverse deflection and stress peaks in comparison to homogeneous plates. Aydogdu and Taskin [7] studied and proposed closed-form solutions for the free vibrations of FGM beams using the Euler-Bernoulli and Timoshenko beam theories, having concluded that the materials' gradual mixture effectively influences the dynamic behaviour, and ceramic-rich beams will be stiffer. Sina et al. [48] developed an analytical approach for FGM beam vibrations, using the power-law material phases' distribution. The influence of the gradation coefficient in the dynamic response was found to be significant. Carvalho et al. [9] studied the influence of variability in material and geometrical properties on the static and free vibration behaviour of FGM plates. The authors also considered the constitution of regression models to predict these responses and analyse how each model parameter contributes to explaining the response variability. Tornabene et al. [52] proposed a theoretical framework to perform the dynamic analysis of doubly-curved shells through the generalized displacement field of the Carrera Unified Formulation and using the Zig-Zag effect from Murakami's function.

The use of carbon nanotubes as a reinforcement agent in the context of composite materials has also deserved the attention of many researchers, who conducted experimental, numerical or analytical studies in the context of material characterization and structures' simulation. Recently, Islam et al. [4], in a comprehensive review work, highlighted CNT applications referring specifically to their mechanical strength, electrical conductivity, and thermal properties. Among a diversity of areas, these nanoparticles enhance electronics, composites, energy storage, and biomedicine. The authors also mention challenges that must continue to be addressed, such as chirality control, biocompatibility, and scalable production. Zhu et al. [63] carried out finite element analyses to analyse the behaviour of CNT plates, using the first-order shear deformation theory. The authors performed mesh sensitivity studies, with proved convergence for vibration and buckling. Bhardwaj et al. [8] analysed the nonlinear flexural dynamic response of CNT laminates, using von Kármán strain theory. The authors considered the possibility of geometric imperfections, confirming their influence on the load capacity reduction. Cinefra et al. [12] investigated the free vibration response of double-walled carbon nanotubes, using a continuum approach and refined shell models obtained through Carrera's Unified Formulation (CUF). The carbon nanotubes are assumed to be a multilayer cylindrical shell with an equivalent thickness, and Young's modulus and van der Waals interactions are estimated by using the Lennard-Jones model. Penna and Feo [34] studied the nonlinear free vibration performance of FGM porous Euler-Bernoulli nano-beams resting on an elastic foundation. To this purpose, these authors considered a stress-driven nonlocal elasticity model, taking into account von Kármán-type nonlinearity and initial geometric imperfections.

In the context of the functionally graded materials domain, but considering their interaction with other fields, such as thermal and piezoelectric, it is relevant to refer that several studies have been developed, such as the studies by Rafiee et al. [36], Alibeigloo and Liew [3], Chakraverty and Pradhan [10] among others. Accordingly, a coupled thermo-electro-aero-elastic model for piezoelectric CNT shells was derived by Rafiee et al. [36], taking into account nonlinearities from temperature and pressure. The authors considered its application to hypersonic applications. Alibeigloo and Liew [3] developed a thermoelasticity model for CNT FGM plate structures, where they showed that the graded mixture of material effectively mitigates the thermal stresses. These results were verified against 3D elasticity solutions, confirming their accuracy in high-temperature environments. Chakraverty and Pradhan [10] addressed the thermal vibration problem in exponential FGMs with arbitrary boundary conditions. The Rayleigh-Ritz method was used to determine how natural frequencies would vary when material properties depend on temperature. The authors considered a set of applications that included heat-resistant panels in hypersonic vehicles.

The integration of nanoparticles, e.g. carbon nanotubes, into composite materials has been considered by many researchers aiming at improving the mechanical performance of structures in a wide number of domains. To illustrate some of the works developed in this context, Wattanasakulpong and Ungbhakorn [57] derived the exact solutions for CNT-reinforced beams supported on Winkler-Pasternak foundations, considering uniform and graded distributions of carbon nanotubes. The authors concluded that these nanoparticles' inclusion enhances the buckling resistance and vibration damping. Ansari et al. [5] studied the nonlinear vibrations of CNT-FGM beams under harmonic loads, concluding on the hardening/softening behaviours dependent on CNT content. Later, in Ansari et al. [6] a semi-analytical approach was proposed to study the vibration behaviour of CNT-reinforced elliptical plates. The conclusions achieved pointed out that higher CNT content increases the plates' fundamental frequencies. An exact analytical solution for free vibration in CNT-FGM beams was proposed by Shi et al. [47], foreseeing different boundary conditions. According to the results, the authors confirmed that the nanotubes' alignment improves stiffness-to-weight ratios. The free vibration performance of composite plates integrating different types of CNT distributions, in a thermal environment, was investigated by Selim et al. [44]. To this purpose, the authors used Reddy's higher-order shear deformation theory in association with an element-free kp-Ritz method. Esen et al. [17] studied CNT-reinforced FGM beams under moving loads, using finite element simulations. The results achieved showed that the inclusion of CNT according to a graded distribution enhances the structural damping and reduces the amplitude of the dynamic displacements. The authors also concluded that higher CNT content close to the beam's tension side provides a better performance. Madenci et al. [28] developed an experimental and analytical study on the buckling response of functionally graded CNT composite sandwich beams. The debonding, interlaminar cracks, and fibre breakage were also observed and analysed.

Despite the enhancements achieved through the inclusion of CNT as reinforcement agents, it is known that the agglomeration phenomena produce an undesirable effect on the mechanical performance of structures, thus being an aspect that should be taken into account and, whenever possible, minimized. Some numerical studies focused on this subject were also developed. For example, Tornabene et al. [53] developed higher-order shear deformation models to investigate the mechanical behaviour of CNT-reinforced doubly-curved shells and demonstrated that the carbon nanotubes' agglomeration affected mode shapes. The authors considered submarine hull design applications. Craveiro and Loja [13] examined the CNT agglomeration effects in the free vibration of nanocomposite functionally graded beams and plates, having concluded that CNT clusters can reduce the structures' stiffness by up to 30% in comparison to a baseline of uniformly dispersed CNTs. In a subsequent work [14] the authors extended the study to thicker structures. The effect of CNT agglomeration in conical shell structures was addressed and quantified by Maleki et al. [51] which used the generalized differential quadrature method. The authors confirmed that the agglomeration of CNT into clusters leads to lower buckling resistance, although this can be minimized via graded distributions.

In a complementary domain, in concrete, the field of metamaterials, such as the auxetic ones, has drawn the attention of many researchers in the more recent years. This is a natural consequence of their notable characteristics and potential applications, and in the synergic approach that can be established with other materials. Several researchers devoted their efforts to studying these materials and structures, such as Al-Furjan et al. [2] who investigated the wave propagation on sandwich beams with an auxetic honeycomb and outer layers of piezoelectric and 2D FGM. The authors concluded that the wave velocity increases by about two and a half times in comparison to the classic theory predictions. Also, the phase velocity is visibly influenced by the auxetic geometrical characteristics. Zhang et al. [62] assessed the possibility of an auxetic material enhancing the buckling capacity of rectangular plates under uniaxial compression. The authors concluded that the buckling performance of those plates can be significantly improved by considering a negative Poisson's ratio material when there is in-plane translation restraint along the unloaded edges. Zhu et al. [64] investigated the behaviour of auxetic honeycomb sandwich plates integrating CNT, finding that these plates revealed enhanced natural frequencies and energy absorption. The authors conducted experimental and numerical studies and concluded on the potential synergy between auxetic geometries and CNTs for lightweight structures. Lim et al. [23] conducted an experimental study on the effect of auxeticity towards impact loads and analysed the auxetic behaviour and the limiting strength under high impact situations.

Referring to some lack of guidance on how to design auxetic structures, Wallbanks et al. [56] highlight commonly used unit-cell types, exploring the design approaches used in the literature for different applications. The authors propose a workflow comprising design, simulation and testing of auxetic structures. Valle et al. [54] proposed a 3D asymmetric auxetic structure, extending the structure from a 2D unit design, using topology analysis and Timoshenko beam theory, to achieve direction-dependent properties, and thus tuning directional stiffness. Srivastava et al. [50] exploited the geometric nonlinearities in re-entrant honeycombs to enhance material auxcticity. Parametric studies were conducted to determine which cell angles would allow for maximum strain recovery. The nonlinear constitutive behaviour and mechanical properties of an auxetic honeycomb structure under large deformations were studied by Ma and Zhang [27]. The authors performed experimental and numerical studies and concluded that these structures presented strain-dependent stiffness evolution and improved energy absorption due to the negative Poisson's ratio. A computational framework integrating topology optimization with experimental validation, for 3D-printed auxetic FGMs was proposed by Pascoa et al. [33]. In their study, it was particularly analysed how tailored microstructures can be considered and designed to achieve materials whose Poisson's ratio and overall stiffness may vary in space. The mechanical properties of 2D re-entrant graded structures produced via additive manufacturing were investigated by Mohammadi Ghalehney et al. [30], having conducted experimental and numerical analyses, which allowed them to conclude that the graded auxetic geometry contributed to improved energy absorption and tunable stiffness. Loja and Barbosa [26] studied the influence of the material and geometric parameters of re-entrant hexagonal honeycomb auxetic structures on shear correction factors and maximum transverse static deflection of beams whose material composition varies through the thickness, according to an exponentially graded profile. The study developed by Xu et al. [60] examined the bending behaviour of an auxetic composite honeycomb sandwich structure. This behaviour was assessed by conducting experimental three-point bending tests, which allow concluding on the clear influence of the core on shear resistance and delayed failure. In a recent review work developed by Siniauskaya et al. [49], the most used mechanical metamaterials are presented, and their applications in vibration isolation, energy absorption, and vibration suppression are discussed. Future developments in the field of applied engineering are also considered.

In a broad sense, data analysis has as a major goal transforming data into meaningful insights that can support decisions or solve problems, by identifying patterns and relationships within datasets. As far as it is possible to see from the literature survey, the connection of exploratory data analysis with structural analysis, related fields, is increasing in different scientific and technical fields. Considering clustering in particular, it is possible to encounter methods of this unsupervised machine learning technique in different fields, namely in data mining, pattern recognition, image segmentation, customer segmentation and anomaly detection, among others. In the wide domain of structural mechanics, several researchers have considered clustering, with a particular focus on anomaly detection, for example, Wu et al. [59] which investigated the damage mechanisms in glass fibre reinforced polymeric sandwich composites under four-point bending using acoustic emission monitoring, and used a clustering approach on the three key parameters (amplitude, peak frequency and duration) to distinguish and characterise different damage mechanisms. A related work is due to Oskouei [41] focusing on glass-polyester composites. Sevillano et al. [46] explored the electromechanical impedance method for structural health monitoring of fibre reinforced polymeric composite beams, and proposed a hierarchical clustering approach to classify sensor data based on damage patterns. In another context, Liu et al. [24] combined clustering and Bayesian optimization to handle crashworthiness design of multimaterial plates and thin-walled structures. Recently, Gaspar et al. [21] proposed a framework procedure based on the prediction, optimization and characterization of optimal solutions for steel frame structures. This dataflow considered the interconnection of the red fox optimization technique with the finite element method and the K-means method to predict, optimize, and group the optimal solutions according to specific affinities. The results obtained by the authors show that this dataflow can provide useful insights and aid in optimal solutions selection.

The present work presents a study on the influence on beams' mechanical response of parameters associated with through-thickness functionally graded beams, whose mixture is ruled by a volume fraction exponential law, while in the plane that contains the beam length and width, the material has a honeycomb auxetic pattern. The material mixture through the thickness involves a metallic phase integrating or not carbon nanotubes and two alternative ceramic phases. The study considers the linear static and free vibration analyses of these beams, using first-order shear deformation theory and accounting for the use of specific shear correction factors. The results obtained in this first part of the study by using the finite element method are subsequently considered in the context of clustering analyses via the K-means method, which enables a complementary characterization of the predicted responses and provides insights into their grouping, resulting from shared behaviour trends. As far as the authors' knowledge, no similar study has been considered previously and published by other researchers.

Following this introductory section, the remainder of this chapter is organized into three additional sections. In section two, the methodology and fundamentals that support the development of the work are presented in a set of sub-sections, each corresponding to a specific topic. The third section presents and discusses the verification studies and the case studies, which are related to the parametric studies performed. The results of these studies are also the object of an exploratory analysis to determine how they can be grouped considering their specific affinity characteristics. In the fourth and last section, some conclusions are drawn based on the results achieved and the previous section discussions.

5.2 Materials and Methods

This section presents the main fundamentals of the topics related to the studies performed in this work.

5.2.1 Functionally Graded Materials

The mixture of material phases, typical of functionally graded materials, is in the present work ruled by an exponential volume fraction function, V_f, [42] which describes the smooth, continuous variation of that mixture through the beam structure thickness (h), associated with the coordinate z

$$V_f = e^{-\delta\left(1-\frac{2z}{h}\right)}. \tag{5.1}$$

The thickness coordinate varies in the interval $[-h/2, h/2]$. Considering the volume fraction expression in (5.1) will be associated with the ceramic phase of a ceramic-metallic FGM, and the ceramic phase is considered to vary from 0% in the lower surface of the plate or beam structure to 100% in the top surface, then a generic property P of this FGM will be given as

$$\delta = \frac{1}{2}\log\left(\frac{P_c}{P_m}\right), \qquad P_{FGM} = P_c V_c, \qquad V_c = V_f. \tag{5.2}$$

In the present case, the subscripts c, and m stand for the ceramic and metallic materials, respectively. Note that in the present exponential function, the gradient of the mixture through the thickness is a function of the relation between the elasticity moduli of the ceramic and metallic phases, in opposition to other volume fraction laws such as the known power law, where a specific parameter adjusts this gradient independently of the materials characteristics. Figure 5.1 illustrates the evolution of the Young's modulus through the thickness for the SiC-Al and the ZrO2-Al composites, respectively. The subscript FGM in (5.2), identifies the resulting functionally graded material, with no pores' assumption, as in Reddy [40].

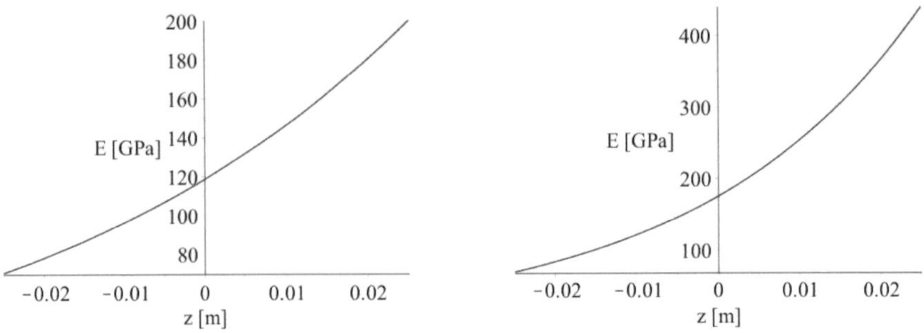

(a) SiC-Al mixture. **(b)** ZrO2-Al mixture.

Fig. 5.1: Young's modulus evolution through thickness.

5.2.2 Carbon Nanotubes Reinforcement

The beams may be constituted by a hierarchic mixture of the metallic phase and single-walled carbon nanotubes graded with the ceramic material, as mentioned in the previous section. As only this type of carbon nanotube was used, there is no need to distinguish them from the multiwalled ones, so, for conciseness, they are referenced throughout the document simply as CNT.

Concerning the mixture of the CNT with the metallic phase, it is assumed that a perfect dispersion is achieved, with no agglomeration occurring. Under this assumption, the Halpin-Tsai equations [1] were considered to estimate the resulting average Young's modulus $E_{m\mathrm{CNT}}$

$$E_{m\mathrm{CNT}} = \frac{E_{\mathrm{m}}}{8}\left[5\left(\frac{1+2\beta_{dd}V_{\mathrm{CNT}}}{1-\beta_{dd}V_{\mathrm{CNT}}}\right)+3\left(\frac{1+2\left(\frac{l_{\mathrm{CNT}}}{d_{\mathrm{CNT}}}\beta_{dl}V_{\mathrm{CNT}}\right)}{1-\beta_{dl}V_{\mathrm{CNT}}}\right)\right], \tag{5.3}$$

where β_{dd} and β_{dl} are given as

$$\beta_{dd} = \frac{\left(\frac{E_{\mathrm{CNT}}}{E_{\mathrm{m}}}\right)-\left(\frac{d_{\mathrm{CNT}}}{4t_{\mathrm{CNT}}}\right)}{\left(\frac{E_{\mathrm{CNT}}}{E_{\mathrm{m}}}\right)+\left(\frac{d_{\mathrm{CNT}}}{2t_{\mathrm{CNT}}}\right)}, \quad \beta_{dl} = \frac{\left(\frac{E_{\mathrm{CNT}}}{E_{\mathrm{m}}}\right)-\left(\frac{d_{\mathrm{CNT}}}{4t_{\mathrm{CNT}}}\right)}{\left(\frac{E_{\mathrm{CNT}}}{E_{\mathrm{m}}}\right)-\left(\frac{l_{\mathrm{CNT}}}{2t_{\mathrm{CNT}}}\right)},$$
$$V_{\mathrm{CNT}} = \frac{w_{\mathrm{CNT}}}{w_{\mathrm{CNT}}+\left(\frac{\rho_{\mathrm{CNT}}}{\rho_{\mathrm{m}}}\right)-\left(\frac{\rho_{\mathrm{CNT}}}{\rho_{\mathrm{m}}}\right)w_{\mathrm{CNT}}} \tag{5.4}$$

The parameters E_{m} and E_{CNT} are the Young's modulus of the metal and the carbon nanotubes. The nanotubes' outer diameter, thickness, and length are respectively represented by d_{CNT}, t_{CNT} and l_{CNT}. The carbon nanotubes' volume fraction is expressed as a function of the corresponding weight fraction and the densities of the phases involved, i.e. in the present case, aluminium and carbon nanotubes. Concerning the remaining material properties for the resulting mixture $P_{m\mathrm{CNT}}$, they are obtained via the Voigt rule of mixtures using the corresponding properties of the materials (subscripts m and CNT)

$$V_{\mathrm{CNT}} + V_{\mathrm{m}} = 1, \qquad P_{\mathrm{mCNT}} = P_{\mathrm{m}} V_{\mathrm{m}} + P_{\mathrm{CNT}} V_{\mathrm{CNT}}. \tag{5.5}$$

5.2.3 Re-Entrant Hexagonal Auxetic Materials

The plate or beam-type structures that will be studied, besides being functionally graded through thickness following the previous exponential volume fraction law through the thickness, possess additionally a cellular composition in the structure plane, defined by the directions x and y. This cellular architecture has a periodic character, in which its elementary unit cell is a re-entrant hexagonal auxetic material, as detailed in [25, 35].

The in-plane elastic properties, considering the material graded character previously presented, are then described as

$$\begin{aligned}
E_1 &= E_{\mathrm{c}} V_{\mathrm{c}} \frac{\eta 3^3(\eta 1 - \sin(\theta))}{\cos^3(\theta)\left[1 + (\tan^2(\theta) + \eta 1 \sec^2(\theta))\eta 3^2\right]},\\
E_2 &= E_{\mathrm{c}} V_{\mathrm{c}} \frac{\eta 3^3}{\cos(\theta)(\eta 1 - \sin(\theta))(\tan^2(\theta) + \eta 3^2)},\\
G_{12} &= E_{\mathrm{c}} V_{\mathrm{c}} \frac{\eta 3^3}{\eta 1\,(1 + 2\eta 1)\, cos\,(\theta)},\\
G_{13} &= G_{\mathrm{c}} V_{\mathrm{c}} \frac{\eta 3}{2\cos(\theta)} \left[\frac{\eta 1 - sin(\theta)}{1 + 2\eta 1} + \frac{\eta 1 + 2\sin^2(\theta)}{2(\eta 1 - \sin(\theta))}\right],\\
G_{23} &= G_{\mathrm{c}} V_{\mathrm{c}} \frac{\eta 3 \cos(\theta)}{\eta 1 - \sin(\theta)},\\
\nu_{12} &= -V_{\mathrm{c}} \frac{\sin(\theta)(1 - \eta 3^2)(\eta 1 - \sin(\theta))}{\cos^2(\theta)\left[1 + (\tan^2(\theta) + \eta 1 \sec^2(\theta))\eta 3^2\right]},\\
\nu_{21} &= -V_{\mathrm{c}} \frac{\sin(\theta)(1 - \eta 3^2)}{(\tan^2(\theta) + \eta 3^2)(\eta 1 - \sin(\theta))},\\
\eta 1 &= \frac{d}{l},\\
\eta 3 &= \frac{t}{l}.
\end{aligned} \tag{5.6}$$

The coefficients E_1, E_2, G_{12}, G_{13}, G_{23}, ν_{12} and ν_{21} are the elastic moduli, the shear moduli and the negative Poisson's ratios. The geometrical aspect ratio parameters $\eta 1$ and $\eta 3$ stand respectively for the re-entrant hexagonal cell aspect ratios d/l and t/l, where the dimensions d and l are the lengths of the horizontal and inclined side members of the hexagonal unit cell, while t is the constant members' thickness of the cell (Fig. 5.2).

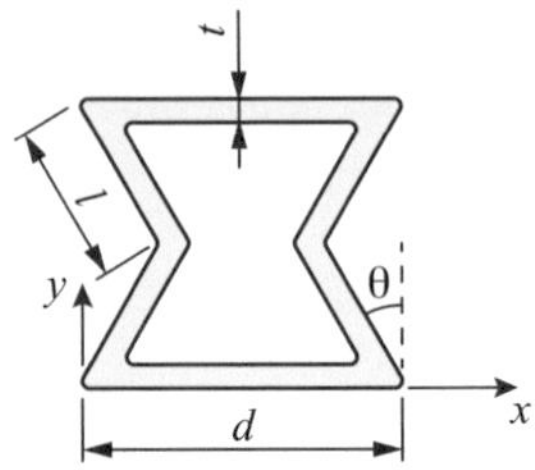

Fig. 5.2 Re-entrant hexagonal unit cell and its relation to the coordinate directions x and y.

When the angle θ is zero, the cell configuration is square or rectangular, depending on the geometrical parameter $\eta 1$ being equal to 2 or different from this value, respectively. In the present study, this angle varies discretely within $\{-60°, -45°, -30°, -15°, 0°, 15°, 30°, 45°, 60°\}$. The extreme values of the angles considered correspond to the flattening of the unit cell along the x and y directions, respectively, although when the angle is positive, the cell suffers a compression and densification in opposition to the deformation suffered when the angles are negative. The values considered for the geometrical parameters $(\eta 1)$ and $(\eta 3)$, correspond respectively to the sets $\{1, 1.25, 1.5, 1.75, 2\}$ and $\{0.01, 0.015, 0.02, 0.025, 0.03\}$ respectively.

5.2.4 Shear Deformation Theory and Constitutive Equations

The displacement field of the first-order shear deformation theory (FSDT), [29], for the analysis of plates, can be written as in (5.4)

$$\begin{aligned} u(x, y, z) &= u^0(x, y) + z\,\theta_x^0(x, y), \\ v(x, y, z) &= v^0(x, y) + z\,\theta_v^0(x, y), \\ w(x, y) &= w^0(x, y). \end{aligned} \tag{5.7}$$

The displacements along the coordinate directions (x, y, z) are described by u, v and w respectively, which describe the kinematics of a generic point of the plate (Fig. 5.3). These displacements are related to the translation and rotation degrees of freedom $\{u^0, v^0, w^0, \theta_x^0, \theta_y^0\}$ referred to the plate mid-surface as denoted by the superscript 0.

The elastic strains are obtained considering the linear kinematical relations from Elasticity Theory, and the constitutive relation is written as

$$\begin{aligned} \begin{Bmatrix} \sigma_{xx} \\ \sigma_{yy} \\ \sigma_{xy} \end{Bmatrix} &= \begin{bmatrix} C_{11} & C_{12} & 0 \\ C_{12} & C_{22} & 0 \\ 0 & 0 & C_{66} \end{bmatrix} \begin{Bmatrix} \epsilon_{xx} \\ \epsilon_{yy} \\ \gamma_{xy} \end{Bmatrix}, \\ \begin{Bmatrix} \sigma_{yz} \\ \sigma_{xz} \end{Bmatrix} &= \begin{bmatrix} k_{23}C_{44} & 0 \\ 0 & k_{13}.C_{55} \end{bmatrix} \begin{Bmatrix} \gamma_{yz} \\ \gamma_{xz} \end{Bmatrix}. \end{aligned} \tag{5.8}$$

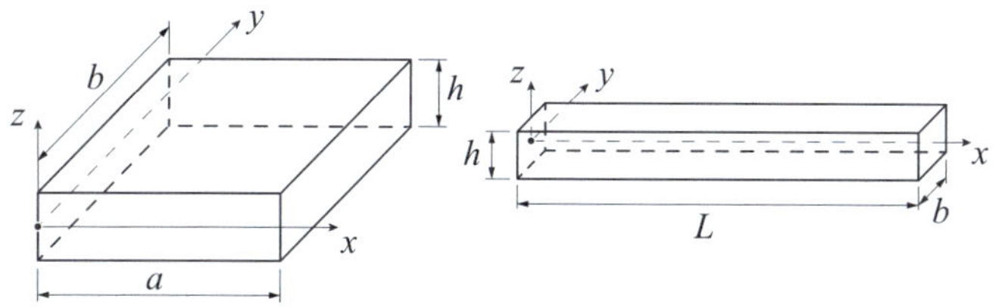

Fig. 5.3: Plate and beam schematic representations and coordinate reference system.

The stress and strain components in the mentioned Cartesian coordinate system are represented as $\{\sigma_{xx}, \sigma_{yy}, \sigma_{xy}, \sigma_{yz}, \sigma_{xz}\}$ and $\{\epsilon_{xx}, \epsilon_{yy}, \gamma_{xy}, \gamma_{yz}, \gamma_{xz}\}$ respectively, while the reduced elastic stiffness coefficients C_{ij} are given in [39]. In the present work, the materials used and the corresponding properties are given in Table 5.1. Concerning the carbon nanotubes, they are additionally characterised by a length of 25 μm, a thickness of 0.34 nm and a diameter of 1.4 nm.

Disregarding for now the inclusion of CNT, two graded mixtures of metal-ceramic were considered, one, constituted by aluminium and silicon carbide (SiC-Al) and a second of aluminium and zirconia (ZrO2-Al). Figures 5.4 and 5.5 illustrate the variation of some stiffness coefficients as a function of different parameters, considering a plate with an aspect ratio of 20 and setting $\eta 3$ to 0.01. Due to different orders of magnitude, the graphical representations for each coefficient of the two materials are separated into two: a first where the distributions at the lower and mid-surface of the composite, and a second at its upper surface.

As can be observed, the distributions of the resulting effective properties of the two materials, are very distinct, which is due to the properties of the ceramic materials. A characteristic that appears along different figures is the effect of the extreme angle values considered, which visibly produce higher stiffnesses. Regarding the effect of the parameter $\eta 1$ on the coefficients, it shows a dependence on the materials involved in the functionally graded material.

Considering the constant shear strain characteristic of the theory used in this work, it is necessary to use shear correction factors [58], which, in the constitutive relation, are denoted by the coefficients k_{mn}. Although the values of the shear correction factors are often set to $5/6$, the typical value for a rectangular homogeneous section, in the present work it was considered adequate to calculate these factors to address

Table 5.1: Materials used and corresponding properties.

	E[GPa]	ν	ρ[kg/m^3]
Silicon carbide (SiC)	440	0.17	3160
Zirconia (ZrO2)	200	0.31	5600
Aluminium (Al)	70.3	0.3	2660
Carbon Nanotubes	640	0.33	1350

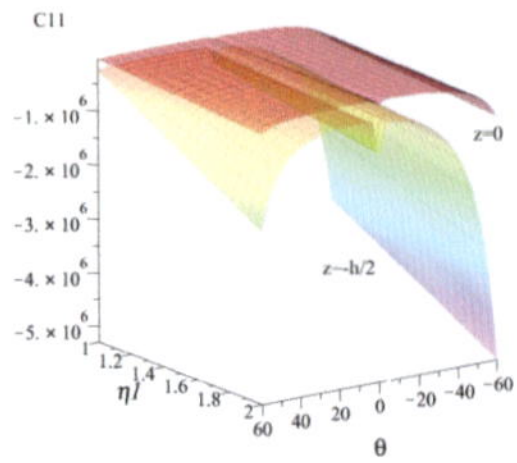

(a) SiC-Al, $z = -h/2$, $z = 0$.

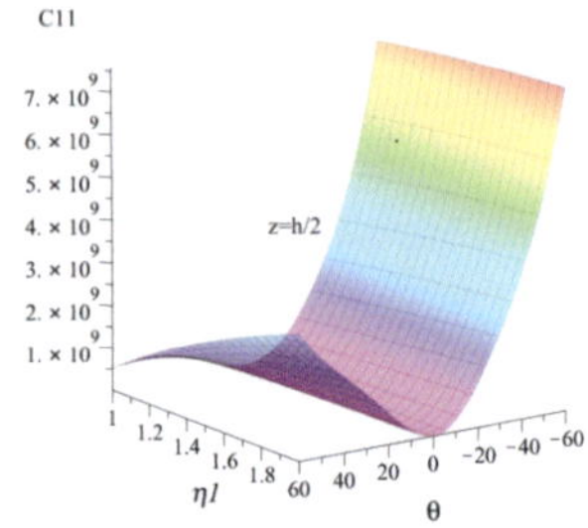

(b) SiC-Al, $z = +h/2$.

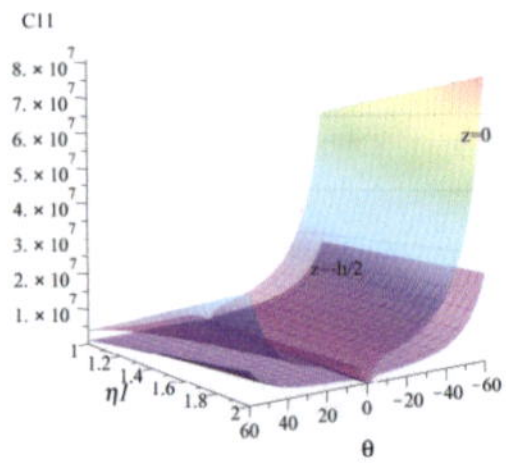

(c) ZrO2-Al, $z = -h/2$, $z = 0$.

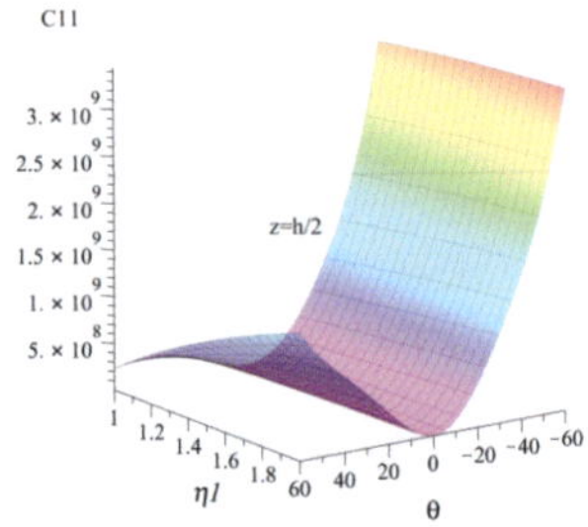

(d) ZrO2-Al, $z = +h/2$.

Fig. 5.4: Reduced stiffness coefficient C_{11} distributions. Parameter $\eta 3$ set to 0.01.

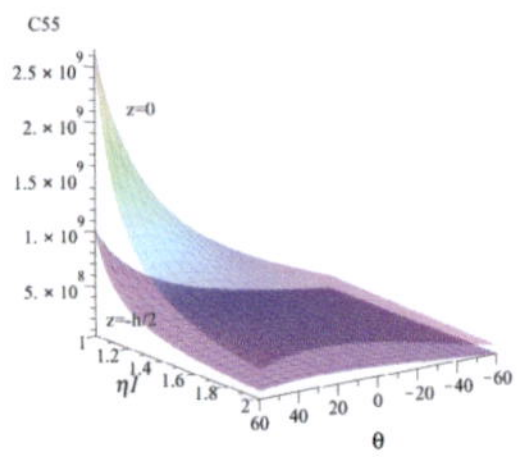

(a) SiC-Al, $z = -h/2$, $z = 0$.

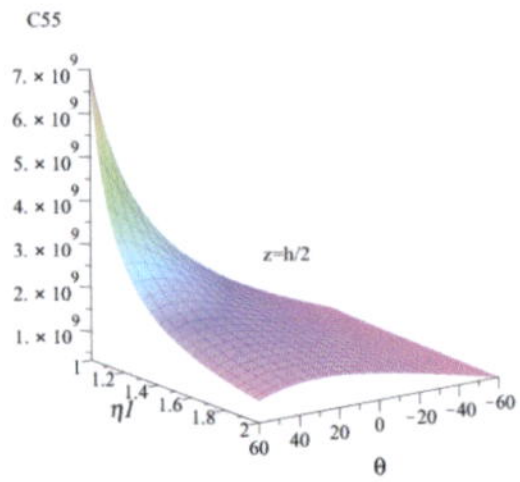

(b) SiC-Al, $z = +h/2$.

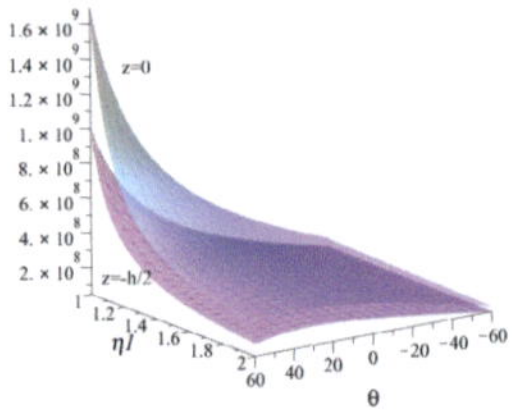

(c) ZrO2-Al, $z = -h/2$, $z = 0$.

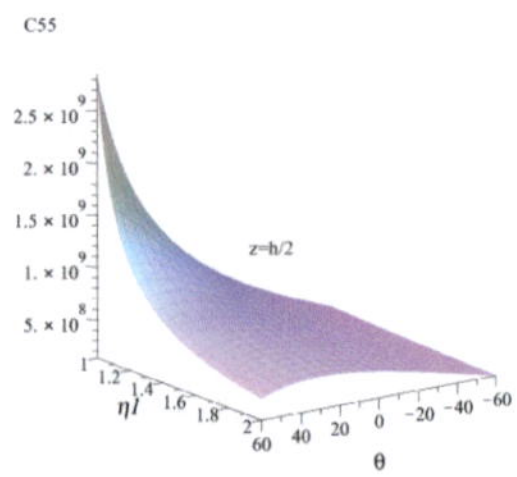

(d) ZrO2-Al, $z = +h/2$.

Fig. 5.5: Reduced stiffness coefficient C_{55} distributions. Parameter $\eta 3$ set to 0.01.

the effective cross-sections non-homogeneity as in [31, 32]. To this purpose, the methodology proposed by Vlachoutsis [55] was considered, which for the calculation of the k_{23} coefficient, for example, yields the following expression

$$k_{23} = \frac{\left[\int_{-h/2}^{h/2} C_{22}(z)(z - z_{n1})^2 \mathrm{d}z \right]^2}{\int_{-h/2}^{h/2} G_{23}(z)\mathrm{d}z \int_{-h/2}^{h/2} \left(\frac{\left(- \int_{-h/2}^{z} C_{22}(z)(z-z_{n1})^2 \mathrm{d}z \right)^2}{G_{23}(z)} \right) \mathrm{d}z}, \tag{5.9}$$

where the neutral surface thickness coordinate is represented by z_{n1} while the mid-surface is associated to the coordinate $z = 0$.

5.2.5 Equilibrium Equations

Considering the aim of predicting and characterising the responses arising from linear elastic static and free vibration analyses for the present auxetic exponentially graded beam/plate structures, the Hamilton's principle [39, 65] was considered. It was considered only, surface-distributed transverse loads applied to the structures.

$$\delta \int_{t_0}^{t_1} (U - T + W)\, \mathrm{d}t = 0, \tag{5.10}$$

where U,T,W are respectively given as

$$\begin{aligned} U &= \frac{1}{2} \int_V (\sigma_x \epsilon_x + \sigma_y \epsilon_y + \sigma_{xy}\gamma_{xy} + \sigma_{xz}\gamma_{xz} + \sigma_{yz}\gamma_{yz})\, \mathrm{d}V, \\ T &= \frac{1}{2} \int_V \rho_{\mathrm{FGM}} (\dot{u}^2 + \dot{v}^2 + \dot{w}^2)\, \mathrm{d}V, \\ W &= \int_A p_z(x, y) w \, \mathrm{d}A. \end{aligned} \tag{5.11}$$

The equilibrium equations obtained after the mathematical manipulations associated to Hamilton's principle can be written as

$$
\begin{aligned}
\frac{\partial N_{xx}}{\partial x} + \frac{\partial N_{xy}}{\partial y} &= I_0 \ddot{u}^0 + I_1 \ddot{\theta}_x^0, \\
\frac{\partial N_{xy}}{\partial x} + \frac{\partial N_{yy}}{\partial y} &= I_0 \ddot{v}^0 + I_1 \ddot{\theta}_y^0, \\
\frac{\partial Q_{xz}}{\partial x} + \frac{\partial Q_{yz}}{\partial y} &= I_0 \ddot{w}^0, \\
\frac{\partial M_{xx}}{\partial x} + \frac{\partial M_{xy}}{\partial y} - Q_{xz} &= I_1 \ddot{u}^0 + I_2 \ddot{\theta}_x^0, \\
\frac{\partial M_{xy}}{\partial x} + \frac{\partial M_{yy}}{\partial y} - Q_{yz} &= I_1 \ddot{v}^0 + I_2 \ddot{\theta}_y^0
\end{aligned}
\tag{5.12}
$$

with

$$
\begin{aligned}
\{N_{ij}, M_{ij}\} &= \int_{-h/2}^{h-2} \sigma_{ij}\{1, z\}\, \mathrm{d}z, \quad i, j = x, y, \\
\{Q_{iz}\} &= \int_{-h/2}^{h-2} \sigma_{iz}\, \mathrm{d}z, \quad i = x, y, \\
\{I_0, I_1, I_2\} &= \int_{-h/2}^{h-2} \rho_{\mathrm{FGM}}\{1, z, z^2\}\, \mathrm{d}z, \\
(\,\dot{}\,) = \frac{\partial}{\partial t}, &\quad (\,\ddot{}\,) = \frac{\partial^2}{\partial t^2}.
\end{aligned}
\tag{5.13}
$$

Assuming either free and harmonic vibrations or a static equilibrium situation, the equilibrium equations at the element level can be written in a compact form as:

$$
\begin{aligned}
([K] - \omega^2 [M])\{q\} &= 0, \\
[K]\{q\} &= \{f\}.
\end{aligned}
\tag{5.14}
$$

These matrices and vectors, $[K]$, $[M]$, $\{f\}$ and $\{q\}$ are respectively the stiffness and mass matrices, and the generalized forces and degrees of freedom vector. ω stand for the kth natural frequency.

These equilibrium equations were implemented at the element level using the finite element method and considering a four-node Lagrange quadrilateral finite element with five degrees of freedom per node (Fig. 5.6). A reduced integration approach was considered for the transverse shear contribution [37, 65]. The whole discretized domain is finally modelled after assembling all the mentioned element matrices and vectors constituting the global equilibrium equations and imposing the boundary conditions.

Fig. 5.6 Linear Lagrange quadrilateral element for first-order shear theory implementation.

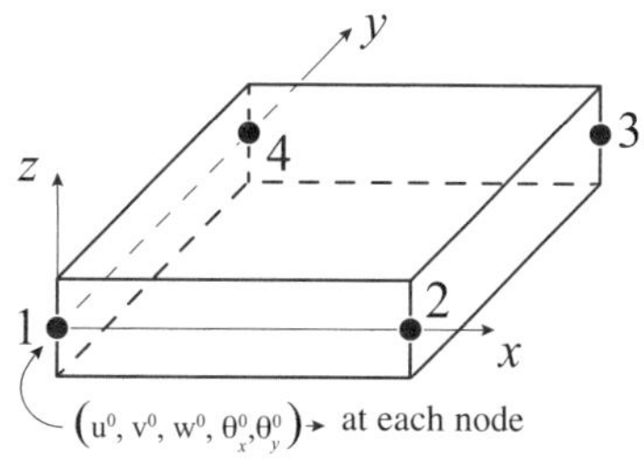

5.2.6 Exploratory Clustering Analysis

For a complementary interpretation of the results achieved, an exploratory clustering analysis was conducted on intermediate and final results, using for this purpose the K-means algorithm [19, 20], an unsupervised machine learning algorithm. K-means method partitions a dataset into non-overlapping clusters, assigning each data point to a certain cluster through an iterative process where the sum of squared distances between the data points $x(i)$ and a cluster centroid is minimized. This minimization may also be assessed through the centroids shift, which also allows assessing convergence, however, with a lower computational cost

$$\begin{aligned} \text{shift} &= \sum_{k=1}^{N_k} ||\mu_k^t - \mu_k^{t-1}|| < \text{tolerance}, \\ \mu_k &= \frac{1}{N_k} \sum_{i=1}^{N_k} x(i). \end{aligned} \tag{5.15}$$

Any cluster (C_k) has a geometric centre, the centroid, which results from the mean (μ_k) of all the data points' values, assigned to that cluster. The centroids' initialization is made using K-means++, an algorithm to select the initial centroids, which spreads out these points, reducing the chances of poor clustering due to bad initialization. The number of clusters can be selected according to prior knowledge of the dataset; however, in this study it was used the Silhouette method which allows the determination of the required and fittest number of clusters. The score given by the method varies within the interval $[-1, 1]$ and the clustering structure will be as fittest as closer to 1

$$\begin{aligned} S(i) &= \frac{1}{N} \sum_{i=1}^{N} s(i), \quad s(i) = \frac{b(i) - a(i)}{\max\{a(i), b(i)\}}, \\ a(i) &= \frac{1}{N_k - 1} \sum_{j \in C_k, j \neq i} d(i, j), \quad b(i) = \min_{C_m \neq C_k} \left(\frac{1}{|C_m|} \sum_{j \in C_m} d(i, j) \right). \end{aligned} \tag{5.16}$$

The scores for a generic $i - th$ point or for the global data set are respectively denoted as $s(i)$ and $S(i)$. The average intra-cluster distance and the average nearest-cluster distance (considering Euclidean distances) are respectively calculated as $a(i)$ and $b(i)$. In the present work, the datasets are organized as matrices, whose columns contain the features' values for each instance, for example, the cell geometrical parameters, $\eta 1$, $\eta 3$ or θ. Each line corresponds to the value obtained for a specific variable of interest, for instance, the shear correction factor or the fundamental frequency, considering a combination of the features' values on that line. In this post-processing stage, and before the clustering analysis, the features are standardized via centring and scaling procedures. This will ensure that each feature will be equally weighted when carrying out the clustering analysis, thus avoiding bias due to possible different orders of magnitude.

5.3 Results and Discussion

This section comprises two verification cases and a set of six case studies. These are parametric studies, with complementary perspectives, to assess the influence of geometrical and material parameters of the re-entrant hexagonal auxetic material in the static and free vibration of structures. The effect these parameters have on the shear correction factors is also considered.

5.3.1 *Verification Cases*

5.3.1.1 Case 1: Maximum Transverse Deflection of Plates

The first verification study considers an isotropic and homogeneous, square, simply supported plate submitted to a uniform transversely distributed load p_z. A set of edge-to-thickness ratios, (a/h), was considered, and the analyses were performed considering double-symmetry with a 4x4 mesh. The results are presented in the non-dimensional form,

$$\overline{w} = \frac{Eh^3 w}{p_z a^4},$$

in Table 5.2, for a set of edge-to-thickness ratios and compared with the results of other authors. Table 5.2 also shows the relative deviations of the present model results from the three reference results. The relative deviations of the present results from different authors are presented within parentheses, and they are calculated as

$$\text{dev} = \frac{\overline{w} - \overline{w}_{\text{ref}}}{\overline{w}_{\text{ref}} \times 100},$$

where $\overline{w}_{\text{ref}}$ stand for the reference non-dimensional deflection.

Table 5.2: Non-dimensional maximum transverse deflection $\overline{w}$ and relative deviations (%).

a/h	$\overline{w}$	Sengupta [45] $\overline{w}$	Rao et al. [38] $\overline{w}$	Salerno and Goldberg [43] $\overline{w}$
4	0.05899	0.05899 (0.44)	0.05631 (4.76)	0.05656 (4.30)
5	0.05369	0.05356 (0.24)	0.05201 (3.23)	0.05360 (0.17)
10	0.04661	0.04666 (-0.11)	0.04627 (0.73)	0.04632 (0.63)
20	0.04484	0.04494 (-0.22)	0.04483 (0.02)	0.04486 (-0.04)

As can be observed, the results present a very good agreement with the alternative authors, particularly with Sengupta [45] whose model is based on Mindlin-Reissner theory and was implemented through a triangular plate bending element with thirty-five degrees of freedom. Regarding the agreement with the remaining references, there is also a good agreement except for the very thick beams, where a greater deviation is observed, although below 5%.

5.3.1.2 Case 2: Shear Correction Factors for Layered Plate with Functionally Graded Outer Layers

This verification study aims to determine the shear correction factors for a layered plate structure with a metallic core and outer layers made of a metal–ceramic FGM. All the materials have a Poisson's ratio of 0.3. The FGM follows a power-law volume fraction distribution. The results are summarized in Table 5.3.

The results of this study align perfectly with those reported by Nguyen et al. [12]. For this reason, this table displays only the obtained values.

Table 5.3: Non-dimensional maximum transverse deflection $\overline{w}$ and relative deviations (%).

	Exponent N							
E_c/E_m	0.00	0.20	0.40	0.60	1.00	2.00	4.00	6.00
1	0.83	0.83	0.83	0.83	0.83	0.83	0.83	0.83
2	0.83	0.81	0.80	0.78	0.77	0.74	0.72	0.71
3	0.83	0.79	0.76	0.73	0.70	0.66	0.62	0.61
4	0.83	0.77	0.72	0.69	0.64	0.59	0.55	0.53
5	0.83	0.75	0.69	0.64	0.59	0.53	0.48	0.47
6	0.83	0.73	0.65	0.61	0.55	0.48	0.44	0.42

5.3.2 Case Studies

In this section, a set of studies is presented to characterise how material and geometrical parameters influence the maximum static deflection and the fundamental frequency of simply supported beams with a length-to-thickness ratio of 20. Besides the influence of these parameters on the beam's mechanical responses, it is also considered the influence in the shear correction factor, of particular importance in the theoretical approach adopted in this work.

The material properties used were presented in Table 5.1 and the meaning of the geometrical features of the re-entrant unit cell are as referred to in Sect. 5.2; $\eta 1$ and $\eta 3$ stand respectively for the re-entrant hexagonal cell aspect ratios d/l and t/l, where the dimensions d and l are the lengths of the horizontal and inclined side members of the hexagonal unit cell, while t is the constant members' thickness of the cell (Fig. 5.2).

5.3.2.1 Case 3: Influence of Parameters $\eta 1$, $\eta 3$ and θ on Shear Correction Factors for Metal-Ceramic Exponentially Graded Beams

In the present case study, two auxetic functionally graded materials are considered, a first one, constituted by aluminium and silicon carbide (SiC-Al) and a second one, whose composition is aluminium and zirconia (ZrO2-Al). In both cases, there is no inclusion of carbon nanotubes. Figure 5.7 illustrates the influence of the geometrical parameters $\eta 1$, $\eta 3$ and θ in the shear correction factor (SCF), in short designated by k, required by the first-order shear deformation theory.

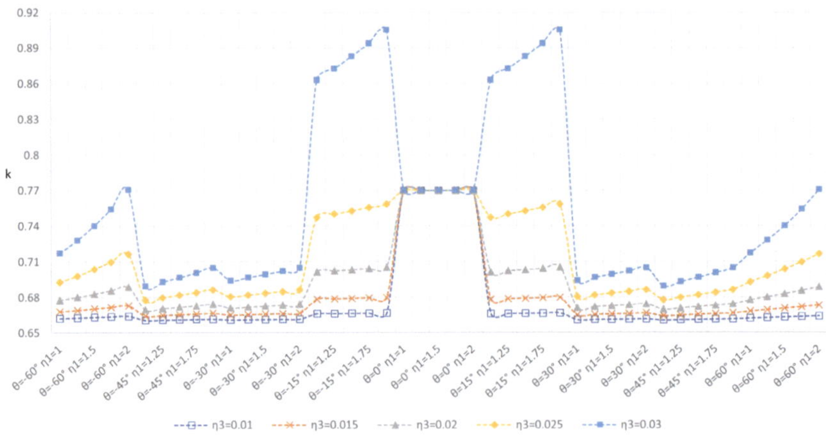

Fig. 5.7: Evolution of the shear correction factor (k) for the SiC-Al auxetic material as a function of θ, $\eta 1$ and $\eta 3$.

The results achieved tell us that the SCF for the SiC-Al auxetic FGM increases with the parameter $\eta 1$ as it goes from 1 to 2. This effect is more evident when $\eta 3$ increases, which means that the relative thickness of the cell influences this quantity. Hence, cells with a higher thickness-to-inclined side length' ratio, lead to larger SCF.

When the angle θ is zero, the re-entrant sides become vertical and the SCF values stabilize, meaning that the variation of these cell aspect ratios has no influence on this factor, showing no auxetic effect. However, the highest values for the shear correction factor are found when the angle θ is equal to $\pm 15°$, where a sharp increasing trend of the shear correction factor with $\eta 1$ is visible. Moderate values for this factor can be found in the interval of $\pm(30°$ to $60°)$, although for the $\pm 60°$ angles, a steepest trend is observed, but in a range of moderate values.

To implement the clustering analysis [15], mentioned in Subsect. 5.2.5 in the context of the present work dataflow, the calculation of the Silhouette scores for a set of possible numbers of clusters was conducted and followed by the application of the K-means method, to obtain the clustering structure of the data. The global average Silhouette scores for some numbers of clusters are shown in Table 5.4.

The first cluster comprises the instances with the angle $\theta = 0°$ for which there is no influence of the parameters $\eta 1$ and $\eta 3$. A second cluster includes the instances ranging from $\pm 30°$ to $\pm 60°$ where a low to moderate range of the SCF values is found and where the influence trend of $\eta 1$ and $\eta 3$ is similarly noted. Finally, the third cluster is characterised by the angles $\pm 15°$, where a moderate to high influence trend of these parameters and the highest values of the SCF are registered.

Overall, the clustering structure obtained indicates a parameter independence when the angle is 0°. When the angle assumes the values $\pm 15°$, the SCF demonstrates a greater sensitivity to the geometrical ratios, which tend to progressively stabilize and be less parameter-sensitive for intermediate magnitude angles. The schematic representation of the clustering structure is shown in Fig. 5.8.

Considering now the other auxetic exponentially graded material, the ZrO2-Al structure, with the same aspect ratio as previously, one obtains the set of curves that represent the influence of the geometrical parameters $\eta 1$, $\eta 3$ and θ in the SCF. The results distributions clearly differ from the previous one, as represented in Fig. 5.9. In this case, the SCF in the set of θ values considered, present a symmetrical pattern with respect to 0°. Similarly to the SiC-Al auxetic FGM, when the angle is zero, the variation in the cell geometrical parameters $\eta 1$, $\eta 3$ has no effect on the shear correction factor, hence no auxetic effect is produced.

In contrast with the SiC-Al auxetic FGM, where for the $\pm 15°$ angles, maximum values of SCF were found, for the ZrO2-Al auxetic FGM, one observes lower values, suggesting a lower shear stiffness. Regarding the influence of the relative thickness

Table 5.4: Global average Silhouette scores for 2-8 clusters. Shear correction factor (k) of SiC-Al auxetic material as a function of θ, $\eta 1$ and $\eta 3$.

Clusters	2	3	4	5	6	7	8
Silhouette score	0.755	**0.847**	0.797	0.774	0.719	0.715	0.655

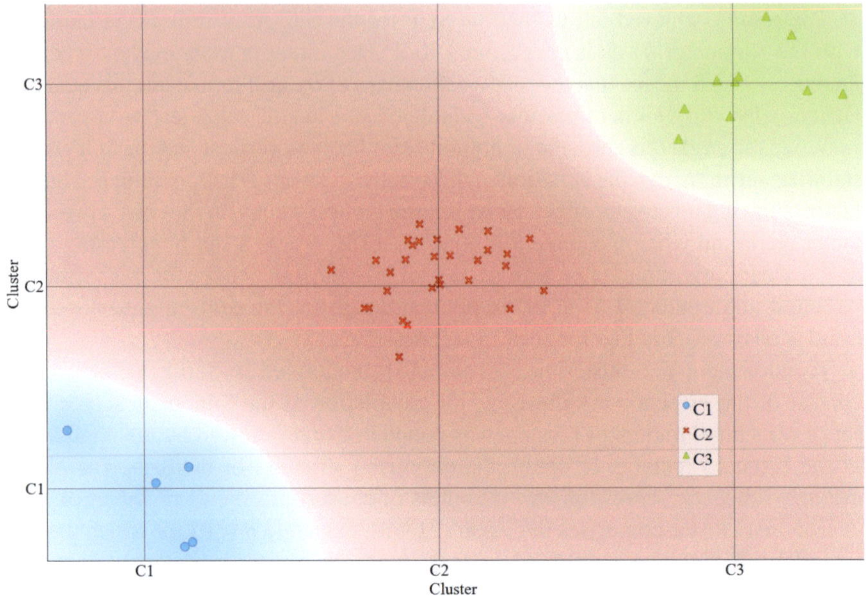

Fig. 5.8: Clustering structure associated with the shear correction factor of SiC-Al auxetic material as a function of θ, $\eta 1$ and $\eta 3$.

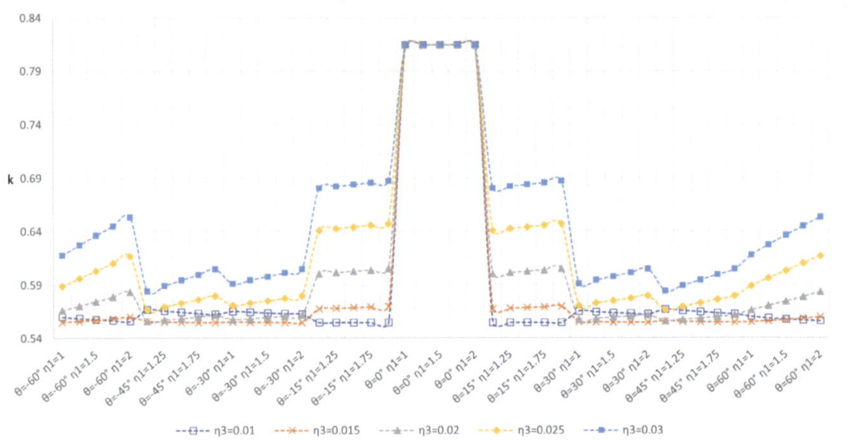

Fig. 5.9: Evolution of the shear correction factor for the ZrO2-Al auxetic material as a function of θ, $\eta 1$ and $\eta 3$.

Table 5.5: Global average Silhouette scores for 2-8 clusters. Shear correction factor (k) of ZrO2-Al auxetic material as a function of θ, $\eta 1$ and $\eta 3$.

Clusters	2	3	4	5	6	7	8
Silhouette score	**0.890**	0.832	0.826	0.704	0.714	0.691	0.667

ratio ($\eta 3$) it is visible that higher values produce an increase in the shear correction factor, the same happening for the influence of $\eta 1$, where a slight increase in that factor is observed.

The results obtained for the global average Silhouette score for different possible clusters are presented in Table 5.5. One of these clusters is characterised by the instances with angle $\theta = 0°$, for which no variation is observed with respect to $\eta 1$ and $\eta 3$. The other cluster comprises the remaining instances where a low and moderate influence of $\eta 1$ and $\eta 3$ respectively, is noticed.

Considering the two cases analysed, it is possible to conclude that both materials demonstrate geometrical parameters' independence when the angle is 0°, showing the ZrO2-Al auxetic material the highest shear correction factor value (for this angle). The SiC-Al auxetic material show a stronger $\eta 3$ sensitivity when $\theta = \pm 15°$. It also shows a greater $\eta 1$ sensitivity at $\theta = +60°$ and particularly at $\pm 15°$, while for the ZrO2 the $\eta 1$ dependence is in general weak except when $\eta = \pm 60°$.

5.3.2.2 Case 4: Influence of Parameters $\eta 1$, θ and Carbon Nanotubes' Weight Fraction on Shear Correction Factors for CNT Reinforced Metal-Ceramic Exponentially Graded Beams

The present case considers that the functionally graded material may include carbon nanotubes with different weight fraction contents (wcn), namely $\{0, 1, 3, 5\}\%$. The aim of this case is now to characterise how this graded reinforcement will affect the shear correction factors. Figure 5.10 illustrates the influence of the parameters, θ , $\eta 1$ and wcn for a fixed value of $\eta 3$ ($\eta 3 = 0.01$) on the shear correction factor (SCF) for the SiC-Al-CNT auxetic FGM.

The results achieved indicate symmetry of the SCF relative to the angle $\theta = 0°$, which presents peak values no matter the ratio $\eta 1$ or the CNT weight content (wcn). This peak trend indicates overall maximum shear stiffnesses for these cases. At the zero angle, the SCF reverses its evolution sign when compared with the other angles. It is also visible that a stronger stiffening effect occurs when increasing the weight content of CNT.

The clustering structure is highly influenced by the wcn factor, which dictates the grouping arrangements achieved. The four clusters (according to the highest Silhouette score in Table 5.6) correspond to the four values assumed by wcn, independently of the $\eta 1$ values, having each cluster three instances related in any case with the $\eta 1 = 1$, 1.5 and 2. For a fixed value of wcn when $\eta 1$ increases, the SCF decreases.

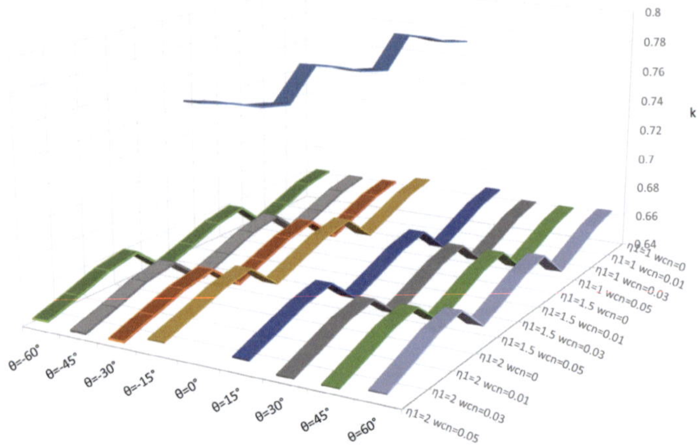

Fig. 5.10: Evolution of the shear correction factor as a function of the parameters, η , $\eta 1$ and wcn for a fixed value of $\eta 3(\eta 3 = 0.01)$ for the SiC-Al-CNT auxetic FGM.

Table 5.6: Global average Silhouette scores for 2-8 clusters. Shear correction factor (k) of SiC-Al auxetic material integrating carbon nanotubes as a function of θ, $\eta 1$ and wcn. $\eta 3 = 0.01$.

Clusters	2	3	4	5	6	7	8
Silhouette score	0.706	0.798	**0.849**	0.688	0.532	0.353	0.173

Figure 5.11 illustrates for the ZrO2-Al-CNT auxetic FGM, the shear correction factor evolution as a function of the parameters, θ , $\eta 1$ and wcn also for the same fixed value of $\eta 3$ ($\eta 3 = 0.01$). As depicted in the figure, the shear correction factor presents symmetry across the angle of 0°, for which it consistently obtains the highest shear correction factor regardless of the values of $\eta 1$ and whether carbon nanotubes are present or not.

Regarding the carbon nanotubes wcn for a fixed $\eta 1$ ($\eta 1 = 1$), and for the non-zero angle, the shear correction factor decreases as this content increases; however, when the angle is zero, the correction factor follows an opposite trend. Hence, a higher carbon nanotubes content when the angle of the auxetic unit cell is zero, represents an increased transverse shear stiffness. If fixed CNT contents is considered, the results show that as the ratio $\eta 1$ increases ($\eta 1 = 1$ to $\eta 1 = 2$), it leads to slightly greater SCF overall, following the trend observed when no CNT were present.

Overall, summarizing Figs. 5.10 and 5.11, it is possible to see that ZrO2-Al-CNT composite presents a higher SCF for the angle $\theta = 0°$, when compared to the SiC-Al-CNT, however, in both cases, the highest SCF occur for this angle, and is dominant over the remaining situations analysed. For the remaining angles, the opposite situation is observed; SiC-Al-CNT shows higher SCF values in comparison with the ones obtained for ZrO2-Al-CNT. The clustering structure and constitution

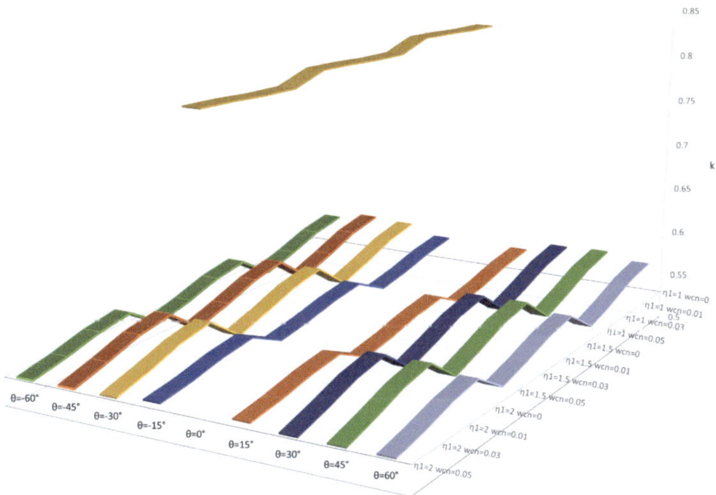

Fig. 5.11: Evolution of the shear correction factor as a function of the parameters, θ , $\eta 1$ and wcn for a fixed value of $\eta 3 (\eta 3 = 0.01)$ for the ZrO2-Al-CNT auxetic FGM.

rationale for this composite follow similar guidelines as those found for the SiC-Al-CNT composite, which, for fixed $\eta 3$, denotes the influence of the CNT content on the shear correction factor.

5.3.2.3 Case 5: Influence of Parameters $\eta 3$, θ and Carbon Nanotubes' Weight Fraction on the Shear Correction Factor for CNT Reinforced Metal-Ceramic Exponentially Graded Beams

Following the previous studies, Fig. 5.12 presents for the SiC-Al-CNT, the evolution of the correction factor as a function of the angle θ, the parameter $\eta 3$ and the CNT wcn for a fixed value of $\eta 1$ $(\eta 1 = 1)$. This figure depicts again the symmetric profiles of the SCF around the angle $\theta = 0°$, now for a fixed value of $\eta 1$ $(\eta 1 = 1)$ and assuming three values for $\eta 3$ $(\eta 3 = 0.01$, 0.02 and 0.03) as well as considering increasing CNT weight contents ($wcn = 0.01, 0.03, 0.05$), for the SiC-Al-CNT case. Globally, for fixed values of $\eta 1$ and θ, increasing values of $\eta 3$ lead to increasing values of the SCF. More specifically, if $\eta 3$ assumes the values 0.01 and 0.02, the SCF maximum values are observed for the angle $\theta = 0°$, however, when $\eta 3$ equals 0.03, a bimodal profile is depicted, showing peak values when θ equals $\pm 15°$.

As an overall trend, it is observed that for any value of $\eta 3$, increasing the CNT weight fraction content, leads to decreasing values of the SCF.

The best clustering structure is the one corresponding to two clusters, as demonstrated by the Silhouette scores presented in Table 5.7. This structure is clearly

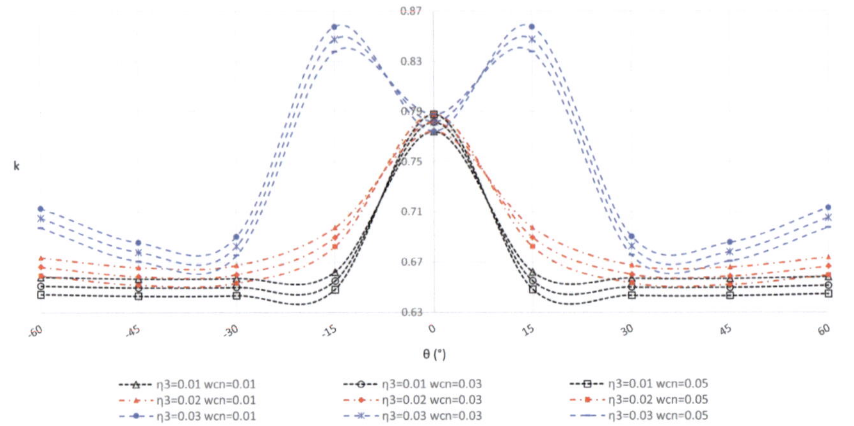

Fig. 5.12: Evolution of the shear correction factor of the SiC-Al-CNT composite material as a function of the angle θ, the parameter $\eta 3$ and the CNT weight fraction (wcn) for a fixed value of $\eta 1$ ($\eta 1 = 1$).

Table 5.7: Global average Silhouette scores for 2-8 clusters. Shear correction factor (k) of SiC-Al auxetic material integrating carbon nanotubes as a function of θ, $\eta 3$ and wcn. $\eta 1 = 1$.

Clusters	**2**	**3**	**4**	**5**	**6**	**7**	**8**
Silhouette score	**0.622**	0.481	0.450	0.349	0.379	0.376	0.373

influenced by the values of $\eta 3$. The first cluster in this case is constituted by all the instances where $\eta 3 = 0.01$ and $\eta 3 = 0.02$, while the second cluster is composed of the instances where $\eta 3 = 0.03$, as clearly evidenced by the trends depicted in Fig. 5.12.

Conducting the same analysis for the ZrO2-Al-CNT composite, one obtains the profiles depicted in Fig. 5.13. Figure 5.13 shows the trend in SCF for the ZrO2-Al-CNT material, considering a fixed value of $\eta 1$ ($\eta 1 = 1$) and varying $\eta 3$, θ and wcn parameters. In the present case, it is observed that despite the value of $\eta 3$, the maximum values for the SCF are always obtained when the angle θ equals zero, maximizing the shear stiffness in this configuration. For thinner cells ($\eta 3 = 0.01$), the SCF presents an abrupt decreasing trend for angles $\pm 15°$ and as progressing to the $\pm 60°$ angles, slightly increases, and almost stabilizes in lower values, contributing to a lower transverse shear stiffness. When $\eta 3$ increases, the drop from the values assumed for $\theta = 0°$ present a progressively smoother variation, starting to increase for angles $\pm 45°$.

Regarding the influence of the CNT weight content (wcn); the results suggest an interaction with the geometrical parameter $\eta 3$, because as depending on this latter value the SCF presents different increasing/decreasing trends and so different shear behaviour. Also, the number of clusters and their constitution are conditioned by the geometrical parameter $\eta 3$ value. Two clusters are the optimal instances' grouping, where the first cluster integrates all the instances where $\eta 3 = 0.01$ and $\eta 3 = 0.02$

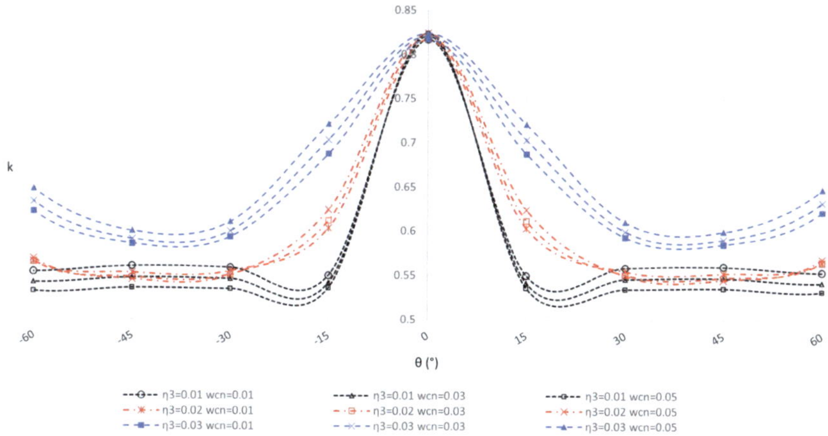

Fig. 5.13: Evolution of the shear correction factor of the ZrO2-Al-CNT composite material as a function of the angle θ, the parameter $\eta 3$ and the CNT weight fraction (wcn) for a fixed value of $\eta 1$ ($\eta 1 = 1$).

independently of the CNT content. This highlights the relative greater influence of this geometrical parameter when compared with the CNT weight fraction.

5.3.2.4 Case 6: Influence of Parameters $\eta 1$, θ and Carbon Nanotubes Weight Fraction on Non-Dimensional Deflection for CNT Reinforced Metal-Ceramic Exponentially Graded Beams

After having analysed the influence of the geometric parameters on the shear correction factor, the present case study is now related to the maximum transverse displacement of beams. A set of numerical analyses was performed, and the results obtained for the SiC-Al-CNT and ZrO2-Al-CNT, respectively, are summarized in Figs. 5.14 and 5.15. The results are presented in a non-dimensional form as

$$w_{\text{adim}} = \frac{E_{\text{m}} w h^3}{10^3 p_z a^4}.$$

Figure 5.14 illustrates how the non-dimensional transverse deflection is influenced by the geometrical parameter $\eta 1$, the CNT weight volume fraction and the angle θ for a fixed value of the parameter $\eta 3 = 0.01$. Overall, it is observed that the non-dimensional transverse deflection decreases as $\eta 1$ increases.

For $\eta 1 = 1$, whether CNT are not present or are included in a weight fraction of 5%, the maximum displacements occur for $\theta = 30°$. However, this changes when $\eta 1$ increases to 1.5 and 2, where the maximum values are observed for $\theta = 15°$. This suggests these configurations provide more flexible structures. For the angles 45°

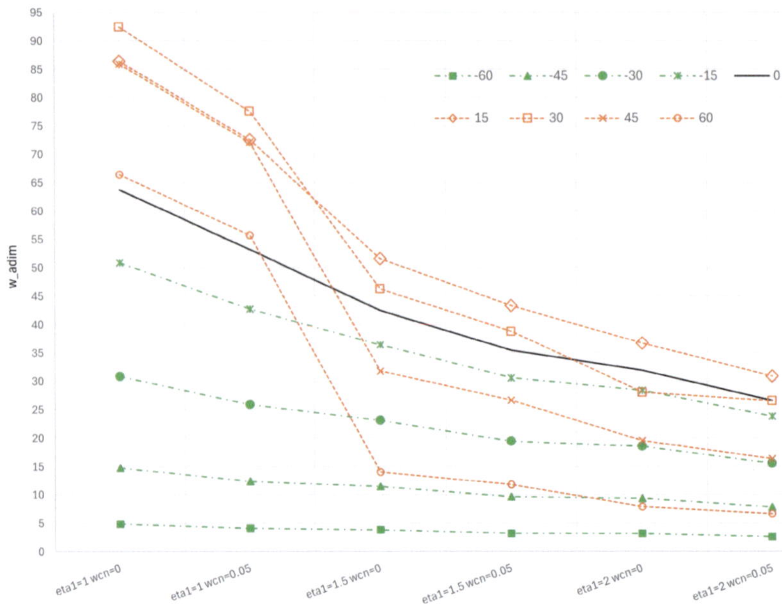

Fig. 5.14: Evolution of the non-dimensional deflection of SiC-Al-CNT beam as a function of the angle θ, the parameter $\eta 1$ and the CNT weight fraction (wcn) for a fixed value of $\eta 3$ ($\eta 3 = 0.01$).

and 60°, a sharp decrease when $\eta 1$ increases to 1.5 is perceived. Globally, for the positive angles, a greater decreasing trend is observed in this situation.

The configurations that lead to lower non-dimensional transverse displacements correspond to the angle $\theta = -60°$, showing a very slight decrease when $\eta 1$ increases, and no significant sensitivity to the inclusion of CNT. This is also noted when $\theta = -45°$. The influence of CNT is increasingly visible when the angle is greater than or equal to $-30°$, and more pronounced for more flexible structures, which happens when $\eta 1 = 1$.

Figure 5.15 presents the evolution of the non-dimensional deflection of the ZrO2-Al-CNT composite beam as a function of the parameters $\eta 1$, wcn and θ for a fixed value of the parameter $\eta 3 = 0.01$.

The non-dimensional maximum transverse displacement for this ZrO2-Al-CNT FGM beam assumes the highest values when $\theta = 0$, independently of the $\eta 1$, and wcn values, denoting that for those parameters' configurations, the less stiff structures are obtained. When $\eta 1 = 1$, high values for the angles $\theta = 15$ and 30° are attained. Lower values are observed when the angles assume the values ±60°, which denote the stiffest structures for these configurations. Overall, stiffer structures will be obtained when the geometrical parameter $\eta 1 > 1$, and the inclusion of CNT shows to be most effective for flexible structures. Analysing these two cases from a clustering perspective, both lead to a two-cluster structure. The best global average Silhouette

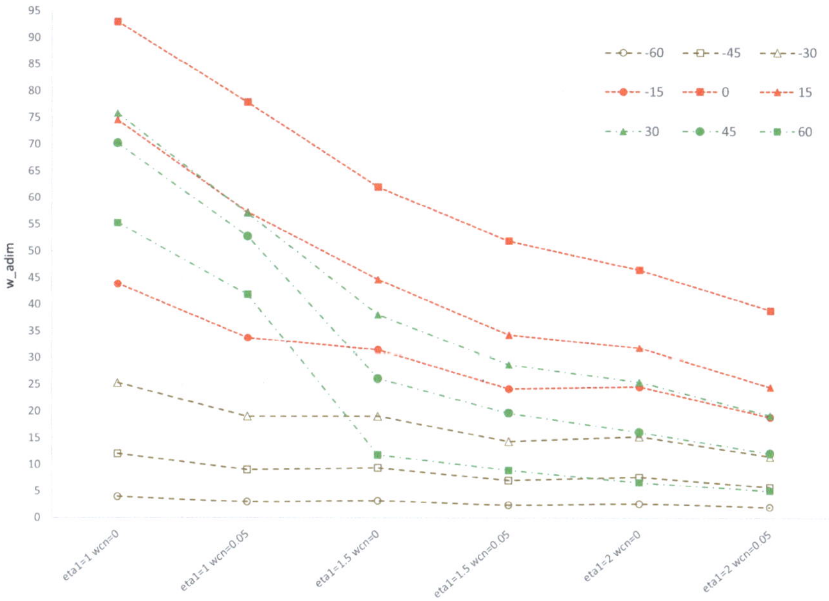

Fig. 5.15: Evolution of the non-dimensional deflection of ZrO2-Al-CNT beam as a function of the angle θ, the parameter $\eta 1$ and the CNT weight fraction (wcn) for a fixed value of $\eta 3$ ($\eta 3 = 0.01$).

scores for the SiC-Al-CNT and ZrO2-Al-CNT composites, for those two-cluster groupings, are respectively 0.749 and 0.687. Both cases follow the same clustering logic; one cluster where the geometrical parameter $\eta 1 = 1$, and is associated with the highest non-dimensional deflections, which decrease with high wcn. And another cluster for $\eta 1 > 1$, where lower non-dimensional deflections are presented.

5.3.2.5 Case 7: Influence of Parameters $\eta 1$, $\eta 3$ and θ for 5% Carbon Nanotubes Weight Fraction on Non-Dimensional Deflection for CNT Reinforced Metal-Ceramic Exponentially Graded Beams

The influence of the geometrical parameters $\eta 1$, $\eta 3$ and θ on the non-dimensional deflection of the SiC-Al-CNT and ZrO2-Al-CNT, when including 5% CNT weight fraction, is now considered in this case study. Figure 5.16 presents the evolution of the SiC-Al-CNT beams' response considering combinations of the different geometrical parameters' sets of values. Figure 5.16 shows that the increase of $\eta 3$ causes a decreasing trend in the non-dimensional maximum deflection, thus correlating with a stiffer behaviour of the beams. For these beams and the selected parameters' values, the maximum transverse non-dimensional displacement is obtained for the

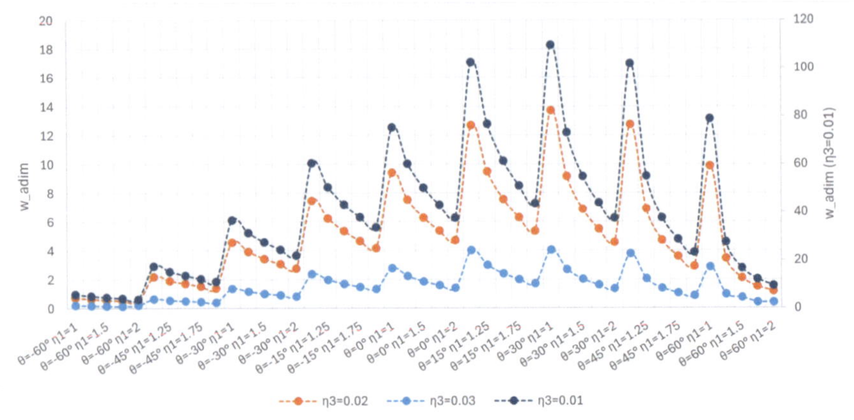

Fig. 5.16: Evolution of the non-dimensional deflection of SiC-Al-CNT beam as a function of the angle θ, the parameters $\eta 1$ and $\eta 3$ for a fixed CNT weight fraction (wcn = 5%).

angle $\theta = 30°$ when $\eta 1 = 1$ and $\eta 3 = 0.01$. The curve associated with $\eta 3 = 0.01$ visibly distinguishes from the others, in terms of magnitude.

Regarding the other composite beams, which instead of SiC involve the other ceramic material, ZrO2, the evolution of the non-dimensional displacement is depicted in Fig. 5.17. As it is possible to observe, the higher non-dimensional deflections are always achieved for $\eta 3 = 0.01$, being the peak value associated with the angle $\theta = 0°$. The higher values for different $\eta 1$ are obtained when the angle is 0°, showing an overall increasing value for negative angles and an overall decreasing trend for pos-

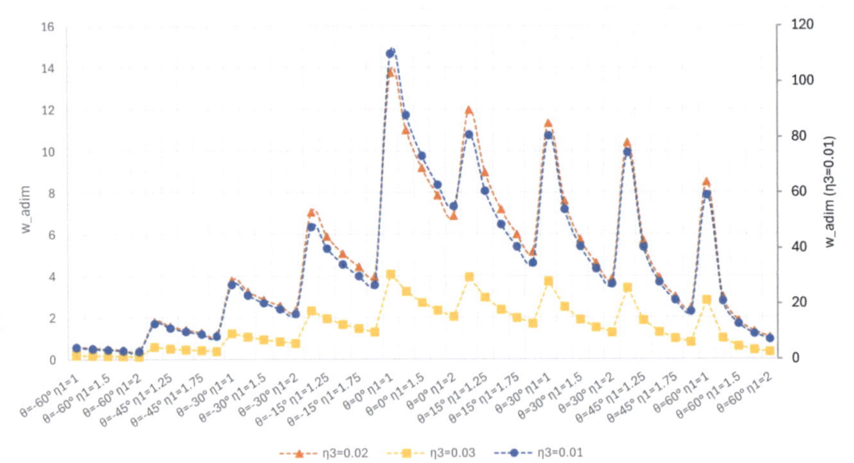

Fig. 5.17: Evolution of the non-dimensional deflection of ZrO2-Al-CNT beam as a function of the angle θ, the parameters $\eta 1$ and $\eta 3$ for a fixed CNT weight fraction (wcn = 5%).

itive angles for fixed $\eta 1$ values. When $\eta 1$ increases, the non-dimensional deflection decreases.

With respect to the clustering structure that best fits the results obtained for the SiC-Al-CNT material, the highest global average Silhouette score is obtained for two clusters, as observed in Table 5.8.

So, using a two-cluster structure, the K-means method leads to a grouping where the cluster separation is determined by a lower or higher non-dimensional transverse deflection pattern. Cluster 1, constituted by lower non-dimensional deflections, is associated with negative-angle configurations, larger ratios $\eta 1$ (>= 1.5), and a wider range of $\eta 3$. Cluster 2, comprising higher non-dimensional deflections, is mainly constituted by positive-angle cases and some $\theta = -15°$, and is associated with smaller $\eta 1$ and larger $\eta 3$ ratios. As an overall perspective, cluster 1 contains the stiffest configurations and the second cluster the more flexible ones. The structures associated to the two materials are represented in Figure 5.18.

For the ZrO2-Al-CNT, the best Silhouette score was also obtained for a two-cluster configuration, with a global average score of 0.605. In this case, the first cluster is characterised by more flexible responses, thus by higher non-dimensional deflections. In opposition, the second cluster possesses stiffer configurations. Regarding the unit cell geometrical parameters, the first cluster is constituted by configurations where $\eta 1$ is low (<= 1.25) and θ >= $-15°$. The instances in the second cluster present

Table 5.8: Global average Silhouette scores for 2-8 clusters. Beam non-dimensional deflection of SiC-Al auxetic material integrating carbon nanotubes as a function of θ, $\eta 1$, $\eta 3$ and wcn.

Clusters	2	3	4	5	6	7	8
Silhouette score	**0.558**	0.546	0.533	0.546	0.537	0.545	0.546

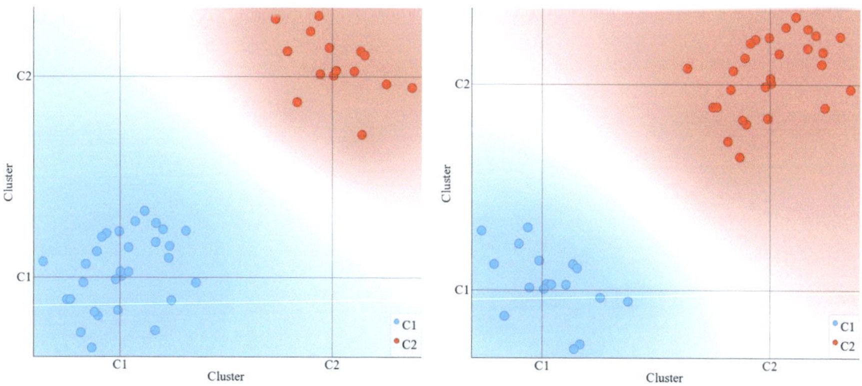

(a) SiC-Al-CNT auxetic composite. (b) ZrO2-Al-CNT auxetic composite.

Fig. 5.18: Schematic representation of the clustering scheme - non-dimensional maximum deflections.

higher $\eta 1$ ratios and angles <= $-30°$. When $\theta = 15°$ and $\theta = 60°$, and for lower values of $\eta 1$, a mixed behaviour is found being present different clusters.

Considering the two materials, it is possible to conclude that the ZrO2-Al FG shows a more flexible behaviour for low $\eta 1$ while SiC-Al FGM shows an opposite trend. In any case, the most influential parameters regarding the cluster aggregation are the $\eta 1$ ratio and the angle θ. Regarding $\eta 3$, although it affects the magnitude of deflection, it does not trigger cluster switching.

5.3.2.6 Case 8: Influence of Parameters $\eta 1$, $\eta 3$, θ and Carbon Nanotubes Weight Fraction on Non-Dimensional Fundamental Frequency for CNT Reinforced Metal-Ceramic Exponentially Graded Beams

Similarly to the parametric studies performed previously regarding the characterization of the corresponding shear correction factor and non-dimensional maximum transverse displacement of beams, in the present and next case studies the geometrical parameters influence on the non-dimensional fundamental frequency are analysed. Therefore, this case study considers the influence of the parameters $\eta 1$, $\eta 3$, θ, and carbon nanotubes' weight fraction on the non-dimensional fundamental frequency, which is obtained through the expression

$$\Omega_{\text{adim}} = \Omega\left(a^2\sqrt{\frac{\rho_{\text{m}} h}{D_{E_{\text{m}}}}}\right).$$

Figure 5.19 presents the evolution of the non-dimensional fundamental frequency of the SiC-Al-CNT auxetic FGM beams against these variables.

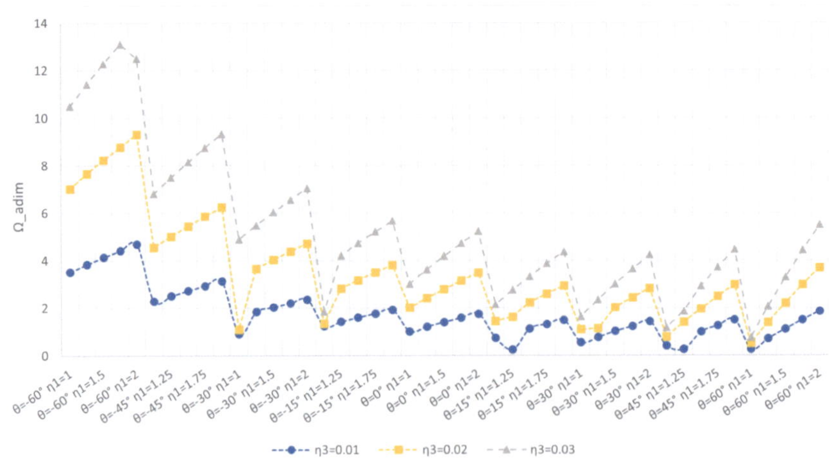

Fig. 5.19: Evolution of the non-dimensional fundamental frequency of SiC-Al-CNT beam as a function of the angle θ, the parameters $\eta 1$ and $\eta 3$ for a fixed CNT weight fraction (wcn = 5%).

As shown in Fig. 5.19, the highest values achieved for the non-dimensional frequencies are observed for the angle $\theta = -60°$, indicating the stiffest parameters' configuration. In opposition, the lower frequencies can be found in the range of angles $\theta = 15°$ to $60°$. Concerning the trend evolution as a function of $\eta 1$; it is possible to observe that as $\eta 1$ increases, the non-dimensional frequency also increases; the same happens when $\eta 3$ increases. The frequencies consistently increase with these parameters, denoting a greater stiffness for these structures. The results achieved for a similar study carried out for the ZrO2-Al-CNT auxetic FGM beams are now presented in Fig. 5.20. For the ZrO2-Al-CNT beams, the largest values are observed for the angle $\theta = -60°$, showing an overall decreasing trend as the angle increases. It is also possible to identify that the non-dimensional frequencies increase with the increase in $\eta 1$ and $\eta 3$, denoting a stiffening effect on the structures.

The best clustering structure in both cases was found to be a two-cluster structure, to which corresponded the highest global average Silhouette scores of 0.655 and 0.664, respectively, for the SiC-Al-CNT and ZrO2-Al-CNT composites. The two-cluster structure for these two cases is schematically represented in Fig. 5.21.

In Fig. 5.21a, Cluster 1 comprises for the SiC-Al-CNT material the non-dimensional natural frequencies for most angles, from −45° to 60°, while the second cluster only appears at $\theta = -60°$ and for some −45°. The second cluster is populated by higher non-dimensional frequencies, when compared with the first cluster, where low to moderate frequency values are grouped, for similar values of the remaining geometrical parameters. The clusters of the ZrO2-Al-CNT material are presented in Figure 5.21b. In this case, the first cluster is constituted by the non-dimensional frequencies associated with angles −60° and −45° and the case $\theta = -30°$ when $\eta 1 = 2$. This cluster groups the higher frequencies within the dataset and the second

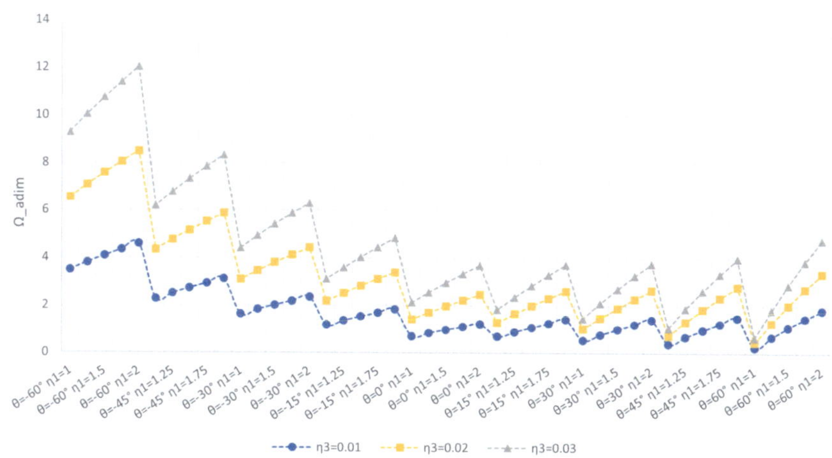

Fig. 5.20: Evolution of the non-dimensional fundamental frequency of ZrO2-Al-CNT beam as a function of the angle θ, the parameters $\eta 1$ and $\eta 3$ for a fixed CNT weight fraction ($wcn = 5\%$).

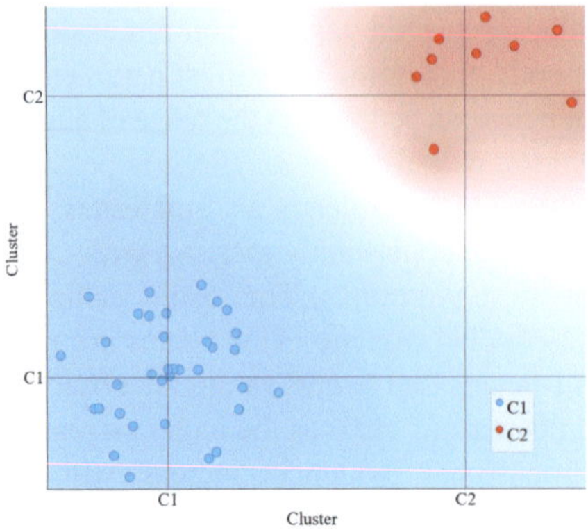

(a) SiC-Al-CNT.

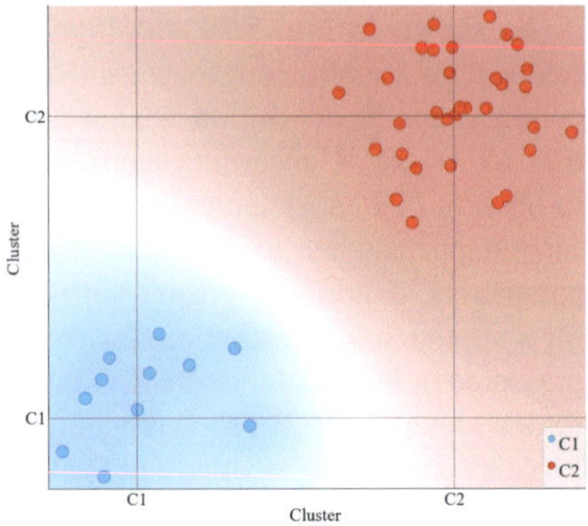

(b) ZrO2-Al-CNT auxetic composites.

Fig. 5.21: Schematic representation of the clustering scheme - non-dimensional fundamental frequencies.

cluster appears for the frequencies where the angle is greater than or equal to 30° (with the previous exception). In this latter case, the frequencies are low to moderate.

5.4 Conclusions

Following the parametric studies performed, it is possible to draw some conclusions about the influence of the selected parameters on the structures analysed:

- In more general terms, the angle θ has shown to be an important driver of the mechanical response, suppressing auxetic effects when $\theta = 0°$. Also, as is known from a wider context, the presence of carbon nanotubes enhances the beams' stiffness, although this effect depends on the unit cell parameters, denoting a parameter-dependent stiffening.
- Regarding the shear correction factor, this factor is influenced by the auxetic unit cell's geometric parameters $\eta 1$, $\eta 3$, and θ. The increase in the relative thickness parameter, $\eta 3$, in general, produces an increase in the shear correction factor, thus indicating a higher transverse shear stiffness. For the specific null angle, the shear correction factor is invariant.
- The structures become stiffer, presenting lower transverse deflections and higher fundamental frequencies, when the geometrical relations $\eta 1$ and $\eta 3$ increase. The stiffest configuration, for both composites, corresponds to the extreme angle values, no matter the specific material involved in the FGM.
- Comparing the two functionally graded composites, it is observed that the SiC-Al composite shows a greater sensitivity to $\eta 3$ and $\eta 1$, especially when the angle θ assumes the values $\pm 15°$, while for the ZrO2-Al composites, lower shear

correction factors occur at these angles. The ZrO2-Al composite presents a higher baseline for lower shear correction factors at the null angle.

- The SiC-Al-CNT beams exhibit greater flexibility when $\theta = 30°$, while for the ZrO2-Al-CNT beams, this occurs when $\theta = 0°$.
- Regarding the clustering schemes, it is overall clear that concerning the static behaviour, the clusters compositions are typified by the separation of configurations into flexible configurations (higher deflections) and stiffer configurations (lower deflections), this being driven by $\eta 1$ and θ. In a dynamic free vibration perspective, these clusters group low-to-moderate frequencies (which occur for most angles) and high frequencies (occurring for extreme angles).
- For the CNT-reinforced beams, clustering is primarily ruled by the CNT weight fraction and the relative thickness ratio parameter $\eta 3$.

As a global conclusion, it is also relevant to note that the clustering analyses performed allow the observation of parameter-driven behavioural trends, providing supported insights, very important in material design and selection for engineering applications.

Acknowledgements In memoriam to Professor Cristóvão Manuel Mota Soares: A very heartfelt posthumous thank you to the Professor and Friend who influenced several generations of students, and who taught us how to fly. The authors acknowledge Fundação para a Ciência e a Tecnologia (FCT) for its financial support via LAETA (project `https://doi.org/10.54499/UID/50022/2025`).

References

[1] Affdl JCH, Kardos JL (1976) The halpin-tsai equations: A review. Polymer Engineering & Science 16(5):344–352, DOI 10.1002/pen.760160512

[2] Al-Furjan MSH, Xu MX, Farrokhian A, Jafari GS, Shen X, Kolahchi R (2025) On wave propagation in piezoelectric-auxetic honeycomb-2d-fgm micro-sandwich beams based on modified couple stress and refined zigzag theories. Waves in Random and Complex Media 35(1):1147–1171, DOI 10.1080/17455030.2022.2030499

[3] Alibeigloo A, Liew KM (2013) Thermoelastic analysis of functionally graded carbon nanotube-reinforced composite plate using theory of elasticity. Composite Structures 106:873–881, DOI 10.1016/j.compstruct.2013.07.002

[4] Aminul Islam M, Hasan M, Rahman M, Hosne Mobarak M, Akter Mimona M, Hossain N (2024) Advances and significances of carbon nanotube applications: A comprehensive review. European Polymer Journal 220:113443, DOI 10.1016/j.eurpolymj.2024.113443

[5] Ansari R, Faghih Shojaei M, Mohammadi V, Gholami R, Sadeghi F (2014) Nonlinear forced vibration analysis of functionally graded carbon nanotube-reinforced composite timoshenko beams. Composite Structures 113:316–327, DOI 10.1016/j.compstruct.2014.03.015

[6] Ansari R, Torabi J, Shakouri AH (2017) Vibration analysis of functionally graded carbon nanotube-reinforced composite elliptical plates using a numerical strategy. Aerospace Science and Technology 60:152–161, DOI 10.1016/j.ast.2016.11.004
[7] Aydogdu M, Taskin V (2007) Free vibration analysis of functionally graded beams with simply supported edges. Materials & Design 28(5):1651–1656, DOI 10.1016/j.matdes.2006.02.007
[8] Bhardwaj G, Upadhyay AK, Pandey R, Shukla KK (2013) Non-linear flexural and dynamic response of cnt reinforced laminated composite plates. Composites Part B: Engineering 45(1):89–100, DOI 10.1016/j.compositesb.2012.09.004
[9] Carvalho A, Silva T, Loja MAR, Damásio FR (2017) Assessing the influence of material and geometrical uncertainty on the mechanical behavior of functionally graded material plates. Mechanics of Advanced Materials and Structures 24(5):417–426, DOI 10.1080/15376494.2016.1191100
[10] Chakraverty S, Pradhan KK (2014) Free vibration of exponential functionally graded rectangular plates in thermal environment with general boundary conditions. Aerospace Science and Technology 36:132–156, DOI 10.1016/j.ast.2014.04.005
[11] Chi SH, Chung YL (2006) Mechanical behavior of functionally graded material plates under transverse load—part i: Analysis. International Journal of Solids and Structures 43(13):3657–3674, DOI 10.1016/j.ijsolstr.2005.04.011
[12] Cinefra M, Carrera E, Brischetto S (2011) Refined shell models for the vibration analysis of multiwalled carbon nanotubes. Mechanics of Advanced Materials and Structures 18(7):476–483, DOI 10.1080/15376494.2011.604601
[13] Craveiro DS, Loja MAR (2020) A study on the effect of carbon nanotubes'distribution and agglomeration in the free vibration of nanocomposite plates. C 6(4):79, DOI 10.3390/c6040079
[14] Craveiro DS, Loja MAR (2021) An assessment of thick nanocomposite plates'behavior under the influence of carbon nanotubes agglomeration. Journal of Composites Science 5(2):41, DOI 10.3390/jcs5020041
[15] Demšar J, Curk T, Erjavec A, Gorup Č, Hočevar T, Milutinovič M, Možina M, Polajnar M, Toplak M, Starič A, Štajdohar M, Umek L, Žagar L, Žbontar J, Žitnik M, Zupan B (2013) Orange: Data mining toolbox in python. Journal of Machine Learning Research 14(71):2349–2353
[16] El-Galy IM, Saleh BI, Ahmed MH (2019) Functionally graded materials classifications and development trends from industrial point of view. SN Applied Sciences 1(11):1378, DOI 10.1007/s42452-019-1413-4
[17] Esen i, Koç MA, Eroğlu M (2024) Effect of functionally graded carbon nanotube reinforcement on the dynamic response of composite beams subjected to a moving charge. Journal of Vibration Engineering & Technologies 12:5203–5218, DOI 10.1007/s42417-023-01192-0
[18] Ferreira AJM, Batra RC, Roque CMC, Qian LF, Martins PALS (2005) Static analysis of functionally graded plates using third-order shear deformation

theory and a meshless method. Composite Structures 69(4):449–457, DOI 10.1016/j.compstruct.2004.08.003

[19] Flach P (ed) (2012) Machine Learning: The Art and Science of Algorithms that Make Sense of Data. Cambridge University Press, Cambridge,

[20] Gan G, Ma C, Wu J (2007) Data Clustering: Theory, Algorithms, and Applications. ASA-SIAM Series on Statistics and Applied Mathematics Vol 20, SIAM, Society for Industrial and Applied Mathematics, Philadelphia, Pa, DOI 10.1137/1.9780898718348

[21] Gaspar JSD, Loja MAR, Barbosa JI (2025) Plane frame structures: Optimization and design solutions clustering. Algorithms 18(7):375, DOI 10.3390/a18070375

[22] Koizumi M (1997) FGM activities in Japan. Composites Part B: Engineering 28(1):1–4, DOI 10.1016/S1359-8368(96)00016-9

[23] Lim TC, Alderson A, Alderson KL (2014) Experimental studies on the impact properties of auxetic materials. physica status solidi (b) 251(2):307–313, DOI 10.1002/pssb.201384249

[24] Liu K, Wu T, Detwiler D, Panchal J, Tovar A (2019) Design for crashworthiness of categorical multimaterial structures using cluster analysis and bayesian optimization. Journal of Mechanical Design 141(121701), DOI 10.1115/1.4044838

[25] Liu Y, Zhao C, Xu C, Ren J, Zhong J (2023) Auxetic meta-materials and their engineering applications: a review. Engineering Research Express 5(4):042003, DOI 10.1088/2631-8695/ad0eb1

[26] Loja MAR, Barbosa JI (2024) Exponentially graded auxetic structures: An assessment of the shear correction factor and static deflection. Applied Sciences 14(20):9356, DOI 10.3390/app14209356

[27] Ma Q, Zhang J (2023) Nonlinear constitutive and mechanical properties of an auxetic honeycomb structure. Mathematics 11(9):2062, DOI 10.3390/math11092062

[28] Madenci E, Özkılıç YO, Bahrami A, Aksoylu C, Asyraf MRM, Hakeem IY, Beskopylny AN, Stel'makh SA, Shcherban EM, Fayed S (2024) Experimental investigation and analytical verification of buckling of functionally graded carbon nanotube-reinforced sandwich beams. Heliyon 10(8):e28388, DOI 10.1016/j.heliyon.2024.e28388

[29] Mindlin RD (2021) Influence of rotatory inertia and shear on flexural motions of isotropic, elastic plates. Journal of Applied Mechanics 18(1):31–38, DOI 10.1115/1.4010217

[30] Mohammadi Ghalehney S, Sadeghi MH, Gharehbaghi H (2024) Mechanical properties of 2d re-entrant gradient structures produced by additive manufacturing. Iranian Journal of Science and Technology, Transactions of Mechanical Engineering 48(3):1395–1404, DOI 10.1007/s40997-023-00724-z

[31] Mota AF, Loja MAR (2023) A study on porous nanocomposite graded plates using equivalent single layer and layerwise models. Mechanics of Advanced Materials and Structures 30(20):4155–4177, DOI 10.1080/15376494.2022.2092665

[32] Mota AF, Loja MAR, Barbosa JI, Rodrigues J (2020) Porous functionally graded plates: An assessment of the influence of shear correction factor on static behavior. Mathematical and Computational Applications 25(2):25, DOI 10.3390/mca25020025

[33] Páscoa FT, Machado CM, Cardoso JO, Borges JP, Velhinho A (2023) Methodology for modelling and simulation of tailored 3d functionally graded auxetic metamaterials. The International Journal of Advanced Manufacturing Technology 125(9):4647–4661, DOI 10.1007/s00170-023-10863-x,

[34] Penna R, Feo L (2020) Nonlinear dynamic behavior of porous and imperfect bernoulli-euler functionally graded nanobeams resting on winkler elastic foundation. Technologies 8(4):56, DOI 10.3390/technologies8040056

[35] Prawoto Y (2012) Seeing auxetic materials from the mechanics point of view: A structural review on the negative Poisson's ratio. Computational Materials Science 58:140–153, DOI 0.1016/j.commatsci.2012.02.012

[36] Rafiee M, Mohammadi M, Sobhani Aragh B, Yaghoobi H (2013) Nonlinear free and forced thermo-electro-aero-elastic vibration and dynamic response of piezoelectric functionally graded laminated composite shells, part i: Theory and analytical solutions. Composite Structures 103:179–187, DOI 10.1016/j.compstruct.2012.12.053

[37] Ramos Loja MA, Mota Soares CM, Mota Soares CA (2002) Modelling and design of adaptive structures using b-spline strip models. Composite Structures 57(1):245–251, DOI 10.1016/S0263-8223(02)00091-0

[38] Rao GV, Venkataramana J, Raju IS (1974) A high precision triangular plate bending element for the analysis of thick plates. Nuclear Engineering and Design 30(3):408–412, DOI 10.1016/0029-5493(74)90225-8

[39] Reddy J (2003) Mechanics of Laminated Composite Plates and Shells: Theory and Analysis, 2nd edn. CRC Press, DOI 10.1201/b12409

[40] Reddy JN (2000) Analysis of functionally graded plates. International Journal for Numerical Methods in Engineering 47(1-3):663–684, DOI 10.1002/(SICI)1097-0207(20000110/30)47:1/3<663::AID-NME787>3.0.CO;2-8

[41] Refahi Oskouei A, Heidary H, Ahmadi M, Farajpur M (2012) Unsupervised acoustic emission data clustering for the analysis of damage mechanisms in glass/polyester composites. Materials & Design 37:416–422, DOI 10.1016/j.matdes.2012.01.018

[42] Rosa RD, Loja MA, Carvalho AC (2018) Toward variability characterization and statistic models'constitution for the prediction of exponentially graded plates'static response. Journal of Composites Science 2(4):59, DOI 10.3390/jcs2040059

[43] Salerno VL, Goldberg MA (1960) Effect of shear deformations on the bending of rectangular plates. Journal of Applied Mechanics 27(1):54–58, DOI 10.1115/1.3643934, URL https://doi.org/10.1115/1.3643934

[44] Selim BA, Zhang LW, Liew KM (2016) Vibration analysis of cnt reinforced functionally graded composite plates in a thermal environment based on reddy's higher-order shear deformation theory. Composite Structures 156:276–290, DOI 10.1016/j.compstruct.2015.10.026

[45] Sengupta D (1991) Stress analysis of flat plates with shear using explicit stiffness matrix. International Journal for Numerical Methods in Engineering 32(7):1389–1409, DOI 10.1002/nme.1620320703

[46] Sevillano E, Sun R, Gil A, Perera R (2016) Interfacial crack-induced debonding identification in frp-strengthened rc beams from pzt signatures using hierarchical clustering analysis. Composites Part B: Engineering 87:322–335, DOI 10.1016/j.compositesb.2015.09.006

[47] Shi Z, Yao X, Pang F, Wang Q (2017) An exact solution for the free-vibration analysis of functionally graded carbon-nanotube-reinforced composite beams with arbitrary boundary conditions. Scientific Reports 7(1):12909, DOI 10.1038/s41598-017-12596-w

[48] Sina SA, Navazi HM, Haddadpour H (2009) An analytical method for free vibration analysis of functionally graded beams. Materials & Design 30(3):741–747, DOI 10.1016/j.matdes.2008.05.015

[49] Siniauskaya V, Wang H, Liu Y, Chen Y, Zhuravkov M, Lyu Y (2024) A review on the auxetic mechanical metamaterials and their applications in the field of applied engineering. Frontiers in Materials Volume 11 - 2024, DOI 10.3389/fmats.2024.1453905

[50] Srivastava C, Bhola L, Mahesh V, Guruprasad PJ, Petrinic N, Scarpa F, Harursampath D, Ponnusami SA (2023) Exploiting nonlinearities through geometric engineering to enhance the auxetic behaviour in re-entrant honeycomb metamaterials. Scientific Reports 13(1):20915, DOI 10.1038/s41598-023-47525-7

[51] Tavakoli Maleki A, Pourseifi M, Zakeri M (2022) Effect of agglomeration of the nanotubes on the vibration frequency of the multi-scale hybrid nanocomposite conical shells: a gdq-based study. Waves in Random and Complex Media 32(1):359–380, DOI 10.1080/17455030.2020.1773007

[52] Tornabene F, Fantuzzi N, Bacciocchi M (2014) Free vibrations of free-form doubly-curved shells made of functionally graded materials using higher-order equivalent single layer theories. Composites Part B: Engineering 67:490–509, DOI 10.1016/j.compositesb.2014.08.012

[53] Tornabene F, Fantuzzi N, Bacciocchi M, Viola E (2016) Effect of agglomeration on the natural frequencies of functionally graded carbon nanotube-reinforced laminated composite doubly-curved shells. Composites Part B: Engineering 89:187–218, DOI 10.1016/j.compositesb.2015.11.016

[54] Valle R, Pincheira G, Tuninetti V (2021) Design of an auxetic cellular structure with different elastic properties in its three orthogonal directions. Proceedings of the Institution of Mechanical Engineers, Part L: Journal of Materials: Design and Applications 235(6):1341–1350, DOI 10.1177/14644207211010209

[55] Vlachoutsis S (1992) Shear correction factors for plates and shells. International Journal for Numerical Methods in Engineering 33(7):1537–1552, DOI 10.1002/nme.1620330712

[56] Wallbanks M, Khan MF, Bodaghi M, Triantaphyllou A, Serjouei A (2022) On the design workflow of auxetic metamaterials for structural applications. Smart Materials and Structures 31(2):023002, DOI 10.1088/1361-665X/ac3f78

[57] Wattanasakulpong N, Ungbhakorn V (2014) Linear and nonlinear vibration analysis of elastically restrained ends fgm beams with porosities. Aerospace Science and Technology 32(1):111–120, DOI 10.1016/j.ast.2013.12.002
[58] Whitney JM (1973) Shear correction factors for orthotropic laminates under static load. Journal of Applied Mechanics 40(1):302–304, DOI 10.1115/1.3422950
[59] Wu Y, Pastor ML, Perrin M, Casari P, Gong X (2023) Characterisation of damage mechanisms of gfrp-balsa sandwich under 4-point bending based on two-step clustering process in acoustic emission analysis. Composites Part B: Engineering 260:110774, DOI 10.1016/j.compositesb.2023.110774
[60] Xu HH, Zhang XG, Han D, Jiang W, Zhang Y, Luo YM, Ni XH, Teng XC, Xie YM, Ren X (2024) Mechanical characteristics of auxetic composite honeycomb sandwich structure under bending. AI in Civil Engineering 3(1):8, DOI 10.1007/s43503-024-00026-6
[61] Yadav S, Liu S, Singh RK, Sharma AK, Rawat P (2024) A state-of-art review on functionally graded materials (fgms) manufactured by 3d printing techniques: Advantages, existing challenges, and future scope. Journal of Manufacturing Processes 131:2051–2072, DOI 10.1016/j.jmapro.2024.10.026
[62] Zhang Y, Li X, Liu S (2016) Enhancing buckling capacity of a rectangular plate under uniaxial compression by utilizing an auxetic material. Chinese Journal of Aeronautics 29(4):945–951, DOI 10.1016/j.cja.2016.06.010
[63] Zhu P, Lei ZX, Liew KM (2012) Static and free vibration analyses of carbon nanotube-reinforced composite plates using finite element method with first order shear deformation plate theory. Composite Structures 94(4):1450–1460, DOI 10.1016/j.compstruct.2011.11.010
[64] Zhu X, Zhang J, Zhang W, Chen J (2019) Vibration frequencies and energies of an auxetic honeycomb sandwich plate. Mechanics of Advanced Materials and Structures 26(23):1951–1957, DOI 10.1080/15376494.2018.1455933
[65] Zienkiewicz OC, Taylor RL, Fox D (eds) (2014) The Finite Element Method for Solid and Structural Mechanics. Butterworth-Heinemann, Oxford, DOI 10.1016/B978-1-85617-634-7.00018-1

Chapter 6
Coupling Asymptotic Homogenization and Physics-Informed Neural Networks for Predicting Effective Properties of One-Dimensional Composites

Boris Mederos-Madrazo, Oscar Luis Cruz-González, Gaizka Farrera-Rivera, Reinaldo Rodríguez-Ramos, Yoanh Espinosa-Almeyda, Jose Merodio, Amaury Alvarez Cruz, and Carlos Quesada-González

Abstract In this contribution, a coupling multiscale approach, based on two-scale Asymptotic Homogenization methods (AHM) and Physics-Informed Neural Networks (PINNs), is implemented to predict the effective elastic properties of one-dimensional periodic composites. The chosen geometrical configuration allows for a comprehensive presentation of the hybrid workflow, explicitly highlighting the stage where the application of PINNs is most relevant. In particular, we utilize PINNs to solve the local problem derived from the asymptotic homogenization approach, the results are then integrated into the calculation of effective coefficients and the formal asymptotic solution. We propose a Dual-Network architecture and optimization choices designed to handle material discontinuities and stiff gradients. Finally, the method is validated through a direct comparison against the solution of the heterogeneous problem. Numerical simulations are presented for a variety of volume fractions, and the influence of material property contrast is analyzed.

Boris Mederos-Madrazo · Gaizka Farrera-Rivera · Yoanh Espinosa-Almeyda
Instituto de Ingeniería y Tecnología, Universidad Autónoma de Ciudad Juárez, Cd. Juárez, Chihuahua, CP 32310, México, e-mail: `boris.mederos@uacj.mx`, `gaizka.farrera@outlook.com`, `yoanh.espinosa@uacj.mx`

Oscar Luis Cruz-González
Aix Marseille Univ, CNRS, Centrale Marseille, IRPHE UMR 7342, Marseille, France, e-mail: `oscar-luis.cruz-gonzalez@univ-amu.fr`

Reinaldo Rodríguez-Ramos (corresponding author)
Departamento de Matemáticas, Facultad de Matemática y Computación, Universidad de La Habana, La Habana, CP 10400, Cuba & PPG-MCCT, Universidade Federal Fluminense, RJ, CP 27255-125, Brasil, e-mail: `reinaldo@matcom.uh.cu`, `reinaldorr@id.uff.br`

Jose Merodio · Carlos Quesada-González
Universidad Politécnica de Madrid, E.T.S.I. de Sistemas Informáticos, Departamento de Matemática Aplicada a las TIC, Madrid, CP 28031, Spain, e-mail: `merodioj@gmail.com`, `carlos.quesada@upm.es`

Amaury Alvarez Cruz
Universidade Federal do Rio de Janeiro, RJ, CP 21941-617, Brasil, e-mail: `amaury@ic.ufrj.br`

H. Altenbach et al. (eds.), *Composite Structures and Technology*,
Advanced Structured Materials 255,
https://doi.org/10.1007/978-3-032-24656-1_6

Key words: Physics-Informed Neural Networks, Asymptotic homogenization method, Effective elastic properties, One-dimensional periodic composites, Dual-Network architecture, SOAP optimizer

6.1 Introduction

The rapid evolution of technology and structural design has increased the demand for composite materials with tailored properties [2, 42]. Predicting the behavior of these materials relies on the integration of physical-mathematical models and numerical simulations, which establish critical structure-property relationships and guide material design [6, 24, 27, 29].

Over the past few decades, analytical and computational micromechanical methods have played a fundamental role within the composite materials research community due to their high ability to describe the relationship between the intrinsic characteristics of materials at the microscale and their effective macroscopic response. They rely on the analysis of the geometrical and physical characteristic effects, such as the geometry and distribution of inclusions, constituents' properties, and the phase interaction effects, on the effective response of the medium. Examples of analytical approaches include Eshelby's inclusion theory [11], the Hashin-Shtrikman bound [21, 35], the generalized self-consistent schemes [10, 38], the well-known Mori-Tanaka [5, 36], and the asymptotic homogenization methods [3, 4, 29, 46], among others, whereas computational micromechanical schemes include finite element-based homogenization, boundary element methods, and fast Fourier transform (FFT)-based approaches [13, 16, 26, 30, 34, 39]. These methods provide closed-form or semi-analytical expressions for the homogenized constitutive equations, offering valuable insights into the influence of microstructural parameters on the macroscopic effective response. However, these conventional homogenization schemes remain limited for modeling multiphase composites with highly heterogeneous or complex microstructures [40].

Recently, research has integrated micromechanical methods with machine learning techniques to tackle these challenges [14, 15, 22]. In particular, the Physics-Informed Neural Networks (PINNs) method has offered solutions to multiscale partial differential equations without conventional discretizations and with little or even no measurement data, relying primarily on the underlying physical laws (see [19, 20, 44, 49]). These contributions highlight the growing potential of PINNs to achieve accurate solutions with high computational efficiency and strong generalization in the modeling of complex multiscale systems. For instance, notable advances have been achieved through Neural Homogenization with PINNs by [32], which combines oversampling and automatic differentiation to enhance accuracy. Further developments in periodic and non-periodic systems are reported. For example, Park and Zhu [40] implement PINNs to calculate homogenized coefficients and approximate solutions of multiscale elliptic problems by leveraging a priori knowledge of the underlying homogenized equations. They also demonstrate that the Pinns-based approaches can obtain approximations that are accurate for homogenized solutions. Jiang et al. [25] proposed a new micromechanics theory called the Physically Informed Deep Homogenization Network (DHN) to model unidirectional multiphase/multi-inclusion thermo-elastic periodic composites. The DHN scheme is based on periodic homogenization theory and deep neural networks. Next, Wu et al. [53] extended the methodology of Jiang et al. [25] to heterogeneous elastic materials with periodic microstructures, adopting a two-scale expansion of the displacement field of a repeating unit cell in terms of averaging and periodic contributions. It establishes the baseline for using Neural Networks to solve the elastic cell problem. In [49], the authors focus the research on Periodic Boundary Conditions (PBCs) and sharp interfaces. They use Fourier features. Moreover, the authors in [17] address the reliability of PINNs for discontinuous coefficients, showing that standard PINNs struggle with interfaces.

In this work, we propose a hybrid computational framework that couples the two-scale Asymptotic Homogenization Method (AHM) with Physics-Informed Neural Networks (PINNs) to address a boundary value problems in heterogeneous elastic media. We adopt a straightforward 1D geometrical configuration, specifically, a two-phase composite with a periodic microstructure, to provide clear insight into the multiscale coupling and the integration of PINNs into the loop. The main goal is to provide a general overview of the procedure and validate the approach through a rigorous comparison. We verify our results by contrasting the solution of the original heterogeneous problem, solved via Finite Differences, against the homogenized problem and the formal asymptotic solution constructed entirely via classical AHM, and the formal asymptotic solution where the local cell problem is solved via our PINN approach.

While the integration of PINNs with asymptotic homogenization theory has gained traction, the present work reveals differences with previous contributions that are worthy to be mentioned. Recent works such as [32] and [40] have primarily focused on scalar diffusion problems or inverse parameter identification from macro-data. In contrast, this work establishes a rigorous forward modeling framework for linear elasticity in two-phase composites with periodic microstructures. We strictly adhere to the Asymptotic Homogenization Method (AHM), utilizing PINNs without any solution data to solve the local problem. In addition, in [7] the authors successfully applied hybrid neural networks to periodic elastic composites. However, while they rely on Fourier feature mappings to enforce periodicity, our work diverges by focusing on optimization stability and a generalized interface formulation within the proposed architecture.

The novelties of the present work lie in the specific Dual-Network architectural and optimization choices designed to handle material discontinuities and stiff gradients. First, we implement a Dual-Network architecture that assigns independent neural networks to the constituents phases, connected strictly via continuity interface conditions. This avoids the smoothing artifacts of single-network approaches in high-contrast media. Second, distinct from the standard optimizers used in [7] and [32], we employ SOAP optimizer [50, 54], a recent second-order optimizer. This choice significantly accelerates convergence in the presence of stiff microstructural gradients compared to the standard Adam [28] or Adam/L-BFGS routines [45] [28] used in previous studies. Finally, we go beyond calculating effective coefficients by reconstructing the full formal asymptotic solution, validating our approach against the Finite Difference solution of the heterogeneous problem to set a robust baseline for future developments.

This manuscript is organized as follows. In Sect. 6.2, we formulate the heterogeneous problem, and present the fundamental equations and bases for the application of the homogenization technique. In Sect. 6.3, the two-scale Asymptotic Homogenization method is introduced, and we briefly explain the standard steps of the procedure application. In Sect. 6.4, we present the PINNs method adapted to the particular scenario under study and we explain its main role within the coupling workflow. In Sect. 6.5, the results are organized by first solving the local problem, then calculating the effective properties, and finally performing a validation by comparing the solution of the heterogeneous problem with the constructed formal asymptotic solution.

6.2 Fundamental Equations

We consider an elastic heterogeneous bar of length L, occupying a bounded region $\Omega = [0, L] \subset \mathbb{R}$ with a boundary $\partial\Omega = \{0, L\}$, see Fig. 6.1. The bar is assumed to be a periodic composite consisting of two constituent phases that are isotropic materials. In the absence of body forces, the equilibrium equation on Ω is given by

$$\frac{d\sigma^{(\epsilon)}(x)}{dx} = 0, \quad x \in (0, L), \tag{6.1}$$

with the corresponding constitutive relation (Hooke's law)

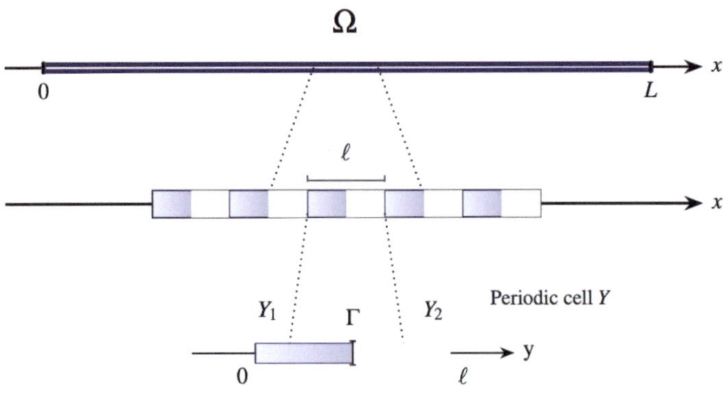

Fig. 6.1: One-dimensional heterogeneous bar with domain Ω. The first zoom reveals the periodic structure with two distinct materials, and the second zoom shows the periodic cell.

$$\sigma^{(\epsilon)}(x) = E^{(\epsilon)}(x)\,\frac{du^{(\epsilon)}(x)}{dx}, \tag{6.2}$$

where $\sigma^{(\epsilon)}(x)$ and $u^{(\epsilon)}(x)$ denote the stress and displacement fields, respectively, and $E^{(\epsilon)}(x)$ is a rapidly-oscillating and piecewise constant coefficient, referred to as Young's modulus. Here, we assume *macroscopic uniformity* [23], implying that the periodic unit cell and constituent properties are independent of the macroscopic variable x. Consequently, the Young's modulus is given by

$$E^{(\epsilon)}(x) = \begin{cases} E_1, & \text{if } x \in \text{mat}_1, \\ E_2, & \text{if } x \in \text{mat}_2, \end{cases} \tag{6.3}$$

where mat_1 and mat_2 denote the constituents.

Substituting Eq. (6.2) into Eq. (6.1) yields the governing differential equation

$$\frac{d}{dx}\left(E^{(\epsilon)}(x)\,\frac{du^{(\epsilon)}(x)}{dx}\right) = 0, \quad x \in (0, L), \tag{6.4}$$

subject to the boundary conditions

$$u^{(\epsilon)}(0) = 0, \qquad u^{(\epsilon)}(L) = 1, \tag{6.5}$$

which represents the boundary value problem associated with the linear elasticity of a heterogeneous bar. In addition, perfect contact conditions between the constituent phases are considered, ensuring the continuity of displacements and tractions across the internal interface points, which are denoted by the set Γ_ϵ, that is,

$$[\![u^{(\epsilon)}(x)]\!] = 0, \qquad [\![\sigma^{(\epsilon)}(x)]\!] = 0, \quad x \in \Gamma_\epsilon, \tag{6.6}$$

where the jump operator is defined as follows

$$[\![f^{(\epsilon)}(x_j)]\!] = \lim_{x \to x_j^+} f^{(\epsilon)}(x) - \lim_{x \to x_j^-} f^{(\epsilon)}(x). \tag{6.7}$$

6.2.1 Separation of Scales

The asymptotic homogenization technique relies on a clear separation between the characteristic length of the periodic micro-structure ℓ, and that of the whole composite L. We therefore introduce the small, dimensionless parameter

$$\epsilon := \frac{\ell}{L} \ll 1, \tag{6.8}$$

which is assumed to be constant (see [3]).

To make explicit the two-scale nature of a given field $\Phi^{(\epsilon)}(x)$, we let it depend on the dimensionless variables

$$\bar{x} := \frac{x}{L} \quad \text{and} \quad \bar{y} := \frac{x}{\ell}, \tag{6.9}$$

where x denotes the physical spatial coordinate, while $\bar{x}$ and $\bar{y}$ are the macroscopic and microscopic dimensionless variables, respectively. Accordingly, we write

$$\Phi^{(\epsilon)}(x) = \breve{\phi}(x; \ell, L) = \bar{\phi}(x/L, x/\ell) = \phi(\bar{x}, \bar{y}), \tag{6.10}$$

with $\bar{x}$ and $\bar{y}$ related, by Eq. (6.8), through

$$\bar{y} = \frac{x}{\ell} = \frac{x/L}{\ell/L} = \frac{\bar{x}}{\epsilon}, \tag{6.11}$$

thus, $\Phi^{(\epsilon)}$ is recast as a function of two formally independent variables, $\bar{x}$ and $\bar{y}$.

Using Eqs. (6.8)–(6.11) and applying the chain rule, the spatial derivative of $\Phi^{(\epsilon)}$ becomes

$$\frac{d\Phi^{(\epsilon)}(x)}{dx} = \frac{1}{L}\frac{\partial\phi(\bar{x}, \bar{y})}{\partial\bar{x}} + \frac{1}{\ell}\frac{\partial\phi(\bar{x}, \bar{y})}{\partial\bar{y}} = \frac{1}{L}\left(\frac{\partial\phi(\bar{x}, \bar{y})}{\partial\bar{x}} + \frac{1}{\epsilon}\frac{\partial\phi(\bar{x}, \bar{y})}{\partial\bar{y}}\right). \tag{6.12}$$

Within this formalism, all equations are expressed in terms of the dimensionless variables. For notational convenience, we henceforth drop the 'bar' notation, so that Eq. (6.10) is rewritten as

$$\Phi^{(\epsilon)}(x) = \phi(x, y), \tag{6.13}$$

with $y = x/\epsilon$, and consider that $L = 1$.

6.3 Two-Scales Asymptotic Homogenization Method

Most engineering problems require multiscale analysis, often within the framework of continuum mechanics and composite materials. In particular, the two-scale asymptotic homogenization method (AHM) [1, 43, 48] takes advantage of the periodic microstructure and the scales separation assumption to calculate the effective properties of the medium at its larger scales. This method reduces the complexity of the heterogeneous model specified in Eqs. (6.1)-(6.6) by finding an "equivalent" homogenized problem through the construction of a formal asymptotic solution. During this process, a problem defined on the unit cell and known as the local problem is derived, and the fine scale information is encoded into the effective coefficients (see Fig. 6.2 for a diagrammatic workflow).

Let us quickly review the method. The solution of the heterogeneous problem (Eqs. (6.4)-(6.6)) can be proposed into a formal asymptotic expansion as follows,

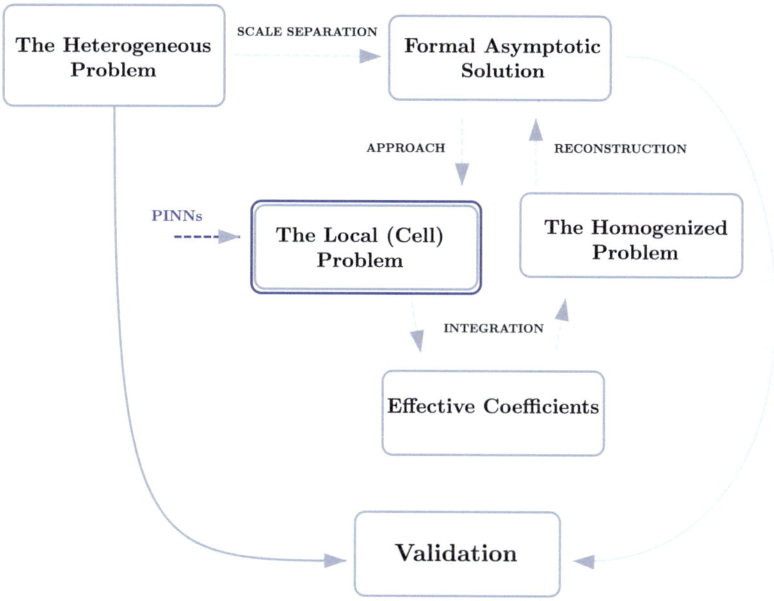

Fig. 6.2: Workflow: Coupling asymptotic homogenization method and PINNs.

$$u^{(\epsilon)}(x) = \epsilon^0 u^{(0)}(x, y) + \epsilon^1 u^{(1)}(x, y) + \epsilon^2 u^{(2)}(x, y) + \cdots, \tag{6.14}$$

where each $u^{(i)}(x, y)$ $(i = 0, 1, \ldots)$ is an infinitely differentiable and 1-periodic function with respect to y.

By incorporating the asymptotic expansion into the governing equation and collecting terms of the same power of ϵ, one derives the local problem, the effective coefficient, and the equivalent homogenized problem. In practice, it suffices to retain terms up to order ϵ^2 for this purpose. In addition, a Lemma is used here that provides a necessary and sufficient condition for the existence of such problems. We shall not go into the technical details of its application, see [8, 9] for further insights.

So then, substituting Eq. (6.14) into Eq. (6.4), applying the chain rule (6.12), and grouping by powers of ϵ gives,

$$\epsilon^{-2}: \quad \left[\frac{\partial}{\partial y}\left(E(y)\frac{\partial}{\partial y}u^{(0)}(x,y)\right)\right] = 0, \tag{6.15}$$

$$\epsilon^{-1}: \quad \left[E(y)\frac{\partial}{\partial x\partial y}u^{(0)}(x,y) + \frac{\partial}{\partial y}\left(E(y)\frac{\partial}{\partial x}u^{(0)}(x,y)\right)\right.$$
$$\left. + \frac{\partial}{\partial y}\left(E(y)\frac{\partial}{\partial y}u^{(1)}(x,y)\right)\right] = 0, \tag{6.16}$$

$$\epsilon^{0}: \quad \left[E(y)\frac{\partial^2}{\partial x^2}u^{(0)}(x,y) + E(y)\frac{\partial^2}{\partial x\partial y}u^{(1)}(x,y) + \frac{\partial}{\partial y}\left(E(y)\frac{\partial}{\partial x}u^{(1)}(x,y)\right)\right.$$
$$\left. + \frac{\partial}{\partial y}\left(E(y)\frac{\partial}{\partial y}u^{(2)}(x,y)\right)\right] = 0. \tag{6.17}$$

From order ϵ^{-2}, Eq. (6.15) has the trivial solution, and together with the periodicity of $u^{(0)}$ in y implies,

$$u^{(0)}(x,y) = u^{(0)}(x), \tag{6.18}$$

showing that the leading-order displacement $u^{(0)}$ is independent of the microscale variable y.

From order ϵ^{-1}, it is assumed that

$$u^{(1)}(x,y) = N(y)\frac{d}{dx}u^{(0)}(x), \tag{6.19}$$

where $N(y)$ is a 1-periodic function, called *local function*, which leads to the *local problem*

$$\frac{d}{dy}\left(E(y)\frac{dN(y)}{dy} + E(y)\right) = 0, \tag{6.20}$$

with uniqueness condition, periodic condition and the continuity of displacement and stress at the interface y_Γ

$$N(0) = 0, \qquad N(1) = 0, \tag{6.21}$$

$$[\![N(y_\Gamma)]\!] = 0, \qquad \left[\!\!\left[E(y_\Gamma)\frac{dN(y_\Gamma)}{dy} + E(y_\Gamma)\right]\!\!\right] = 0. \tag{6.22}$$

Similarly, using the above results in order ϵ^0, Eq. (6.17) leads to the homogenized problem

$$E^*\frac{d^2u^{(0)}(x)}{dx^2} = 0, \tag{6.23}$$

where $u^{(0)}(x) = x$ is the solution. In addition, E^* is called the effective coefficient, having the following expression

$$E^* = \left\langle E(y) + E(y)\frac{dN(y)}{dy}\right\rangle_Y, \tag{6.24}$$

where the angular brackets $\langle \cdot \rangle$ denote the spatial averaging operator over the microscopic periodic cell Y.

It should be noted that, in practice, the heterogeneous problem is prohibited and the goal is, in fact, to skip it entirely to save computation time. For this purpose, the methodology prioritizes the local problem as the central constitutive characterization phase (see Fig. 6.2).

As a general approach, the solution to the local problem (Eqs. (6.20)-(6.22)) is substituted into Eq. (6.24) to derive the effective coefficient. This coefficient is subsequently utilized to define the homogenized problem (Eq. (6.23)) for global structural analysis, thereby enabling the simulation of macroscopic behavior.

In the specific context of this research, we restrict the analysis to a straightforward one-dimensional geometric configuration. This simplification renders the entire homogenization pipeline transparent, providing accessible reference solutions for all governing components. The heterogeneous problem is resolved via the Finite Difference method (see A.1), the local problem admits an analytical solution (see A.2), the effective coefficient is determined through a closed-form expression (see A.3), and the homogenized problem allows for a direct solution.

Several computational frameworks have been established to solve the local problem in homogenization, including the Finite element method (FEM) [34], Voronoi cell methods (VC) [18], Fast Fourier Transform (FFT) solvers [37], and Semi-analytical finite element methods (SAFEM) [39]. Despite the efficacy of traditional approaches, this study investigates Physics-Informed Neural Networks (PINNs) to leverage their emerging capabilities and significant potential in solving complex microstructural problems. Utilizing this fully characterized scenario as a rigorous benchmarking environment, we apply PINNs to solve the local problem (see Fig. 6.2) and validate their predictive capability against the established analytical ground truth.

6.4 Physics-Informed Neural Networks for Solving Local Problems

Physics-Informed Neural Networks (PINNs) are a recent neural-network-based approach for approximating solutions to problems governed by differential equations [31, 44]. Unlike classical machine learning, which relies on data, PINNs integrate physical knowledge directly into the training process, that is, PINNs embed the governing equations, expressed as PDEs, into the loss function of the neural network [12, 47]. The loss typically includes the PDE residuals, initial conditions, boundary conditions, and some or no observational data, ensuring that the predicted solution remains physically consistent across the domain.

In this work, the local problem (Eqs. (6.20)-(6.22)) is solved using the PINNs approach as a direct solver without any observational data. Here, the local functions $N_i(y)$, with $i = 1, 2$ corresponding to each phase of the material, are estimated by satisfying the PDE residuals, the uniqueness condition, the periodic condition, and the interface conditions.

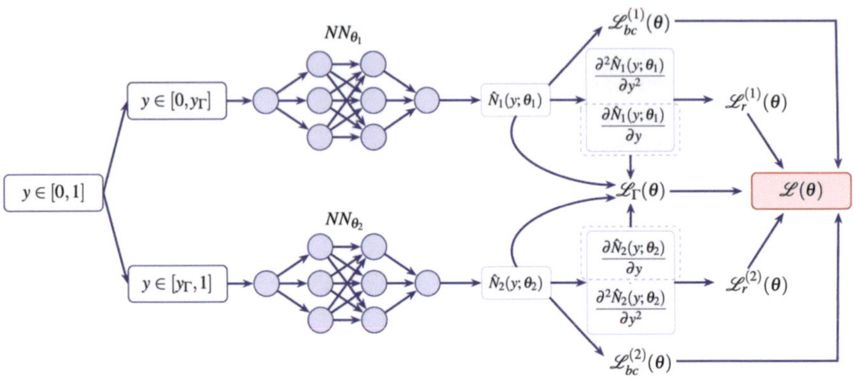

Fig. 6.3: Schematic representation of the domain decomposition strategy for PINNs in solving local problems.

In particular, the local problem (Eqs. (6.20)-(6.22)) is rewritten as the following equivalent system,

$$\frac{d}{dy}\left(E_1 \frac{dN_1(y)}{dy} + E_1\right) = 0, \quad \text{on} \quad Y_1, \tag{6.25}$$

$$N_1(0) = 0 \tag{6.26}$$

$$\frac{d}{dy}\left(E_2 \frac{dN_2(y)}{dy} + E_2\right) = 0, \quad \text{on} \quad Y_2, \tag{6.27}$$

$$N_2(1) = 0 \tag{6.28}$$

$$N_1(y) = N_2(y), \quad \text{on} \quad \Gamma, \tag{6.29}$$

$$E_1 \frac{dN_1(y)}{dy} + E_1 = E_2 \frac{dN_2(y)}{dy} + E_2, \quad \text{on} \quad \Gamma \tag{6.30}$$

For this purpose, we propose the network architecture displayed in Fig. 6.3. Starting from the decomposition of the domain into two subdomains, Y_1 and Y_2, respectively, the proposed model consists of two multi-layer perceptron (MLPs) networks (NN_{θ_i}, $i = 1, 2$), one for each subdomain Y_i. Each network takes as input the independent variables $y \in Y_i$, corresponding to spatial coordinates, and outputs the predicted local displacements $\hat{N}_i(y; \theta_i)$, unknown solutions to the respective local problem and its derivatives $\partial \hat{N}_i(y; \theta_i)/\partial y$, computed via automatic differentiation [41]. Then, the total loss function $\mathcal{L}(\theta)$ is determined. Here, the term $\theta = \theta_1 \cup \theta_2$ represents the neural network parameters to be adjusted by minimizing $\mathcal{L}(\theta)$. The proposed architecture for the MLP networks NN_{θ_i}, $i = 1, 2$ is identical, consisting of three layers with 100 neurons each. It is presented in Appendix B, along with the training parameters.

Based on the above, the total loss function $\mathcal{L}(\theta)$ can be defined as follows

$$\mathcal{L}(\theta) = \mathcal{L}_r(\theta) + \mathcal{L}_\Gamma(\theta) + \mathcal{L}_{bc}(\theta), \tag{6.31}$$

where $\mathcal{L}_r(\theta)$ is the residual loss term that enforces the fulfillment of the governing equations (6.25) and (6.27), $\mathcal{L}_\Gamma$ is the interface loss term that enforces the interface conditions given in Eqs. (6.29) and (6.30), and $\mathcal{L}_{bc}(\theta)$ is the loss term that guarantees the uniqueness and periodic conditions (6.26) and (6.28), respectively.

The residual loss $\mathcal{L}_r(\theta)$ is defined as the sum of the contributions from the two phases, Y_1 and Y_2, and is given by

$$\mathcal{L}_r(\theta) - \mathcal{L}_r^{(1)}(\theta_1) + \mathcal{L}_r^{(2)}(\theta_2), \tag{6.32}$$

where $\mathcal{L}_r^{(1)}(\theta_1)$ and $\mathcal{L}_r^{(2)}(\theta_2)$ are the loss components associated with the neural networks NN_{θ_1} and NN_{θ_2}, respectively. These components are computed as the mean squared error of the residuals evaluated at a finite set of collocation points,

$$\mathcal{L}_r^{(1)}(\theta_1) = \frac{1}{k_1}\sum_{i=1}^{k_1}\left|R_1\left(y_1^{(i)};\theta_1\right)\right|^2, \quad \mathcal{L}_r^{(2)}(\theta_2) = \frac{1}{k_2}\sum_{i=1}^{k_2}\left|R_2\left(y_2^{(i)};\theta_2\right)\right|^2. \tag{6.33}$$

Here, $\left\{y_1^{(1)},\ldots,y_1^{(k_1)}\right\} \subset Y_1$ and $\left\{y_2^{(1)},\ldots,y_2^{(k_2)}\right\} \subset Y_2$ serve as the inputs to the networks during training, with k_1 and k_2 denoting the total number of points in each set. Finally, denoting the network predictions as $\hat{N}_1(y;\theta_1)$ and $\hat{N}_2(y;\theta_2)$, the residuals R_1 and R_2 are defined as

$$R_1(y;\theta_1) = \frac{\partial}{\partial y}\left(E_1\frac{\partial \hat{N}_1(y;\theta_1)}{\partial y} + E_1\right), \quad \text{on } Y_1, \tag{6.34}$$

$$R_2(y;\theta_2) = \frac{\partial}{\partial y}\left(E_2\frac{\partial \hat{N}_2(y;\theta_2)}{\partial y} + E_2\right), \quad \text{on } Y_2. \tag{6.35}$$

The interface loss term $\mathcal{L}_\Gamma(\theta)$ is computed as the sum of two terms: $\mathcal{L}_{\text{disp}}(\theta)$, which ensures the continuity of pseudo-displacement, $\mathcal{L}_{\text{stress}}(\theta)$, which enforces the continuity of local stresses

$$\mathcal{L}_\Gamma(\theta) = \mathcal{L}_{\text{disp}}(\theta) + \mathcal{L}_{\text{stress}}(\theta), \tag{6.36}$$

where

$$\mathcal{L}_{\text{disp}}(\theta) = \left(\hat{N}_1(y_\Gamma;\theta_1) - \hat{N}_2(y_\Gamma;\theta_2)\right)^2, \tag{6.37}$$

$$\mathcal{L}_{\text{stress}}(\theta) = \left(E_1 + E_1\frac{\partial \hat{N}_1(y_\Gamma;\theta_1)}{\partial y} - E_2 - E_2\frac{\partial \hat{N}_2(y_\Gamma;\theta_2)}{\partial y}\right)^2. \tag{6.38}$$

The boundary condition loss term $\mathcal{L}_{bc}(\theta)$ is expressed as

$$\mathcal{L}_{\text{bc}}(\theta) = \mathcal{L}_{\text{bc}}^{(1)}(\theta) + \mathcal{L}_{\text{bc}}^{(2)}(\theta), \tag{6.39}$$

where

$$\mathcal{L}_{\text{bc}}^{(1)}(\theta) = (\hat{N}_1(0;\theta_1))^2, \quad \mathcal{L}_{\text{bc}}^{(2)}(\theta) = (\hat{N}_2(1;\theta_2))^2. \tag{6.40}$$

Once the local functions $\hat{N}_1$ and $\hat{N}_2$ have been obtained, the predicted effective coefficient ($\hat{E}^*$) can be computed using the averaged equation,

$$\hat{E}^* = \left\langle E_1(y) + E_1(y)\frac{d\hat{N}_1(y)}{dy}\right\rangle_{Y_1} + \left\langle E_2(y) + E_2(y)\frac{d\hat{N}_2(y)}{dy}\right\rangle_{Y_2}, \tag{6.41}$$

which represents an approximation of the effective coefficient E^* defined in Eq. (6.24).

In summary, by employing a dual-network architecture, where a dedicated neural network is assigned to each subdomain, we can embed the specific governing equations into separate loss functions. This ensures that each network independently satisfies the physical constraints of its respective subdomain. The unified solution is then achieved by coupling these networks through interface conditions enforced at the common boundary.

6.5 Results

This section presents an assessment of the proposed PINNs formulation. The results are organized to evaluate three aspects of the model:

- Its ability to predict the solution of the local problem (Eqs. (6.20)-(6.30)) was examined from two complementary directions. The first involves varying the volume fraction V_1 while keeping the material properties of the constituents fixed at. $E_1 = 45$GPa and $E_2 = 12$GPa. The second explores the influence of material contrast by systematically changing the elastic modulus ratio E_1/E_2 over the range of 1.5 to 10 keeping the volume fraction fixed at $V_1 = 0.6$. As ground trust, we use the analytical solution of the local problem determined in A.2. This analysis aims to validate the dual-network architecture of Sect. 6.4.
- Once the local problem is solved, we evaluate the integration of the predicted local functions $\hat{N}_1$ and $\hat{N}_2$ into the expression of the effective coefficient (6.41). We then perform a direct comparison against the closed-form expression in A.3. For consistency, the predicted effective response is evaluated along the same two directions: (i) varying the volume fraction V_1 with constituent properties fixed at $E_1 = 45$GPa and $E_2 = 12$GPa, and (ii) varying the elastic modulus ratio E_1/E_2 from 1.5 to 10 with volume fraction fixed at $V_1 = 0.6$.
- Finally, following the diagram in Fig. 6.2, we close the loop by including a validation of the whole process. For this purpose, we perform a comparison between the numerical solution of the heterogeneous problem, obtained via finite differences (see A.1), the direct solution of the homogenized problem (see Eq. (6.23)), and the built-in formal asymptotic solution following two variants: one

entirely with AHM (see Eq. (6.42)) and the other integrating PINNs into the process (Eq. (6.43)).

6.5.1 Local Problem

The left panel of Fig. 6.4 depicts the local displacement solution across the unit cell coordinate y for different volume fractions V_1, i.e., 0.2, 0.25, 0.5, 0.55, 0.8, and 0.85, comparing the proposed model predictions, denoted with continuous lines, against the analytical solutions, in dashed lines, of the local problem (see Eqs. (6.25)-(6.30)). Both curve families exhibit good agreement throughout the spatial domain, demonstrating that the proposed PINN architecture faithfully reproduces the shapes of the ground trust. It also shows that the PINN solution captures the uniqueness condition, the periodic condition and the continuities at the interface. Additionally, the right panel of Fig. 6.4 shows the absolute error $e(y) = |N(y) - \hat{N}(y)|$ between the analytical (N) and the PINNs-predicted ($\hat{N}$) local problem solutions for the same volume fractions. As shown, there is an absolute error of order 10^{-7}.

For validating the influence of the material contrast, the left panel of Fig. 6.5 depicts the local problem solutions for several elastic modulus ratios, i.e, $E_1/E_2 =$ 1.5, 3.0, 5.0, 6.0, 8.5 and 10. In all scenarios, there is good agreement between the PINNs predicted solution, denoted with solid lines, and the analytical solution, in dashed lines. The predicted curves reproduce the characteristic V-shaped of the piecewise linear analytical solution. Similarly, the right panel shows the absolute error $e(y) = |N(y) - \hat{N}(y)|$ computed for each modulus ratios. The error curves remain on the order of 10^{-6} for the higher ratios $E_1/E_2 = 8.5$ and 10.0, while for the smaller ratios $E_1/E_2 = 1.5$ and 3.0 the errors are of the order of 10^{-7}. It can be seem that the higher fluctuations occurs at the interface.

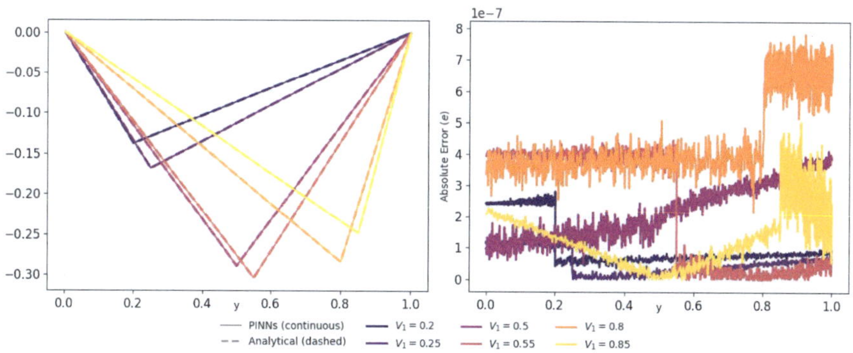

Fig. 6.4: Comparison between the analytical solution (N) of the local problem (6.20)-(6.30) and the predicted solution ($\hat{N}$) via PINNs for several volume fractions V_1.

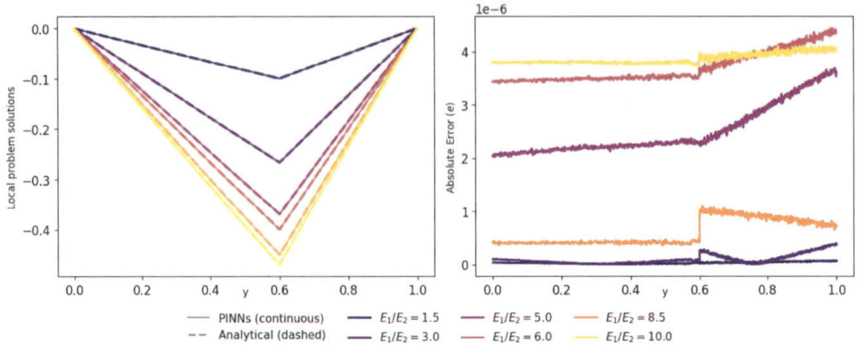

Fig. 6.5: Comparison between the analytical solution (N) of the local problem (6.20)-(6.30) and the predicted solution ($\hat{N}$) via PINNs for several material contrasts.

6.5.2 *Effective Coefficients*

Now, a comparison between the effective Young's modulus E^* and the predicted one $\hat{E}^*$ is presented in Fig. 6.6, for different volume fractions V_1. As observed, the results exhibit a good agreement. This can be confirmed in Table 6.1, where a more comprehensive report is presented for values of V_1 ranging from 0.1 to 0.95. In addition, the relative error R_E in percentage is included, such as $R_E = |\hat{E}^* - E^*|/E^* \times 100\%$. From the Table 6.1, the maximum relative error obtained is $1.34 \times 10^{-4}\%$ for $V_1 = 0.95$, which is suitable from a practical point of view. These

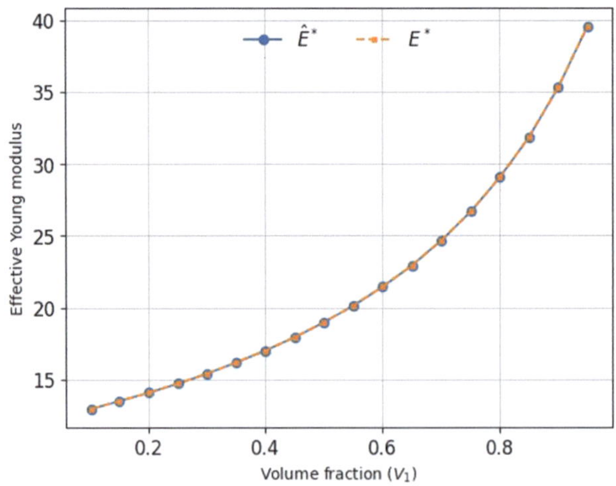

Fig. 6.6: Comparison between the effective Young modulus (E^*) obtained via Eq. (6.60), and the predicted one ($\hat{E}^*$) Eq. (6.41) as a function of the volume fraction (V_1).

Table 6.1: Corresponding data from Fig 6.6 showing the comparison between the effective Young modulus (E^*) and the predicted one ($\hat{E}^*$).

V_1	$\hat{E}^*$	E^*	$R_E\,(\%)$
0.10	12.949641	12.949640	7.23e-06
0.15	13.483148	13.483146	1.15e-05
0.20	14.062500	14.062500	0
0.25	14.693878	14.693877	4.24e-06
0.30	15.384604	15.384615	7.72e-05
0.35	16.143500	16.143497	1.00e-05
0.40	16.981133	16.981132	2.51e-06
0.45	17.910440	17.910447	4.08e-05
0.50	18.947361	18.947368	3.92e-05
0.55	20.111732	20.111731	3.18e-06
0.60	21.428577	21.428571	2.80e-05
0.65	22.929935	22.929936	3.71e-06
0.70	24.657540	24.657534	2.08e-05
0.75	26.666683	26.666666	6.20e-05
0.80	29.032257	29.032258	3.40e-06
0.85	31.858389	31.858407	5.71e-05
0.90	35.294147	35.294117	8.46e-05
0.95	39.560387	39.560439	1.34e-04

results indicate that the PINNs model accurately estimates the effective coefficients, demonstrating its potential for this type of application.

Additionally, to assess the proposed methodology over different material contrasts, Fig. 6.7 presents the effective Young's modulus E^* together with the predicted one $\hat{E}^*$ for a range of elastic modulus contrasts E_1/E_2. The curves indicate that the PINN prediction accurately recovers the effective coefficients. This observation is also supported by the data in Table 6.2, where the relative errors R_E, expressed as percentages, remain mostly on the order of $10^{-5}\%$ – $10^{-6}\%$, with only two isolated cases of order of $10^{-4}\%$. The results confirm that our PINN approach is able to capture the effective response of the heterogeneous medium across a wide range of material contrasts.

6.5.3 Comparison Between the Approaches

Finally, in this section, a comparison is performed among the numerical solution of the heterogeneous problem (see A.1), denoted by $u_{\text{het}}^{(\epsilon)}(x)$, the direct solution of the homogenized problem $u^{(0)}(x)$ (see Eq. (6.23)), and the formal asymptotic solutions $u^{(\epsilon)}_{\text{twoscale}}(x)$ and $u^{(\epsilon)}_{\text{pinns}}(x)$, corresponding respectively to the pure AHM approach and the PINN-integrated approach, defined as follows,

$$u^{(\epsilon)}_{\text{twoscale}}(x) = u^{(0)}(x) + \epsilon N(x/\epsilon), \tag{6.42}$$

$$u^{(\epsilon)}_{\text{pinns}}(x) = u^{(0)}(x) + \epsilon \hat{N}(x/\epsilon). \tag{6.43}$$

Notice that Eq. (6.42) is the truncated expression (6.14) that arises from a two-scale analysis and Eq. (6.43) is the equivalent expression but considering the predicted local function $\hat{N}$ by means of the PINNs approach.

Based on that, Fig. 6.8 illustrates the displacement field for different values of $\epsilon \in \{0.5, 0.1, 0.01\}$, a fixed volume fraction $V_1 = 0.6$, and the material properties set at $E_1 = 45$GPa and $E_2 = 12$GPa. As observed, the heterogeneous solution $u^{(\epsilon)}_{\text{het}}$ exhibits the typical piecewise-linear oscillations induced by the geometrical structure of the layered medium. Both $u^{(\epsilon)}_{\text{twoscale}}$ and $u^{(\epsilon)}_{\text{pinns}}$ accurately reproduce these oscillations and closely following the heterogeneous solution along the entire domain.

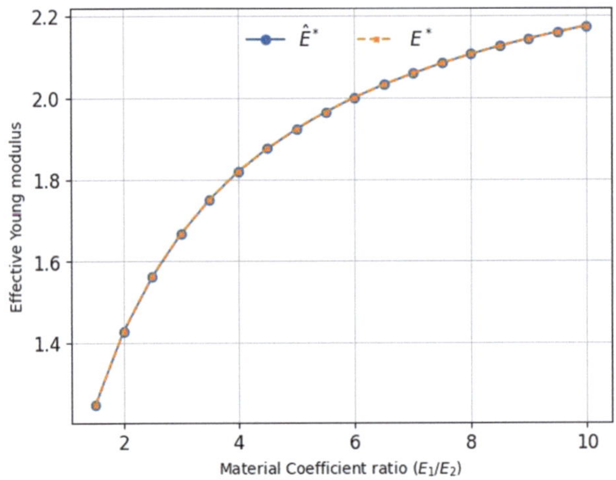

Fig. 6.7: Comparison between the effective Young modulus (E^*) and the predicted one ($\hat{E}^*$) as a function of the material contrast.

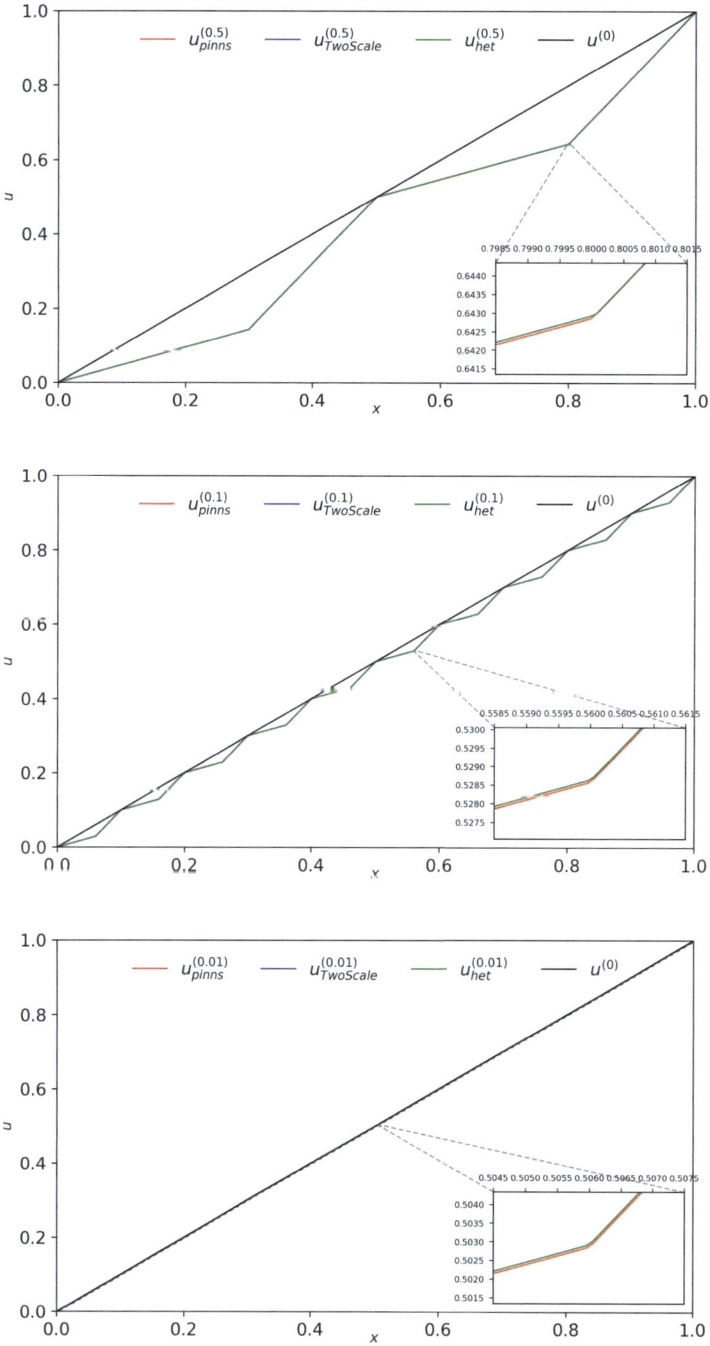

Fig. 6.8: Visual comparison between the solutions of the heterogeneous problems, the homogenized problem and the two formal asymptotic expansions given in Eqs. (6.42)-(6.43) for $\epsilon \in \{0.5, 0.1, 0.01\}$, material properties $E_1 = 45$GPa and $E_2 = 12$GPa and for a fixed volume fraction $V_1 = 0.6$.

Table 6.2: Corresponding data from Fig 6.7 showing the comparison between the effective Young modulus (E^*) and the predicted one ($\hat{E}^*$).

E_1/E_2	$\hat{E}^*$	E^*	R_E (%)
1.5	1.2499998	1.2499999	1.91e-05
2.0	1.4285715	1.4285714	2.38e-06
2.5	1.5625001	1.5625000	7.63e-06
3.0	1.6666679	1.6666666	7.63e-05
3.5	1.7499999	1.7500000	6.81e-06
4.0	1.8181825	1.8181818	3.58e-05
4.5	1.8750000	1.8750000	0
5.0	1.9230797	1.9230769	1.46e-04
5.5	1.9642854	1.9642857	1.73e-05
6.0	2.0000014	1.9999999	7.15e-05
6.5	2.0312500	2.0312500	0
7.0	2.0588243	2.0588235	3.75e-05
7.5	2.0833333	2.0833333	3.81e-06
8.0	2.1052628	2.1052632	1.91e-05
8.5	2.1250002	2.1250000	1.12e-05
9.0	2.1428623	2.1428571	2.42e-04
9.5	2.1590900	2.1590909	4.02e-05
10.0	2.1739135	2.1739130	2.00e-05

Note that we have added some small panels that zoom in on specific regions where discontinuities occur in the material coefficients for each value of ϵ, in particular, the intervals containing an interface point. Under this zoom, the solutions $u^{(\epsilon)}_{\text{twoscale}}$ and $u^{(\epsilon)}_{\text{pinns}}$ remain very close and seem nearly identical, matching the heterogeneous solution with only minor deviations. As expected, the discrepancy with the homogenized solution $u^{(0)}$ is more pronounced for larger ϵ, while for smaller ones, the heterogeneous solution approaches the homogenized limit, as well as $u^{(\epsilon)}_{\text{twoscale}}$ and $u^{(\epsilon)}_{\text{pinns}}$, in agreement with homogenization theory.

Before going further, lets define the absolute root mean square errors $e^{(\epsilon)}_{\text{twoscale}}$ and $e^{(\epsilon)}_{\text{pinns}}$ as follows,

$$e^{(\epsilon)}_{\text{twoscale}} = e\left(u^{(\epsilon)}_{\text{het}}, u^{(\epsilon)}_{\text{twoscale}}\right) = \sqrt{\frac{1}{n}\sum_{i=1}^{n}\left(u^{(\epsilon)}_{\text{het}}(x_i) - u^{(\epsilon)}_{\text{twoscale}}(x_i)\right)^2}, \tag{6.44}$$

$$e^{(\epsilon)}_{\text{pinns}} = e\left(u^{(\epsilon)}_{\text{het}}, u^{(\epsilon)}_{\text{pinns}}\right) = \sqrt{\frac{1}{n}\sum_{i=1}^{n}\left(u^{(\epsilon)}_{\text{het}}(x_i) - u^{(\epsilon)}_{\text{pinns}}(x_i)\right)^2}, \tag{6.45}$$

where $\{x_i\}_{i=1}^{n} \subset [0,1]$. So then, we complement Fig. 6.8, by presenting the errors $e^{(\epsilon)}_{\text{twoscale}}$ and $e^{(\epsilon)}_{\text{pinns}}$ in Fig. 6.9, for the same values of epsilon, i.e., $\epsilon \in \{0.5, 0.1, 0.01\}$ and material properties but varying the volume fractions. As shown, the errors remain within the order of 10^{-4} for all analyzed volume fractions. In summary, all results show that the PINN-integrated approach predict the effective response of 1D elastic laminated composites with high fidelity, achieving accuracy remarkably close to the analytical solution and very close to the finite difference solution of the heterogeneous problem.

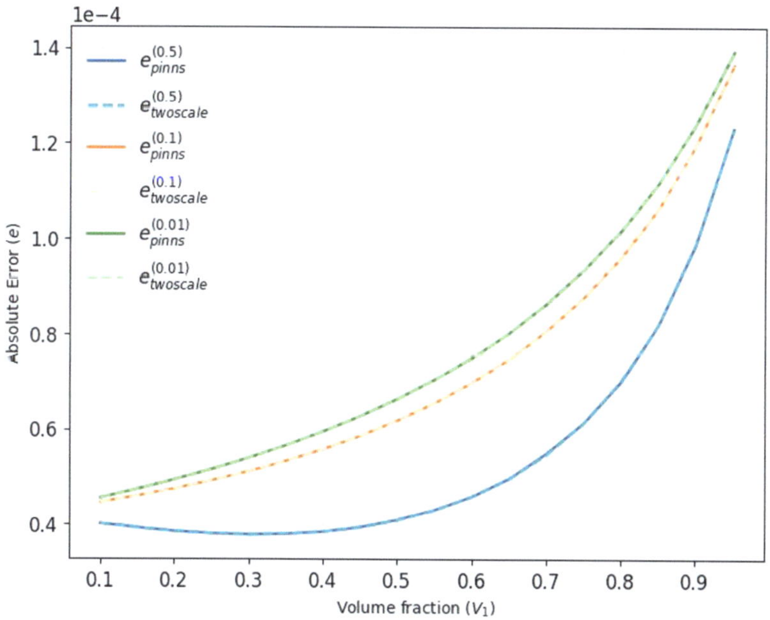

Fig. 6.9: Root mean square errors Eqs. (6.44) - (6.45) evaluated across different volume fractions V_1 for $\epsilon \in \{0.5, 0.1, 0.01\}$ and material properties $E_1 = 45$GPa and $E_2 = 12$GPa.

6.6 Conclusions

In this work, we presented a hybrid multiscale computational framework that couples the Asymptotic Homogenization Methods (AHM) with the flexibility of Physics-Informed Neural Networks (PINNs). The main goal was to evaluate the capability of PINNs to solve the local problem and accurately predict the effective elastic properties of periodic composites. By focusing on a one-dimensional framework, we were able to fully detail the coupling workflow. We proposed a Dual-Network architecture and optimization choices designed to handle material discontinuities and stiff gradients. The PINN was trained to minimize a composite loss function comprising the equilibrium equations ($\mathcal{L}_r(\theta)$), the uniqueness and periodic conditions ($\mathcal{L}_{bc}(\theta)$), and the interface continuity constraints ($\mathcal{L}_\Gamma(\theta)$), without relying on labeled training data. The resulting local field solutions were successfully integrated into the calculation of the effective elastic coefficients and the formal asymptotic solution.

The choice of a one-dimensional configuration was purposely made. While 1D elasticity admits closed-form analytical solutions, it serves as a rigorous benchmark for validating the neural network approximations. Our results demonstrate that the proposed PINN framework converges to the exact analytical solution of the local problem with high accuracy across a wide range of volume fractions and material contrast scenarios.

The main strengths of integrating PINNs into the workflow are its meshless nature, which eliminates possible discretization issues, and its tailored Dual-Network architecture, which facilitates seamless generalization to complex 2D and 3D microstructures. While the initial computational cost of training the PINN model to achieve this high fidelity is substantial, rendering it slower than instantaneous analytical methods for this specific 1D benchmark, this expense represents an essential initial investment. The resulting trained model, providing the high-accuracy local function ($\hat{N}$) and the effective property (E^*), serves as an accurate, integrated, and rapidly inferable solver for future applications. This investment is justified because the efficiency gap is expected to narrow significantly in higher dimensions and complex geometries, where the scaling properties of neural networks become advantageous and conventional meshing techniques become the limiting bottleneck. Furthermore, the resulting differentiable framework is inherently suitable for challenging inverse homogenization problems.

Having validated the workflow against the exact 1D solution, several extensions are now suitable. The immediate next step is the application of this framework to address imperfect contact conditions. Then, the extension to 2D and 3D scenarios, where the mesh generation for complex geometries becomes a non-trivial bottleneck for FEM. Furthermore, due to the differentiable nature of neural networks, this framework is particularly well-suited for inverse homogenization problems, identifying the optimal microstructural topology required to achieve a target effective property, which remains a challenging frontier in materials design.

Acknowledgements R. Rodríguez-Ramos thanks the National Council for Scientific and Technological Development, CNPq No. 09/2023, the Research on Productivity at CNPq - level 2 process No. 307188/2023-0, and the call Office of the Vice-Rector for Research, Postgraduate Studies and Innovation (PROPPI), Federal Fluminense University (UFF) (No. 05/2022) Postgraduate Program in Computational Modeling in Science and Technology (PPG-MCCT). Y. Espinosa-Almeyda acknowledges SECIHTI for the postdoctoral scholarship "Estancias Postdoctorales por México para la Formación y Consolidación de Investigadores por México" at IIT, UACJ, 2024-2026. All authors thank to SECIHTI, México.

Appendix A Supplementary Derivations

A.1 Finite Difference Method for Solving the Heterogeneous Problem

To solve the heterogeneous contour problem Eqs. (6.4) - (6.6), the approach proposed in [33] is adopted. Following this methodology, a uniform mesh $\{x_i\}_{i=1}^{M+1}$ is built within the interval $[0, 1]$ with a mesh size $h = 1/M$. Based on this, an approximation of Eqs. (6.4) is computed using the central finite difference method as follows,

$$\frac{E_{i+\frac{1}{2}}(u_{i+1} - u_i) - E_{i-\frac{1}{2}}(u_i - u_{i-1})}{h^2} = 0, \tag{6.46}$$

where $u_i = u(x_i)$ and $i = 1, 2, \ldots, M - 2$. The Eq. (6.46) leads to a homogeneous system of equations

$$E_{i-\frac{1}{2}}u_{i-1} - (E_{i-\frac{1}{2}} + E_{i+\frac{1}{2}})u_i + E_{i+\frac{1}{2}}u_{i+1} = 0, \tag{6.47}$$

which can be rewritten as a linear system $Au = b$, where the boundary conditions are added to obtain a system of M equations.
The matrix $A_{M\times M}$ and the vectors $b_{M\times 1}$ and $u_{M\times 1}$ are given by

$$A = \begin{bmatrix} 1 & 0 & 0 & 0 & \cdots & 0 \\ a_1 & b_1 & c_1 & 0 & \ddots & \vdots \\ 0 & a_2 & b_2 & c_2 & \ddots & 0 \\ \vdots & \ddots & \ddots & \ddots & \ddots & \vdots \\ 0 & \cdots & 0 & a_{M-2} & b_{M-2} & c_{M-2} \\ 0 & \cdots & 0 & 0 & 0 & 1 \end{bmatrix}, \quad b = \begin{bmatrix} 0 \\ 0 \\ 0 \\ \vdots \\ 0 \\ 1 \end{bmatrix}, \quad \text{and} \quad u = \begin{bmatrix} u_1 \\ u_2 \\ u_3 \\ \vdots \\ u_{M-1} \\ u_M \end{bmatrix}, \tag{6.48}$$

where the nonzero entries of the 3-band structure of A are defined as

$$a_i = \frac{E_{i-1/2}}{h^2}, \quad b_i = -\frac{E_{i-1/2} + E_{i+1/2}}{h^2}, \quad c_i = \frac{E_{i+1/2}}{h^2}. \tag{6.49}$$

Then, according to [33], the coefficients $E_{i+1/2}$ and $E_{i-1/2}$ are computed as

$$E_{i+1/2} = \left[\frac{1}{h} \int\limits_{x_i}^{x_{i+1}} E^{-1}(x)dx \right]^{-1} \quad \text{and} \quad E_{i-1/2} = E_{i-1+1/2}, \tag{6.50}$$

where, in our particular case, the final expression of these coefficients depends on whether a point in $P = \{k\epsilon\}_{k=1}^{M}$, with $M = \lfloor 1/\epsilon \rfloor$ or in the interface $\Gamma = \{\gamma_k\}_{k=1}^{M}$ is between two consecutive points of the mesh. The mesh $\{x_i\}_{i=1}^{N}$ is considered fine enough, so that for each i, there is no more than one point of $P \cup \Gamma$ in the interval (x_i, x_{i+1}).

Three cases are considered to compute $E_{i+1/2}$. The first case occurs when $\gamma_k \in (x_i, x_{i+1})$ for some k. In this situation, $E(x) = E_{\text{mat}_1}$ for $x \in [x_i, \gamma_k]$ and $E(x) = E_{\text{mat}_2}$ for $x \in [\gamma_k, x_{i+1}]$ (see Eq. (6.3)). Therefore,

$$E_{i+1/2} = \left[\frac{1}{h} \left(\frac{\gamma_k - x_i}{E_{\text{mat}_1}} + \frac{x_{i+1} - \gamma_k}{E_{\text{mat}_2}} \right) \right]^{-1} = \frac{E_{\text{mat}_1} E_{\text{mat}_2} h}{E_{\text{mat}_1}(x_{i+1} - \gamma_k) + E_{\text{mat}_2}(\gamma_k - x_i)}. \tag{6.51}$$

The second case corresponds to $k\epsilon \in (x_i, x_{i+1})$ for some integer k. This leads to

$$E_{i+1/2} = \left[\frac{1}{h} \left(\frac{k\epsilon - x_i}{E_{\text{mat}_1}} + \frac{x_{i+1} - k\epsilon}{E_{\text{mat}_2}} \right) \right]^{-1} = \frac{E_{\text{mat}_1} E_{\text{mat}_2} h}{E_{\text{mat}_1}(x_{i+1} - k\epsilon) + E_{\text{mat}_2}(k\epsilon - x_i)}. \tag{6.52}$$

Finally, the third case applies when $E(x)$ is constant throughout the interval (x_i, x_{i+1}). Thus,

$$E_{i+1/2} = \left[\frac{1}{h} \left(\frac{x_{i+1} - x_i}{E(x_i + h/2)} \right) \right]^{-1} = E(x_i + h/2). \tag{6.53}$$

Once the values of the $E_{i+1/2}$ are found, the matrix A is completed, and the system $Au = b$ can be solved, resulting in the approximate solution to the heterogeneous problem.

A.2 Analytical Solution of the Local Problem

Consider a two-phase composite where the elastic modulus $E(y)$ is piecewise constant. The unit cell is divided into two domains,

$$Y_1 = [0, y_\Gamma) \quad \text{and} \quad Y_2 = (y_\Gamma, 1],$$

separated by the interface $\Gamma = \{y_\Gamma\}$. Consequently, the modulus takes the values E_1 in Y_1 and E_2 in Y_2. Since $E(y)$ is constant within each phase, its spatial derivative vanishes

$$\frac{dE}{dy} = 0 \tag{6.54}$$

everywhere except at the interface Γ. To obtain an analytical solution of the local problem, we differentiate Eq. (6.20) with respect to y, which yields,

$$\frac{dE(y)}{dy}\frac{dN(y)}{dy} + E(y)\frac{d^2N(y)}{dy^2} + \frac{dE(y)}{dy} = 0. \tag{6.55}$$

By applying the condition (6.54) within the phases, the equation simplifies to,

$$\frac{d^2N(y)}{dy^2} = 0, \quad y \in Y_1 \cup Y_2. \tag{6.56}$$

Integrating twice (6.56) with respect to the variable y for each phase leads to

$$N(y) = \begin{cases} N_1(y) = A_1\, y + B_1, & y \in Y_1 = [0, y_\Gamma), \\ N_2(y) = A_2\, y + B_2, & y \in Y_2 = (y_\Gamma, 1], \end{cases} \tag{6.57}$$

where A_i and B_i, with $i = 1, 2$ are unknown constants.

The boundary conditions $N_1(0) = 0$ and $N_2(1) = 0$ imply

$$B_1 = 0, \qquad B_2 = -A_2. \tag{6.58}$$

Applying the continuity of displacement and stress across the interface Γ and using Eq. (6.58) gives

$$A_1 = \frac{(E_2 - E_1)(1 - y_\Gamma)}{(1 - y_\Gamma)E_1 + E_2 y_\Gamma}, \qquad A_2 = \frac{(E_1 - E_2)\, y_\Gamma}{(1 - y_\Gamma)E_1 + E_2 y_\Gamma}. \tag{6.59}$$

A.3 Closed Form Expression for the Effective Coefficient

Using these expressions in Eq. (6.24), the effective modulus becomes

$$\begin{aligned} E^* &= \int_0^{y_\Gamma} \left(E_1 + E_1 A_1\right) dy + \int_{y_\Gamma}^{1} \left(E_2 + E_2 A_2\right) dy \\ &= \frac{E_1 E_2}{E_1(1 - y_\Gamma) + E_2 y_\Gamma} = \left[\frac{V_1}{E_1} + \frac{V_2}{E_2}\right]^{-1} = \left\langle \frac{1}{E(y)} \right\rangle^{-1}, \end{aligned} \tag{6.60}$$

where $V_1 = |[0, y_\Gamma)| = y_\Gamma$ and $V_1 + V_2 = 1$.

Appendix B Training Methodology

The training procedure is essential to obtain accurate and stable PINN solutions, where the loss terms may be strongly unbalanced and the optimization landscape is poorly conditioned. The training utilized the SOAP optimizer [50, 54], a sharpness-aware first-order method recently introduced to improve stability and generalization

Table 6.3: PINN configuration.

Parameter	Value
Dual MLP Architecture:	Both networks share the same architecture
Depth	3
Neuron	100
Activation	Tanh
Xavier init	TRUE
Optimization	
Optimizer	SOAP
- Base learning rate	10^{-3}
Exponential decay	TRUE
- Decay rate	0.9
- Decay steps	1,000
Training	
Iterations	10^5
Batch size	2000
Loss weighting	Self-adaptive loss balancing
- Updated every	100 iters

in deep models. The architecture of the neural networks and the parameters for their training are given in Table 6.3. Minibatches were randomly selected from each region $Y_i, i = 1, 2$. We also employ self-adaptive loss balancing [51, 52] and an exponential learning-rate decay. Therefore, the total loss takes the form

$$\mathcal{L} = w_r\, \mathcal{L}_r + w_{bc}\, \mathcal{L}_{bc} + w_\Gamma\, \mathcal{L}_\Gamma, \tag{6.61}$$

where the weights w_r, w_{bc}, and w_Γ are dynamically updated throughout training at a constant frequency. This adaptive weighting mitigates imbalances between the different loss terms by increasing the contribution of components with smaller gradient values. The adaptive weights are computed as

$$\begin{aligned} w_\Gamma &= \frac{\|\nabla_\theta \mathcal{L}_\Gamma\| + \|\nabla_\theta \mathcal{L}_{bc}\| + \|\nabla_\theta \mathcal{L}_r\|}{\|\nabla_\theta \mathcal{L}_\Gamma\|}, \\ w_{bc} &= \frac{\|\nabla_\theta \mathcal{L}_\Gamma\| + \|\nabla_\theta \mathcal{L}_{bc}\| + \|\nabla_\theta \mathcal{L}_r\|}{\|\nabla_\theta \mathcal{L}_{bc}\|}, \\ w_r &= \frac{\|\nabla_\theta \mathcal{L}_\Gamma\| + \|\nabla_\theta \mathcal{L}_{bc}\| + \|\nabla_\theta \mathcal{L}_r\|}{\|\nabla_\theta \mathcal{L}_r\|}. \end{aligned} \tag{6.62}$$

The learning rate η_k follows an exponential decay schedule. At the k-th update, η_k is computed as $\eta_k = \eta_0 e^{-\gamma k}$, where γ denotes the decay rate. This strategy enables the modification of the network parameters by taking large steps at the beginning and progressively smaller ones as training advances. The decay is applied at fixed update intervals throughout the training epochs. The initial learning rate $\eta_0 = 10^{-3}$ and a constant decay rate of $\gamma = 0.9$. Training losses for different values E_1/E_2 over 100000 epochs are shown in Fig. 6.10, exhibiting low losses despite late oscillations.

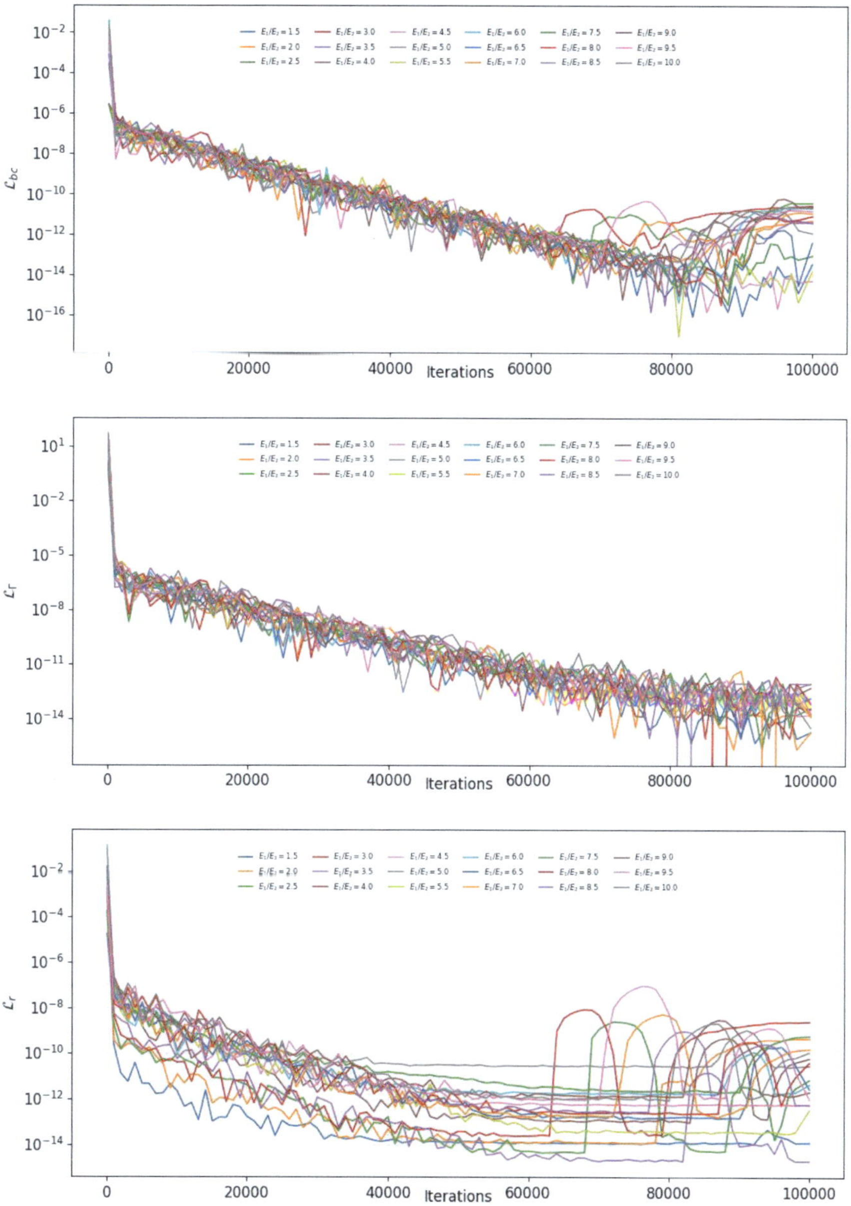

Fig. 6.10: Training losses for different values of the elastic modulus contrast E_1/E_2, showing the residual loss $\mathcal{L}_r$, the boundary loss $\mathcal{L}_{bc}$, and the data loss $\mathcal{L}_\Gamma$, computed over a total of 100 000 training epochs. The training is performed using the SOAP optimizer [50] with adaptive weights and an exponential learning rate schedule. Low values despite late oscillations.

References

[1] Allaire G (1992) Homogenization and two-scale convergence. SIAM Journal on Mathematical Analysis 23(6):1482-1518, DOI 10.1137/0523084

[2] Arabnejad S, Pasini D (2013) Mechanical properties of lattice materials via asymptotic homogenization and comparison with alternative homogenization methods. International Journal of Mechanical Sciences 77:249-262, DOI 10.1016/j.ijmecsci.2013.10.003

[3] Bakhvalov NS, Panasenko GP (1989) Homogenisation: Averaging Processes in Periodic Media. Mathematical Problems in the Mechanics of Composite Materials, Mathematics and its Applications, Vol 36. Springer Science & Business Media, Dordrecht, DOI 10.1007/978-94-009-2247-1

[4] Bensoussan A, Lions JL, Papanicolaou G. (1978) Asymptotic Analysis for Periodic Structures. North-Holland, Amsterdam

[5] Benveniste Y (1987) A new approach to the application of Mori-Tanaka's theory in composite materials. Mechanics of Materials 6(2):147-157, DOI 10.1016/0167-6636(87)90005-6

[6] Chen Q, Chatzigeorgiou G, Meraghni F (2023) Extended mean-field homogenization of unidirectional piezoelectric nanocomposites with generalized Gurtin-Murdoch interfaces. Composite Structures 307:116639, DOI 10.1016/j.compstruct.2022.116639

[7] Chen Q, Du X, Chatzigeorgiou G, Meraghni F, Zhao G, Yang Z (2025) Hybrid homogenization neural networks for periodic composites. International Journal of Solids and Structures 284:112521, DOI 10.1016/j.ijsolstr.2025.113622

[8] Cruz-González OL, Rodríguez-Ramos R, Bravo-Castillero J, Martínez-Rosado R, Guinovart-Díaz R, Otero JA, Sabina FJ (2017) Effective viscoelastic properties of one-dimensional composites. American Research Journal of Physics 3(1):1-17, DOI 10.21694/2380-5714.17002

[9] Cruz-González OL, Ramírez-Torres A, Rodríiguez-Ramos R, Penta R, Lebon F (2022) Hierarchical heterogeneous one-dimensional problem in linear viscoelastic media. European Journal of Mechanics - A/Solids 95:104617, DOI 10.1016/j.euromechsol.2022.104617

[10] Dong XN, Zhang X, Huang YY, Guo XE (2005) A generalized self-consistent estimate for the effective elastic moduli of fiber-reinforced composite materials with multiple transversely isotropic inclusions. International Journal of Mechanical Sciences 47(6):922-940, DOI 10.1016/j.ijmecsci.2005.01.008

[11] Eshelby JD (1957) The determination of the elastic field of an ellipsoidal inclusion, and related problems. Proceedings of the Royal Society of London. Series A. Mathematical and Physical Sciences 241(1226):376-396, DOI 10.1098/rspa.1957.0133

[12] Es'kin VA, Davydov DV, Gur'eva JV, Malkhanov AO, Smorkalov ME (2024) Separable physics-informed neural networks for the solution of elasticity problems. arXiv preprint arXiv:2401.13486, DOI 10.48550/arXiv.2401.13486

[13] Fish J, Wagiman A (1993) Multiscale finite element method for a locally nonperiodic heterogeneous medium. Computational Mechanics 12(3):164-180, DOI 10.1007/BF00371991

[14] Gajek S, Schneider M, Böhlke T (2020) On the micromechanics of deep material networks. Journal of the Mechanics and Physics of Solids 142:103984, DOI 10.1016/j.jmps.2020.103984

[15] Gajek S, Schneider M, Böhlke T (2021) An FE-DMN method for the multiscale analysis of short fiber reinforced plastic components. Computer Methods in Applied Mechanics and Engineering 384:113952, DOI 10.1016/j.cma.2021.113952

[16] Geers MGD, Kouznetsova VG, Brekelmans WAM (2010) Multi-scale computational homogenization: Trends and challenges. Journal of Computational and Applied Mathematics 234(7):2175-2182, DOI 10.1016/j.cam.2009.08.077

[17] Gaynutdinova L, Doškář M, Rokoš O, Pultarová I (2025) Homogenization with guaranteed bounds via primal-dual physically informed neural networks. arXiv preprint arXiv:2509.07579, DOI 10.48550/arXiv.2509.07579

[18] Ghosh S, Moorthy S (1995) Elastic-plastic analysis of arbitrary heterogeneous materials with the Voronoi Cell finite element method. Computer Methods in Applied Mechanics and Engineering 121(1-4):373-409, DOI 10.1016/0045-7825(94)00687-I

[19] Han J, Lee Y (2022) A neural network approach for homogenization of multiscale problems. arXiv preprint arXiv:2206.02032, DOI 10.48550/arXiv.2206.02032

[20] Han J, Lee Y (2023) A neural network approach for homogenization of multiscale problems. Multiscale Modeling & Simulation 21(2):716-734, DOI 10.1137/22M1500903

[21] Hashin Z, Shtrikman S (1962) A variational approach to the theory of the effective magnetic permeability of multiphase materials. Journal of Applied Physics 33(10):3125-3131, DOI 10.1063/1.1728579

[22] Henkes A, Wessels H, Mahnken R (2022) Physics-informed neural networks for continuum micromechanics. Computer Methods in Applied Mechanics and Engineering 393:114790, DOI 10.1016/j.cma.2022.114790

[23] Holmes MH (2012) Introduction to Perturbation Methods. Texts in Applied Mathematics Vol 20. Springer Science & Business Media, New York, DOI 10.1007/978-1-4614-5477-9

[24] Huang Z, Xing Y, Gao Y (2022) Two-scale asymptotic homogenization method for composite Kirchhoff plates with in-plane periodicity. Aerospace 9(12):751, DOI 10.3390/aerospace9120751

[25] Jiang J, Wu J, Chen Q, Chatzigeorgiou G, Meraghni F (2023) Physically informed deep homogenization neural network for unidirectional multiphase/multi-inclusion thermoconductive composites. Computer Methods in Applied Mechanics and Engineering 409:115972, DOI 10.1016/j.cma.2023.115972

[26] Kamiński M (2005) Homogenization-based finite element analysis of unidirectional composites by classical and multiresolutional techniques. Computer

Methods in Applied Mechanics and Engineering 194(18):2147-2173, DOI 10.1016/j.cma.2004.07.030
[27] Katouzian M, Vlase S, Itu C, Scutaru ML (2024) Calculation of homogenized mechanical coefficients of fiber-reinforced composite using finite element method. Materials 17(6):1334, DOI 10.3390/ma17061334
[28] Kingma DP, Ba J (2015) Adam: A method for stochastic optimization. Proceedings of the International Conference on Learning Representations (ICLR), arXiv:1412.6980, DOI 10.48550/arXiv.1412.6980
[29] Klusemann B, Svendsen B (2010) Homogenization methods for multiphase elastic composites: comparisons and benchmarks. Technische Mechanik 30(4):374-386
[30] Kouznetsova VG, Geers MGD, Brekelmans WAM (2001) Multi-scale constitutive modelling of heterogeneous materials with a gradient-enhanced computational homogenization scheme. International Journal for Numerical Methods in Engineering 54(8):1235-1260, DOI 10.1002/nme.541
[31] Lagaris IE, Likas A, Fotiadis DI (1998) Artificial neural networks for solving ordinary and partial differential equations. IEEE Transactions on Neural Networks 9(5):987-1000, DOI 10.1109/72.712178
[32] Leung WT, Lin G, Zhang Z (2022) NH-PINN: Neural homogenization-based physics-informed neural network for multiscale problems. Journal of Computational Physics 470:111539, DOI 10.1016/j.jcp.2022.111539
[33] LeVeque RJ, Li Z (1994) The immersed interface method for elliptic equations with discontinuous coefficients and singular sources. SIAM Journal on Numerical Analysis 31(4):1019-1044, DOI 10.1137/0731054
[34] Michel JC, Moulinec H, Suquet P (1999) Effective properties of composite materials with periodic microstructure: A computational approach. Computer Methods in Applied Mechanics and Engineering 172(1-4):109-143, DOI 10.1016/S0045-7825(98)00227-8
[35] Milton GW (2002) The Theory of Composites. Cambridge University Press, Cambridge, DOI 10.1137/1.9781611977486
[36] Mori T, Tanaka K (1973) Average stress in matrix and average elastic energy of materials with misfitting inclusions. Acta Metallurgica 21(5):571-574, DOI 10.1016/0001-6160(73)90064-3
[37] Moulinec H, Suquet P (1998) A numerical method for computing the overall response of nonlinear composites with complex microstructure. Computer Methods in Applied Mechanics and Engineering 157(1-2):69-94, DOI 10.1016/S0045-7825(97)00218-1
[38] Nemat-Nasser S, Hori M (1999) Micromechanics: Overall Properties of Heterogeneous Materials, 2nd ed. Elsevier, Amsterdam
[39] Otero JA, Rodríguez-Ramos R, Bravo-Castillero J, Guinovart-Díaz R, Sabina FJ, Monsivais G (2013) Semi-analytical method for computing effective properties in elastic composite under imperfect contact. International Journal of Solids and Structures 50(3):609-622, DOI 10.1016/j.ijsolstr.2012.11.001

[40] Park JSR, Zhu X (2022) Physics-informed neural networks for learning the homogenized coefficients of multiscale elliptic equations. Journal of Computational Physics 467, 111420, DOI 10.1016/j.jcp.2022.111420
[41] Paszke A, Gross S, Massa F, Lerer A, Bradbury J, Chanan G, Killeen T, Lin Z, Gimelshein N, Antiga L, Desmaison A, Köpf A, Yang E, DeVito Z, Raison M, Tejani A, Chilamkurthy S, Steiner B, Fang L, Bai J, Chintala S (2019) PyTorch: an imperative style, high-performance deep learning library. In: Proceedings of the 33rd International Conference on Neural Information Processing Systems (NeurIPS) Article No 721:8026-8037, Curran Associates Inc., Red Hook, NY, USA
[42] Penta R, Gerisch A (2025) Investigation of the potential of asymptotic homogenization for elastic composites via a three-dimensional computational study. Computing and Visualization in Science 17(4):185-201, DOI 10.1007/s00791-015-0257-8
[43] Pobedrya BE (1984) Mekhanika kompozitsionnykh materialov (in Russ., Mechanics of Composite Materials). Izdatel'stvo MGU, Moscow
[44] Raissi M, Perdikaris P, Karniadakis GE (2019) Physics-informed neural networks: A deep learning framework for solving forward and inverse problems involving nonlinear partial differential equations. Journal of Computational Physics 378:686-707, DOI 10.1016/j.jcp.2018.10.045
[45] Rathore P, Lei W, Frangella Z, Lu L, Udell M (2024) Challenges in training PINNs: A loss landscape perspective. arXiv preprint arXiv:2402.01868, DOI 10.48550/arXiv.2402.01868
[46] Rodríguez-Ramos R, Sabina FJR, Guinovart-Díaz R, Bravo-Castillero J. (2001) Closed-form expressions for the effective coefficients of a fiber-reinforced composite with transversely isotropic constituents - I. Elastic and square symmetry. Mechanics of Materials 33(4):223-235, DOI 10.1016/S0167-6636(00)00059-4
[47] Roy AM, Bose R, Sundararaghavan V, Arróyave R (2023) Deep learning-accelerated computational framework based on physics-informed neural networks for the solution of linear elasticity. Neural Networks 162:472–489, DOI 10.1016/j.neunet.2023.03.014
[48] Sanchez-Palencia E (1980) Non-Homogeneous Media and Vibration Theory. Lecture Notes in Physics, Vol 127. Springer-Verlag, Berlin, Heidelberg, DOI 10.1007/3-540-10000-8
[49] Soyarslan C, Pradas M (2024) Physics-informed machine learning in asymptotic homogenization of elliptic equations. Computer Methods in Applied Mechanics and Engineering 427:117043, DOI 10.1016/j.cma.2024.117043
[50] Vyas N, Morwani D, Zhao R, Kwun M, Shapira I, Brandfonbrener D, Janson L, Kakade S (2025) SOAP: Improving and stabilizing Shampoo using Adam. arXiv preprint arXiv:2409.11321, DOI 10.48550/arXiv.2409.11321
[51] Wang S, Sankaran S, Wang H, Perdikaris P (2023) An expert's guide to training physics-informed neural networks arXiv preprint arXiv:2308.08468, DOI 10.48550/arXiv.2308.08468

[52] Wang S, Teng Y, Perdikaris P (2021) Understanding and mitigating gradient flow pathologies in physics-informed neural networks. SIAM Journal on Scientific Computing 43(5):A3055-A3081, DOI 10.1137/20M1318043
[53] Wu J, Jiang J, Chen Q, Chatzigeorgiou G, Meraghni F (2023) Deep homogenization networks for elastic heterogeneous materials with two- and three-dimensional periodicity. International Journal of Solids and Structures 284:112521, DOI 10.1016/j.ijsolstr.2023.112521
[54] Wang S, Bhartari AK, Li B, Perdikaris P (2025) Gradient alignment in physics-informed neural networks: A second-order optimization perspective. arXiv preprint arXiv:2502.00604, DOI 10.48550/arXiv.2502.00604

Chapter 7
Higher-Order Magneto-Electro-Elastic Theory for Laminated Shell Structures

Francesco Tornabene, Matteo Viscoti, and Rossana Dimitri

Abstract The manuscript presents a refined two-dimensional (2D) formulation, based on higher-order theories, for the coupled magneto-electro-elastic analysis of laminated, doubly-curved shells. The geometry of the structure is described using curvilinear principal coordinates, and the unified formulation is adopted to represent the variation of the unknown field variables along the thickness direction. Based on the Equivalent Layer-Wise (ELW) methodology, the model allows for the enforcement of displacement fields as well as electric and magnetostatic potentials at the top and bottom surfaces. The constitutive relations account for both piezoelectric and piezomagnetic coefficients and enable a full coupling between the electric and magnetic fields and flux components through magneto-electric coefficients. The governing equations are derived under quasistatic conditions from the Master Balance principle by evaluating the stationary configuration of the total energy of the system. A semi-analytical solution is, then, obtained using the Navier approach. In the post-processing, an efficient recovery procedure is adopted to reconstruct the through-the-thickness distributions of both primary and secondary variables. A systematic comparison with numerical predictions obtained from three-dimensional (3D) commercial finite element software is carried out through some numerical examples. In this way, the results highlight both the accuracy and the computational efficiency of the proposed methodology for multifield analysis.

Key words: Anisotropic materials, Doubly-curved shells, Magneto-electro-elastic materials, Recovery procedure

---

Francesco Tornabene (corresponding author) · Matteo Viscoti · Rossana Dimitri
University of Salento, Edificio La Stecca - S.P. 6, Lecce - Monteroni, Lecce, Italy, e-mail: `francesco.tornabene@unisalento.it`, `matteo.viscoti@unisalento.it`, `rossana.dimitri@unisalento.it`

H. Altenbach et al. (eds.), *Composite Structures and Technology*,
Advanced Structured Materials 255,
https://doi.org/10.1007/978-3-032-24656-1_7

7.1 Introduction

Recent advancements in many engineering fields increasingly involve structures with complex geometries and advanced smart materials [10, 11, 17]. These structures are employed not only in classical structural applications but also in severe environmental conditions, where both thermal and hygrometric effects play a significant role [1, 11]. Furthermore, many applications rely on the response of structures under coupled electro-mechanical fields, such as in smart piezoelectric sensors and actuators [22, 33]. For this reason, an accurate evaluation of the structural response in these contexts requires advanced modeling strategies [28]. Conventional numerical approaches commonly rely on a staggered solution [16] of the fundamental governing equations, solving each physical problem independently and subsequently evaluating the stresses induced by each field [4, 5]. The overall mechanical response is, then, obtained by applying the principle of superimposition. This approach neglects the mutual interactions between the different physical fields. In the case of materials that simultaneously respond to electric and magnetic fields, such as in piezoelectric and piezomagnetic analyses, magneto-electro coupling effects modify the global response of the structure [32, 50]. For this reason, a monolithic solution of the governing equations is required to properly account for these couplings. This approach introduces additional Degrees of Freedom (DOFs) within a single solution step; therefore the computational cost can significantly increase. For this reason, in the case of laminated panels, two-dimensional (2D) formulations are often adopted in place of fully three-dimensional (3D) modes to reduce the computational effort [3, 23]. Among 2D theories, two main classes are commonly employed: the Equivalent Single Layer (ESL) and the Layer-Wise (LW) approaches [12, 15]. Furthermore, a hybrid approach, known as the Equivalent Layer-Wise (ELW) theory, combines the advantages of both ESL and LW methodologies and allows for the enforcement of prescribed values of unknown variables at the top and bottom surfaces of the shell [36]. Classical ESL theories include well-established models such as the Classical Plate Theory (CPT) and the First-order Shear Deformation Theory (FSDT) [25, 27]. When adopting 2D approaches, the accuracy of the solution can be limited due to the axiomatic assumption introduced along the thickness direction. To improve computational efficiency without compromising accuracy, higher-order polynomials or trigonometric functions are commonly used to expand the unknown variables through the thickness [7]. In particular, the Third-order Shear Deformation Theory (TSDT) employs third-order polynomials for the in-plane displacement components, while the out-of-plane component is assumed to be uniformly distributed along the thickness [24]. A significant enhancement of these theories is the adoption of the unified formulation [26, 52], which allows the governing equations to be fully derived independently of the specific expressions of the thickness functions used to expand the unknown variables. This approach yields a comprehensive model that embeds in itself several classical theories, including CPT, FSDT, and TSDT [51]. Furthermore, it allows for the introduction within the kinematic model of zigzag functions to capture the interactions between adjacent laminae [8, 18]. Higher-order theories have extensively been applied to mechanical simulations of laminated structures [41, 42].

On the other hand, ELW higher-order theories, originally developed for mechanical analyses, have found wide application in refined 2D multiphysics simulations. In particular, Refs. [35, 39] present a comprehensive model based on higher-order theories for hygro-thermo-magneto-electro-mechanical analysis of laminated shells. Other papers address single-physics problems, including thermo-mechanical [46], thermo-electro-mechanical [38], and thermo-magneto-mechanical analyses [48]. In these cases, temperature variations generate additional electric and magnetic fields due to pyroelectric and pyromagnetic effects. When thermal and hygrometric fields are simultaneously considered, additional mass flux arises due to temperature gradients [43, 47], and the temperature itself is influenced by mass flux within the solid [30, 31]. These phenomena, known as thermophoretic and thermal diffusion effects [6, 9], are computed using appropriate phenomenological relations in combination with the Onsager reciprocity theorem [19, 20]. Another important aspect to consider for improving computational efficiency is the approach used to solve the governing equations. Indeed, these equations can be solved either in the strong or the weak form [37]. In the strong form, boundary conditions involve both prescribed unknown variables and secondary variables [49]. On the other hand, in the weak formulation, only prescribed variables at the boundary of the domain are considered [40]. In some specific cases, closed-form solutions can be derived, such as Navier-type solutions, where a 2D harmonic expansion of the unknown variables is introduced. Numerical solutions in the strong form can be efficiently computed using spectral collocation methods. Among these, the Generalized Differential Quadrature (GDQ) method has proven to be particularly effective for solving partial differential equations [34]. GDQ approximates derivatives of a given function by applying a quadrature formula based on higher-order interpolation of the solution over the entire computational domain. The weighting coefficients can be determined using different approaches [2], depending on the selection of discrete sample points. Starting from the GDQ method, the Generalized Integral Quadrature (GIQ) method is derived to perform numerical integration [29]. On the other hand, solutions in weak form are commonly obtained using the well-known Finite Element Method (FEM), which approximate the solution through shape functions. In this way, derivatives are evaluated numerically from the properties of the shape functions. Within classical FEM, when domain decomposition is employed, continuity of the solution is ensured through the assembly procedure [53]. However, smoothness of the solution is not automatically guaranteed, except in FEM-based formulations that include derivatives of the solution among the unknown variables [21, 42]. Accordingly, the development of analytical solutions plays a crucial role in the validation of advanced numerical models for multifield analysis. These solutions serve as benchmark and allow for investigation of the interactions among the various physical phenomena without introducing numerical issues. For this reason, this chapter presents an ELW theory for the analysis of laminated shell structures under thermodynamic equilibrium conditions, subjected simultaneously to mechanical, electric, and magnetic loads. The governing equations are derived in the principal coordinate system from the Master Balance principle in the strong form, accounting for the complete coupling among the involved fields, including piezoelectric, piezomagnetic, and magneto-electric effects. The constitutive properties are obtained via an analytical homogenization approach based on the Mori-Tanaka scheme [13, 14]. A semi-analytical Navier-type solution is derived for simply-supported shells. In the post-processing, an efficient recovery procedure, based on the 3D equilibrium and balance equations, is adopted to reconstruct the actual multifield response of the structure [44, 45]. The model is applied to some numerical examples, focusing on structures with different curvatures and lamination schemes. Systematic comparisons with 3D reference solutions developed with FEM-based commercial software highlight both the accuracy and computational efficiency of the proposed theory. It is shown that when electric and magnetic loads are applied simultaneously, electric and magnetic fields interact with each other, resulting in deviations compared to those simulations with only a single field applied. The formulation provides a reliable tool for investigating the coupled magneto-electro-mechanical response of laminated structures.

7.2 Higher-Order Equivalent Layer-Wise Magneto-Electro-Elastic Theory for Laminated Shell Structures

According to the ELW theory, the position vector of an arbitrary point of a doubly-curved shell structure within the 3D Euclidean space is expressed in principal coordinates [35] according to the relation reported below, which is defined over the rectangular 2D domain $\left[\alpha_1^0, \alpha_1^1\right] \times \left[\alpha_2^0, \alpha_2^1\right]$

$$\mathbf{R}\left(\alpha_1, \alpha_2, \zeta\right) = \mathbf{r}\left(\alpha_1, \alpha_2\right) + \frac{h}{2} z \mathbf{n}\left(\alpha_1, \alpha_2\right). \tag{7.1}$$

In the previous equation, h denotes the total thickness of the shell, obtained as the sum of the thicknesses $h_k = \zeta_{k+1} - \zeta_k$ of each k-th layer in the stacking sequence, with $k = 1, ..., l$, where l is the total number of layers in the laminate. The vector $\mathbf{r}$ represents the reference surface position vector, described using the principal coordinate set α_1, α_2. The variable $z = 2\zeta/h \in [-1, 1]$ is a dimensionless thickness coordinate that identifies the points within the solid along the thickness direction. On the other hand, the physical thickness coordinate is defined over a closed interval, so that $\zeta \in [-h/2, h/2]$. Finally, $\mathbf{n}$ denotes the outward normal unit vector, defined as follows

$$\mathbf{n}\left(\alpha_1, \alpha_2\right) = \frac{\mathbf{r}_{,1} \times \mathbf{r}_{,2}}{\left|\mathbf{r}_{,1} \times \mathbf{r}_{,2}\right|}, \tag{7.2}$$

where $\mathbf{r}_{,i} = \partial \mathbf{r}/\partial \alpha_i$ with $i = 1, 2$ is the first-order partial derivative of the reference surface position vector. If the second-order partial derivatives are instead denoted by $\mathbf{r}_{,ij} = \partial^2 \mathbf{r}/\partial \alpha_i \partial \alpha_j$, the principal radii of curvature R_i with $i = 1, 2$ can be evaluated as shown below

$$R_i\left(\alpha_1, \alpha_2\right) = -\frac{\mathbf{r}_{,i} \cdot \mathbf{r}_{,i}}{\mathbf{r}_{,ii} \cdot \mathbf{n}}. \tag{7.3}$$

In this way, the thickness scaling parameters can be introduced, defined as $H_i = 1 + \zeta/R_i$ for $i = 1, 2$. Finally, the Lamé parameters of the reference surface, denoted by A_i for $i = 1, 2$, can be evaluated as follows [35]

$$A_i\left(\alpha_1, \alpha_2\right) = \sqrt{\mathbf{r}_i \cdot \mathbf{r}_i}. \tag{7.4}$$

At this point, a generalized kinematic model is introduced. This model expands the multifield unknown variables along the thickness direction using $N + 1$ arbitrary shape functions, denoted by $F_\tau^{\alpha_i(k)}$ for $i = 1, ..., 5$ [35]. The unknown vector $\mathbf{\Delta}^{(k)}$ of multifield 3D variables include the displacement field components $U_1^{(k)}$, $U_2^{(k)}$, and $U_3^{(k)}$, the electric potential $\Delta\phi^{(k)}$, and the magnetostatic potential $\Delta\psi^{(k)}$. This model introduces a set of arbitrary unknown variables for each τ-th kinematic expansion order, each of them collected in the vector $\boldsymbol{\delta}^\tau$

$$\mathbf{U}^k = \sum_{\tau=0}^{N+1} \mathbf{F}_\tau^{(k)} \boldsymbol{\delta}^{(\tau)}. \tag{7.5}$$

Using an expanded notation, Eq.(7.5) can be expressed as follows

$$\begin{bmatrix} U_1^{(k)} \\ U_2^{(k)} \\ U_3^{(k)} \\ \Delta\phi^{(k)} \\ \Delta\psi^{(k)} \end{bmatrix} = \sum_{\tau=0}^{N+1} \begin{bmatrix} F_\tau^{\alpha_1(k)} & 0 & 0 & 0 & 0 \\ 0 & F_\tau^{\alpha_2(k)} & 0 & 0 & 0 \\ 0 & 0 & F_\tau^{\alpha_3(k)} & 0 & 0 \\ 0 & 0 & 0 & F_\tau^{\alpha_4(k)} & 0 \\ 0 & 0 & 0 & 0 & F_\tau^{\alpha_5(k)} \end{bmatrix} \begin{bmatrix} u_1^{(\tau)} \\ u_2^{(\tau)} \\ u_3^{(\tau)} \\ \phi^{(\tau)} \\ \psi^{(\tau)} \end{bmatrix}. \tag{7.6}$$

Each generalized thickness function $F_\tau^{\alpha_i(k)}$ introduced in Eq. (7.6) takes the following form for any $\tau = 0, ..., N + 1$ [35]

$$F_\tau^{\alpha_i(k)} = \begin{cases} \dfrac{1-\tilde{z}}{2} & \tau = 0, \\ \tilde{z}^{\tau+1} - \tilde{z}^{\tau-1} & \tau = 1, ..., N, \\ \dfrac{1+\tilde{z}}{2} & \tau = N. \end{cases} \tag{7.7}$$

Finally, the $\tau = N + 1$ kinematic expansion order is associated with the zigzag function, which is expressed as follows for any $i = 1, ..., 5$

$$F_{N+1}^{\alpha_i(k)} = \begin{cases} (-1)^k z_k = (-1)^k \dfrac{\zeta - \zeta_k}{\zeta_{k+1} - \zeta_k} & k = 1, \\ (-1)^k z_k = (-1)^k \dfrac{2\left(\zeta - \zeta_k\right)}{\zeta_{k+1} - \zeta_k} & k = 2, \ldots l - 1, \\ (-1)^k z_k = (-1)^k \dfrac{\zeta - \zeta_{k+1}}{\zeta_{k+1} - \zeta_k} & k = l. \end{cases} \tag{7.8}$$

The thickness functions associated with $\tau = 0$ and $\tau = N$ in Eq. (7.7) are defined so that the corresponding generalized unknown variables coincide with the 3D variables at the top ($\zeta = h/2$) and bottom ($\zeta = -h/2$) surfaces of the laminated structure

$$\Delta_i = \left(\alpha_1, \alpha_2, \zeta = -\tfrac{h}{2}\right) = \delta_i^{(0)}\left(\alpha_1, \alpha_2\right),$$
$$\Delta_i = \left(\alpha_1, \alpha_2, \zeta = \tfrac{h}{2}\right) = \delta_i^{(N)}\left(\alpha_1, \alpha_2\right). \tag{7.9}$$

In this way, arbitrary values of the unknown variables can be enforces as external boundary condition [35]. To identify the adopted kinematic model in the simulation, the acronym ELDZL-N is used when the zigzag function of Eq. (7.8) is employed, otherwise, the acronym ELD-N is adopted. Once the higher-order ELW kinematic model is introduced, the definition equations for a laminated doubly-curved shell are derived from the corresponding ones in the 3D theory in principal coordinates [35], as reported below in compact matrix form

$$\boldsymbol{\pi}^{(k)} = \mathbf{D}\boldsymbol{\Delta}^{(k)} = \mathbf{D}_\zeta \mathbf{D}_\Omega \boldsymbol{\Delta}^{(k)} = \mathbf{D}_\zeta \left(\sum_{i=1}^{5} \mathbf{D}_\Omega^{\alpha_i}\right)\boldsymbol{\Delta}^{(k)}. \tag{7.10}$$

The vector of primary variables is denoted by

$$\boldsymbol{\pi}^{(k)} = \left[\boldsymbol{\varepsilon}^{(k)T}\ \boldsymbol{\mathcal{E}}^{(k)T}\ \boldsymbol{\mathcal{H}}^{(k)T}\right]^T.$$

The vector

$$\boldsymbol{\varepsilon}^{(k)} = \left[\varepsilon_1\ \varepsilon_2\ \gamma_{12}\ \gamma_{13}\ \gamma_{23}\ \varepsilon_3\right]^T$$

contains the 3D strain components, while

$$\boldsymbol{\mathcal{E}}^{(k)} = \left[\mathcal{E}_1\ \mathcal{E}_2\ \mathcal{E}_3\right]^T$$

and

$$\boldsymbol{\mathcal{H}}^{(k)} = \left[\mathcal{H}_1\ \mathcal{H}_2\ \mathcal{H}_3\right]^T$$

are the vectors of the 3D electric and magnetic fields, respectively. On the other hand, the differential operator $\mathbf{D}$ is split into the operators $\mathbf{D}_\zeta$ and $\mathbf{D}_\Omega$, which contain the first-order derivatives with respect to the thickness coordinate ζ and the in-plane coordinates α_1 and α_2, respectively. By substituting the generalized kinematic model of Eq. (7.5) into the 3D definition equations of Eq. (7.10), the following expression is obtained

$$\boldsymbol{\pi}^{(k)} = \sum_{\tau=0}^{N+1}\sum_{i=1}^{5} \mathbf{Z}^{(k\tau)\alpha_i}\mathbf{D}_\Omega^{\alpha_i}\boldsymbol{\delta}^{(\tau)} = \sum_{\tau=0}^{N+1}\sum_{i=1}^{5} \mathbf{Z}^{(k\tau)\alpha_i}\boldsymbol{\pi}^{(\tau)\alpha_i}. \tag{7.11}$$

In expanded form, the previous equation can be expressed as follows

$$\begin{bmatrix} \boldsymbol{\varepsilon}^{(k)} \\ \boldsymbol{\mathcal{E}}^{(k)} \\ \boldsymbol{\mathcal{H}}^{(k)} \end{bmatrix} = \sum_{\tau=0}^{N+1} \sum_{i=1}^{5} \begin{bmatrix} \mathbf{Z}_1^{(k\tau)\alpha_i} & \mathbf{0} & \mathbf{0} \\ \mathbf{0} & \mathbf{Z}_2^{(k\tau)\alpha_i} & \mathbf{0} \\ \mathbf{0} & \mathbf{0} & \mathbf{Z}_2^{(k\tau)\alpha_i} \end{bmatrix} \begin{bmatrix} \boldsymbol{\varepsilon}^{(\tau)\alpha_i} \\ \boldsymbol{\mathcal{E}}^{(\tau)\alpha_i} \\ \boldsymbol{\mathcal{H}}^{(\tau)\alpha_i} \end{bmatrix}. \tag{7.12}$$

As shown in Eq. (7.11), the vector of 3D primary variables is expressed in terms of the generalized primary variables vector, denoted by $\boldsymbol{\pi}^{(\tau)\alpha_i}$ for $\tau = 0, ..., N+1$ and $i = 1, ..., 5$. In accordance with Eq. (7.12), this vector is obtained by assembling the vectors $\boldsymbol{\varepsilon}^{(\tau)\alpha_i}$, $\boldsymbol{\mathcal{E}}^{(\tau)\alpha_i}$ and $\boldsymbol{\mathcal{H}}^{(\tau)\alpha_i}$, which contain the generalized strain components, electric field components, and magnetic field components, respectively. These vectors are listed below in expanded form

$$\begin{aligned} \boldsymbol{\varepsilon}^{(\tau)\alpha_i} &= \left[\varepsilon_1^{(\tau)\alpha_i} \; \varepsilon_2^{(\tau)\alpha_i} \; \gamma_1^{(\tau)\alpha_i} \; \gamma_2^{(\tau)\alpha_i} \; \gamma_{13}^{(\tau)\alpha_i} \; \gamma_{23}^{(\tau)\alpha_i} \; \omega_{13}^{(\tau)\alpha_i} \; \omega_{23}^{(\tau)\alpha_i} \; \varepsilon_3^{(\tau)\alpha_i} \right]^T, \\ \boldsymbol{\mathcal{E}}^{(\tau)\alpha_i} &= \left[\mathcal{E}_1^{(\tau)\alpha_i} \; \mathcal{E}_2^{(\tau)\alpha_i} \; \mathcal{E}_3^{(\tau)\alpha_i} \right]^T, \\ \boldsymbol{\mathcal{H}}^{(\tau)\alpha_i} &= \left[\mathcal{H}_1^{(\tau)\alpha_i} \; \mathcal{H}_2^{(\tau)\alpha_i} \; \mathcal{H}_3^{(\tau)\alpha_i} \right]^T \end{aligned} . \tag{7.13}$$

At this point, the 3D constitutive relation is introduced for an arbitrary point within the doubly-curved solid. This relation accounts for the full coupling between the primary and secondary variables of mechanical, electrical, and magnetic fields [35]. Denoting by $\boldsymbol{\chi}^{(k)}$ the vector of 3D secondary variables, one obtains

$$\boldsymbol{\chi}^{(k)} = \bar{\boldsymbol{\Gamma}}^{(k)} \boldsymbol{\pi}^{(k)} \tag{7.14}$$

or equivalently in expanded form

$$\begin{bmatrix} \boldsymbol{\sigma}^{(k)} \\ \boldsymbol{\mathcal{D}}^{(k)} \\ \boldsymbol{\mathcal{B}}^{(k)} \end{bmatrix} = \begin{bmatrix} \bar{\boldsymbol{\Gamma}}_{\mathrm{C}}^{(k)} & -\bar{\boldsymbol{\Gamma}}_{\mathrm{P}}^{(k)T} & -\bar{\boldsymbol{\Gamma}}_{\mathrm{Q}}^{(k)T} \\ \bar{\boldsymbol{\Gamma}}_{\mathrm{P}}^{(k)} & \bar{\boldsymbol{\Gamma}}_{\mathrm{L}}^{(k)} & \bar{\boldsymbol{\Gamma}}_{\mathrm{D}}^{(k)} \\ \bar{\boldsymbol{\Gamma}}_{\mathrm{Q}}^{(k)} & \bar{\boldsymbol{\Gamma}}_{\mathrm{D}}^{(k)} & \bar{\boldsymbol{\Gamma}}_{\mathrm{M}}^{(k)} \end{bmatrix} \begin{bmatrix} \boldsymbol{\varepsilon}^{(k)} \\ \boldsymbol{\mathcal{E}}^{(k)} \\ \boldsymbol{\mathcal{H}}^{(k)} \end{bmatrix}, \tag{7.15}$$

Here, $\bar{\boldsymbol{\Gamma}}_{\mathrm{C}}^{(k)}$ is the 3D elastic stiffness matrix, while $\bar{\boldsymbol{\Gamma}}_{\mathrm{L}}^{(k)}$ and $\bar{\boldsymbol{\Gamma}}_{\mathrm{M}}^{(k)}$ denote the dielectric and magnetic permittivity matrices, respectively, expressed in the geometric reference system. Furthermore, $\bar{\boldsymbol{\Gamma}}_{\mathrm{P}}^{(k)}$ accounts for the direct and converse piezoelectric effects, and $\bar{\boldsymbol{\Gamma}}_{\mathrm{Q}}^{(k)}$ represents the piezomagnetic and magnetostrictive effects. Finally, $\bar{\boldsymbol{\Gamma}}_{\mathrm{D}}^{(k)}$ is the magneto-electric coupling matrix. The vectors of magneto-electro-elastic secondary variables, namely the 3D stress components, electric field components, and magnetic field components, are reported below in expanded form [35]

$$\begin{aligned} \boldsymbol{\sigma}^{(k)} &= \left[\sigma_1^{(k)} \; \sigma_2^{(k)} \; \tau_{12}^{(k)} \; \tau_{13}^{(k)} \; \tau_{23}^{(k)} \; \sigma_3^{(k)} \right]^T, \\ \boldsymbol{\mathcal{D}}^{(k)} &= \left[\mathcal{E}_1^{(k)} \; \mathcal{E}_2^{(k)} \; \mathcal{E}_3^{(k)} \right]^T, \\ \boldsymbol{\mathcal{B}}^{(k)} &= \left[\mathcal{H}_1^{(k)} \; \mathcal{H}_2^{(k)} \; \mathcal{H}_3^{(k)} \right]^T. \end{aligned} \tag{7.16}$$

From the computation of the virtual variation of the internal energy of the system, and after some mathematical manipulations, a generalized constitutive relation is obtained based on the higher-order kinematic expansion. The following relation is, thus, derived for any $\tau = 0, ..., N + 1$ and $i = 1, ..., 5$ [35]

$$\boldsymbol{\Sigma}^{(\tau)\alpha_i} = \sum_{\eta=0}^{N+1}\sum_{j=1}^{5}\left(\sum_{k=1}^{l}\int_{\zeta_k}^{\zeta_{k+1}}\left(\mathbf{Z}^{(k\tau)\alpha_i}\right)^T \bar{\boldsymbol{\Gamma}}^{(k)}\mathbf{Z}^{(k\eta)\alpha_j} H_1 H_2 d\zeta\right)\boldsymbol{\pi}^{(\eta)\alpha_j}$$

$$= \sum_{\eta=0}^{N+1}\sum_{j=1}^{5}\mathbf{A}^{(\tau\eta)\alpha_i\alpha_j}\boldsymbol{\pi}^{(\eta)\alpha_j}. \quad (7.17)$$

The previous equation can be expressed in expanded form as follows

$$\boldsymbol{\Sigma}^{(\tau)\alpha_i} = \begin{bmatrix}\mathbf{S}^{(\tau)\alpha_i}\\ \mathbf{D}^{(\tau)\alpha_i}\\ \mathbf{B}^{(\tau)\alpha_i}\end{bmatrix} = \sum_{\eta=0}^{N+1}\sum_{j=1}^{5}\begin{bmatrix}\mathbf{A}_{\varepsilon\varepsilon}^{(\tau\eta)\alpha_i\alpha_j} & \mathbf{A}_{\varepsilon\phi}^{(\tau\eta)\alpha_i\alpha_j} & \mathbf{A}_{\varepsilon\psi}^{(\tau\eta)\alpha_i\alpha_j}\\ \mathbf{A}_{\phi\varepsilon}^{(\tau\eta)\alpha_i\alpha_j} & \mathbf{A}_{\phi\phi}^{(\tau\eta)\alpha_i\alpha_j} & \mathbf{A}_{\phi\psi}^{(\tau\eta)\alpha_i\alpha_j}\\ \mathbf{A}_{\psi\varepsilon}^{(\tau\eta)\alpha_i\alpha_j} & \mathbf{A}_{\psi\phi}^{(\tau\eta)\alpha_i\alpha_j} & \mathbf{A}_{\psi\psi}^{(\tau\eta)\alpha_i\alpha_j}\end{bmatrix}\begin{bmatrix}\boldsymbol{\varepsilon}^{(\eta)\alpha_j}\\ \boldsymbol{\mathcal{E}}^{(\eta)\alpha_j}\\ \boldsymbol{\mathcal{H}}^{(\eta)\alpha_j}\end{bmatrix}. \quad (7.18)$$

The arbitrary term of the matrix $\mathbf{A}^{(\tau\eta)\alpha_i\alpha_j}$ in Eq. (7.17) is evaluated using the relation reported below [35]

$$A_{rsnm(pq)}^{(\tau\eta)[fg]\alpha_i\alpha_j} = \sum_{k=1}^{l}\int_{\zeta_k}^{\zeta_{k+1}}\bar{\Gamma}_{nm}^{(k)}\frac{\partial^f F_\tau^{(k)\alpha_i}}{\partial\zeta^f}\frac{\partial^g F_\eta^{(k)\alpha_j}}{\partial\zeta^g}\frac{H_1 H_2}{H_1^p H_2^q}d\zeta \quad (7.19)$$

for $\tau, \eta = 0, ..., N + 1$, $n, m = 1, ..., 6$, $r, s = \varepsilon, \phi, \psi$, $p, q = 0, 1, 2$, $f, g = 0, 1$, and $i, j = 1, ..., 5$, setting $\bar{\Gamma}_{nm}^{(k)}$ an arbitrary element of the multifield 3D constitutive matrix. Finally, the vectors $\mathbf{S}^{(\tau)\alpha_i}$, $\mathbf{D}^{(\tau)\alpha_i}$, and $\mathbf{B}^{(\tau)\alpha_i}$ defined for $\tau = 0, \ldots, N + 1$ and $i = 1, \ldots, 5$ assume the following expanded form

$$\mathbf{S}^{(\tau)\alpha_i} = \begin{bmatrix} N_1^{(\tau)\alpha_i} & N_2^{(\tau)\alpha_i} & N_{12}^{(\tau)\alpha_i} & N_{21}^{(\tau)\alpha_i} & T_1^{(\tau)\alpha_i} & T_2^{(\tau)\alpha_i} & P_1^{(\tau)\alpha_i} & P_2^{(\tau)\alpha_i} & S_3^{(\tau)\alpha_i}\end{bmatrix}^T,$$
$$\mathbf{D}^{(\tau)\alpha_i} = \begin{bmatrix} D_1^{(\tau)\alpha_i} & D_2^{(\tau)\alpha_i} & D_3^{(\tau)\alpha_i}\end{bmatrix}^T,$$
$$\mathbf{B}^{(\tau)\alpha_i} = \begin{bmatrix} B_1^{(\tau)\alpha_i} & B_2^{(\tau)\alpha_i} & B_3^{(\tau)\alpha_i}\end{bmatrix}^T. \quad (7.20)$$

The generalized magneto-electro-mechanical external actions $q_1^{(\tau)}, q_2^{(\tau)}, q_3^{(\tau)}, q_D^{(\tau)}$, $q_B^{(\tau)}$, applied at the reference middle surface, are obtained through the static equivalence principle, considering the virtual work of the multifield loads acting at the top (+) and bottom (−) surfaces of the shell. These loads include the mechanical surface tractions $q_1^{(\pm)}, q_2^{(\pm)}, q_3^{(\pm)}$, the electric flux $q_D^{(\pm)}$, and the magnetic flux load $q_B^{(\pm)}$. The generalized actions are, thus, determined, for any $\tau = 0, ..., N + 1$, from the expression reported below [35]

$$
\begin{aligned}
q_1^{(\tau)} &= q_1^{(-)} F_\tau^{(1)\alpha_1(-)} H_1^{(-)} H_2^{(-)} + q_1^{(+)} F_\tau^{(l)\alpha_1(+)} H_1^{(+)} H_2^{(+)}, \\
q_2^{(\tau)} &= q_2^{(-)} F_\tau^{(1)\alpha_2(-)} H_1^{(-)} H_2^{(-)} + q_2^{(+)} F_\tau^{(l)\alpha_2(+)} H_1^{(+)} H_2^{(+)}, \\
q_3^{(\tau)} &= q_3^{(-)} F_\tau^{(1)\alpha_3(-)} H_1^{(-)} H_2^{(-)} + q_3^{(+)} F_\tau^{(l)\alpha_3(+)} H_1^{(+)} H_2^{(+)}, \\
q_{\mathrm{D}}^{(\tau)} &= q_{\mathrm{D}}^{(-)} F_\tau^{(1)\alpha_4(-)} H_1^{(-)} H_2^{(-)} + q_{\mathrm{D}}^{(+)} F_\tau^{(l)\alpha_4(+)} H_1^{(+)} H_2^{(+)}, \\
q_{\mathrm{B}}^{(\tau)} &= q_{\mathrm{B}}^{(-)} F_\tau^{(1)\alpha_5(-)} H_1^{(-)} H_2^{(-)} + q_{\mathrm{B}}^{(+)} F_\tau^{(l)\alpha_5(+)} H_1^{(+)} H_2^{(+)}.
\end{aligned}
\tag{7.21}
$$

In the previous equation, the positions $H_1^{(-)} = 1 - h/2R_1$ and $H_2^{(-)} = 1 - h/2R_2$ are introduced. Furthermore, $F_\tau^{(1)\alpha_i(-)}$ and $F_\tau^{(l)\alpha_i(+)}$, for $i = 1, ..., 5$, denote the values of the thickness functions in the kinematic model, evaluated at $\zeta = -h/2$ and $\zeta = h/2$, respectively. These loads are conveniently collected, for any $\tau = 0, ..., N+1$, within the vector

$$
\boldsymbol{q}^{(\tau)} = \left[q_1^{(\tau)} \; q_2^{(\tau)} \; q_3^{(\tau)} \; q_{\mathrm{D}}^{(\tau)} \; q_{\mathrm{B}}^{(\tau)} \right]^T .
$$

Finally, the higher-order two-dimensional magneto-electro-elastic balance equations are derived from the Master Balance principle by integrating over an arbitrary time interval $[t_1, t_2]$ the virtual variation of the total internal energy Γ and the virtual work, denoted by $\delta\mathcal{L}_s$, of the generalized surface loads introduced in Eq. (7.21) [35]

$$
\int_{t_1}^{t_2} (\delta\mathcal{L}_s - \delta\Gamma)\, dt = 0. \tag{7.22}
$$

The virtual work $\delta\mathcal{L}_s$ of the external surface loads, given in Eq. (7.21), is evaluated using the following expression:

$$
\begin{aligned}
\delta\mathcal{L}_s = \int_{\alpha_1}\int_{\alpha_2} \sum_{\tau=0}^{N+1} & q_1^{(\tau)} \delta u_1^{(\tau)} + q_2^{(\tau)} \delta u_2^{(\tau)} + q_3^{(\tau)} \delta u_3^{(\tau)} + \\
& + q_{\mathrm{D}}^{(\tau)} \delta\phi^{(\tau)} + q_{\mathrm{B}}^{(\tau)} \delta\psi^{(\tau)}) A_1 A_2 d\alpha_1 d\alpha_2.
\end{aligned}
\tag{7.23}
$$

Similarly, the virtual variation of the internal energy $\delta\Gamma$ of the system can be computed as:

$$
\begin{aligned}
\delta\Gamma = & \sum_{\tau=0}^{N+1}\sum_{i=1}^{5} \int_{\alpha_1}\int_{\alpha_2} \left(\delta\boldsymbol{\varepsilon}^{(\tau)\alpha_i}\right)^T \boldsymbol{S}^{(\tau)\alpha_i} A_1 A_2 d\alpha_1 d\alpha_2 \\
& - \sum_{\tau=0}^{N+1}\sum_{i=1}^{5} \int_{\alpha_1}\int_{\alpha_2} \left(\delta\boldsymbol{\mathcal{E}}^{(\tau)\alpha_i}\right)^T \boldsymbol{D}^{(\tau)\alpha_i} A_1 A_2 d\alpha_1 d\alpha_2 \\
& - \sum_{\tau=0}^{N+1}\sum_{i=1}^{5} \int_{\alpha_1}\int_{\alpha_2} \left(\delta\boldsymbol{\mathcal{H}}^{(\tau)\alpha_i}\right)^T \boldsymbol{B}^{(\tau)\alpha_i} A_1 A_2 d\alpha_1 d\alpha_2.
\end{aligned}
\tag{7.24}
$$

The application of the integration by parts rule [35] allows the final form of the multifield balance equations to be obtained for any $\tau = 0, ..., N+1$

$$\sum_{i=1}^{5} \boldsymbol{D}_{\Omega}^{*\alpha_i} \boldsymbol{\Sigma}^{(\tau)\alpha_i} + \boldsymbol{q}^{(\tau)} = \boldsymbol{0}. \tag{7.25}$$

In the previous equation, $\boldsymbol{D}_{\Omega}^{*\alpha_i}$ for $i = 1, ..., 5$ represents the balance operators expressed in principal coordinates α_1, α_2, while $\boldsymbol{\Sigma}^{(\tau)\alpha_i}$ is the vector introduced in Eq. (7.18). The integration by parts in Eq. (7.22), together with the balance relation of Eq. (7.25), leads to the definition of the following relations associated with the edges of the 2D physical domain $\left[\alpha_1^0, \alpha_1^1\right] \times \left[\alpha_2^0, \alpha_2^1\right]$

$$\begin{gathered}
\left(\bar{N}_1^{(\tau)\alpha_1} - N_1^{(\tau)\alpha_1}\right) \delta u_1^{(\tau)} = 0 \quad \left(\bar{N}_{21}^{(\tau)\alpha_1} - N_{21}^{(\tau)\alpha_1}\right) \delta u_1^{(\tau)} = 0, \\
\left(\bar{N}_{12}^{(\tau)\alpha_2} - N_{12}^{(\tau)\alpha_2}\right) \delta u_2^{(\tau)} = 0 \quad \left(\bar{N}_2^{(\tau)\alpha_2} - N_2^{(\tau)\alpha_2}\right) \delta u_2^{(\tau)} = 0, \\
\left(\bar{T}_1^{(\tau)\alpha_3} - T_1^{(\tau)\alpha_3}\right) \delta u_3^{(\tau)} = 0 \quad \left(\bar{T}_2^{(\tau)\alpha_3} - T_2^{(\tau)\alpha_3}\right) \delta u_3^{(\tau)} = 0, \\
\left(\bar{D}_1^{(\tau)\alpha_4} - D_1^{(\tau)\alpha_4}\right) \delta \phi^{(\tau)} = 0, \left(\bar{D}_2^{(\tau)\alpha_4} - D_2^{(\tau)\alpha_4}\right) \delta \phi^{(\tau)} = 0, \\
\left(\bar{B}_1^{(\tau)\alpha_5} - B_1^{(\tau)\alpha_5}\right) \delta \psi^{(\tau)} = 0, \left(\bar{B}_2^{(\tau)\alpha_5} - B_2^{(\tau)\alpha_5}\right) \delta \psi^{(\tau)} = 0.
\end{gathered} \tag{7.26}$$

The following relations allow the assessment of the boundary conditions for the multifield problem. In particular, the simply-supported boundary conditions along the four edges of the 2D rectangular physical domain are expressed as follows

$$\begin{cases}
N_1^{(\tau)\alpha_1} = 0, \quad u_2^{(\tau)} = u_3^{(\tau)} = 0, \quad \alpha_1 = \alpha_1^0, \alpha_1^1, \\
N_2^{(\tau)\alpha_2} = 0, \quad u_1^{(\tau)} = u_3^{(\tau)} = 0, \quad \alpha_2 = \alpha_2^0, \alpha_2^1.
\end{cases} \tag{7.27}$$

At this point, a semi analytical solution of the fundamental system of equations in Eq. (7.26) is determined by considering the 2D trigonometric expansion of each generalized displacement field components, for any kinematic expansion order $\tau = 0, ..., N+1$

$$\begin{aligned}
u_1^{(\tau)}(s_1, s_2) &= \sum_{n=1}^{\infty} U_{1nm}^{(\tau)} \cos\left(\frac{n\pi}{L_1} s_1\right) \sin\left(\frac{n\pi}{L_1} s_1\right), \\
u_2^{(\tau)}(s_1, s_2) &= \sum_{n=1}^{\infty} U_{2nm}^{(\tau)} \sin\left(\frac{n\pi}{L_1} s_1\right) \cos\left(\frac{n\pi}{L_1} s_1\right), \\
u_3^{(\tau)}(s_1, s_2) &= \sum_{n=1}^{\infty} U_{3nm}^{(\tau)} \sin\left(\frac{n\pi}{L_1} s_1\right) \sin\left(\frac{n\pi}{L_1} s_1\right), \\
\phi^{(\tau)}(s_1, s_2) &= \sum_{n=1}^{\infty} \Phi_{nm}^{(\tau)} \sin\left(\frac{n\pi}{L_1} s_1\right) \sin\left(\frac{n\pi}{L_1} s_1\right), \\
\psi^{(\tau)}(s_1, s_2) &= \sum_{n=1}^{\infty} \Psi_{nm}^{(\tau)} \sin\left(\frac{n\pi}{L_1} s_1\right) \sin\left(\frac{n\pi}{L_1} s_1\right).
\end{aligned} \tag{7.28}$$

Similarly, the multifield surface loads acting on the top and bottom surfaces are assumed to be expressed using a double trigonometric expansion, in accordance with the definition reported below

$$
\begin{aligned}
q_1^{(\pm)}(s_1, s_2) &= \sum_{n=1}^{\infty} Q_{1nm}^{(\pm)} \cos\left(\frac{n\pi}{L_1} s_1\right) \sin\left(\frac{n\pi}{L_1} s_1\right), \\
q_2^{(\pm)}(s_1, s_2) &= \sum_{n=1}^{\infty} Q_{2nm}^{(\pm)} \sin\left(\frac{n\pi}{L_1} s_1\right) \cos\left(\frac{n\pi}{L_1} s_1\right), \\
q_3^{(\pm)}(s_1, s_2) &= \sum_{n=1}^{\infty} Q_{3nm}^{(\pm)} \sin\left(\frac{n\pi}{L_1} s_1\right) \sin\left(\frac{n\pi}{L_1} s_1\right), \\
q_{\mathrm{D}}^{(\pm)}(s_1, s_2) &= \sum_{n=1}^{\infty} Q_{\mathrm{D}nm}^{(\pm)} \sin\left(\frac{n\pi}{L_1} s_1\right) \sin\left(\frac{n\pi}{L_1} s_1\right), \\
q_{\mathrm{B}}^{(\pm)}(s_1, s_2) &= \sum_{n=1}^{\infty} Q_{\mathrm{B}nm}^{(\pm)} \sin\left(\frac{n\pi}{L_1} s_1\right) \sin\left(\frac{n\pi}{L_1} s_1\right).
\end{aligned}
\tag{7.29}
$$

As a consequence, the generalized surface loads, corresponding to any τ-th kinematic expansion order and applied on the reference surface, are expressed through the following relation

$$
\begin{aligned}
q_1^{(\tau)}(s_1, s_2) &= \sum_{n=1}^{\infty} Q_{1nm}^{(\tau)} \cos\left(\frac{n\pi}{L_1} s_1\right) \sin\left(\frac{n\pi}{L_1} s_1\right), \\
q_2^{(\tau)}(s_1, s_2) &= \sum_{n=1}^{\infty} Q_{2nm}^{(\tau)} \sin\left(\frac{n\pi}{L_1} s_1\right) \cos\left(\frac{n\pi}{L_1} s_1\right), \\
q_3^{(\tau)}(s_1, s_2) &= \sum_{n=1}^{\infty} Q_{3nm}^{(\tau)} \sin\left(\frac{n\pi}{L_1} s_1\right) \sin\left(\frac{n\pi}{L_1} s_1\right), \\
q_{\mathrm{D}}^{(\tau)}(s_1, s_2) &= \sum_{n=1}^{\infty} Q_{\mathrm{D}nm}^{(\tau)} \sin\left(\frac{n\pi}{L_1} s_1\right) \sin\left(\frac{n\pi}{L_1} s_1\right), \\
q_{\mathrm{B}}^{(\tau)}(s_1, s_2) &= \sum_{n=1}^{\infty} Q_{\mathrm{B}nm}^{(\tau)} \sin\left(\frac{n\pi}{L_1} s_1\right) \sin\left(\frac{n\pi}{L_1} s_1\right).
\end{aligned}
\tag{7.30}
$$

The semi-analytical amplitudes $Q_{1nm}^{(\tau)}, Q_{2nm}^{(\tau)}, Q_{3nm}^{(\tau)}, Q_{\mathrm{D}nm}^{(\tau)}, Q_{\mathrm{B}nm}^{(\tau)}$ appearing in Eq. (7.30) are expressed as follows

$$
\begin{aligned}
Q_{1nm}^{(\tau)} &= Q_{1nm}^{(-)} F_\tau^{(1)\alpha_1(-)} H_1^{(-)} H_2^{(-)} + Q_{1nm}^{(+)} F_\tau^{(l)\alpha_1(+)} H_1^{(+)} H_2^{(+)}, \\
Q_{2nm}^{(\tau)} &= Q_{2nm}^{(-)} F_\tau^{(1)\alpha_2(-)} H_1^{(-)} H_2^{(-)} + Q_{2nm}^{(+)} F_\tau^{(l)\alpha_2(+)} H_1^{(+)} H_2^{(+)}, \\
Q_{3nm}^{(\tau)} &= Q_{3nm}^{(-)} F_\tau^{(1)\alpha_3(-)} H_1^{(-)} H_2^{(-)} + Q_{3nm}^{(+)} F_\tau^{(l)\alpha_3(+)} H_1^{(+)} H_2^{(+)}, \\
Q_{\mathrm{D}nm}^{(\tau)} &= Q_{\mathrm{D}nm}^{(-)} F_\tau^{(1)\alpha_4(-)} H_1^{(-)} H_2^{(-)} + Q_{\mathrm{D}nm}^{(+)} F_\tau^{(l)\alpha_4(+)} H_1^{(+)} H_2^{(+)}, \\
Q_{\mathrm{B}nm}^{(\tau)} &= Q_{\mathrm{B}nm}^{(-)} F_\tau^{(1)\alpha_5(-)} H_1^{(-)} H_2^{(-)} + Q_{\mathrm{B}nm}^{(+)} F_\tau^{(l)\alpha_5(+)} H_1^{(+)} H_2^{(+)}.
\end{aligned}
\tag{7.31}
$$

Another important hypothesis concerns the geometric shape of the doubly-curved shell. In particular, the structure is assumed to have a uniform curvature throughout the physical domain. In other words,the following relation must be satisfied

$$\begin{aligned}\frac{\partial^{n+m} A_1}{\partial s_1^n s_2^m} = 0 \quad \frac{\partial^{n+m} A_2}{\partial s_1^n s_2^m} = 0,\\ \frac{\partial^{n+m} R_1}{\partial s_1^n s_2^m} = 0 \quad \frac{\partial^{n+m} R_2}{\partial s_1^n s_2^m} = 0.\end{aligned} \tag{7.32}$$

With regard to the constitutive equations, the semi-analytical solution is derived under the assumption of cross-ply lamination schemes. To this end, the sub-matrices in Eq. (7.15) take the following form [35]

$$\begin{aligned}
\bar{\mathbf{\Gamma}}_{\mathrm{C}}^{(k)} &= \begin{bmatrix} \bar{C}_{11}^{(k)} & \bar{C}_{12}^{(k)} & 0 & 0 & 0 & \bar{C}_{13}^{(k)} \\ \bar{C}_{12}^{(k)} & \bar{C}_{22}^{(k)} & 0 & 0 & 0 & \bar{C}_{23}^{(k)} \\ 0 & 0 & \bar{C}_{66}^{(k)} & 0 & 0 & 0 \\ 0 & 0 & 0 & \bar{C}_{44}^{(k)} & 0 & 0 \\ 0 & 0 & 0 & 0 & \bar{C}_{55}^{(k)} & 0 \\ \bar{C}_{13}^{(k)} & \bar{C}_{23}^{(k)} & 0 & 0 & 0 & \bar{C}_{33}^{(k)} \end{bmatrix}, \\
\bar{\mathbf{\Gamma}}_{\mathrm{L}}^{(k)} &= \begin{bmatrix} \bar{l}_{11}^{(k)} & 0 & 0 \\ 0 & \bar{l}_{22}^{(k)} & 0 \\ 0 & 0 & \bar{l}_{33}^{(k)} \end{bmatrix}, \\
\bar{\mathbf{\Gamma}}_{\mathrm{M}}^{(k)} &= \begin{bmatrix} \bar{m}_{11}^{(k)} & 0 & 0 \\ 0 & \bar{m}_{22}^{(k)} & 0 \\ 0 & 0 & \bar{m}_{33}^{(k)} \end{bmatrix}, \\
\bar{\mathbf{\Gamma}}_{\mathrm{P}}^{(k)} &= \begin{bmatrix} 0 & 0 & 0 & \bar{p}_{14}^{(k)} & 0 & 0 \\ 0 & 0 & 0 & 0 & \bar{p}_{25}^{(k)} & 0 \\ \bar{p}_{31}^{(k)} & \bar{p}_{32}^{(k)} & 0 & 0 & 0 & \bar{p}_{33}^{(k)} \end{bmatrix}, \\
\bar{\mathbf{\Gamma}}_{\mathrm{Q}}^{(k)} &= \begin{bmatrix} 0 & 0 & 0 & \bar{q}_{14}^{(k)} & 0 & 0 \\ 0 & 0 & 0 & 0 & \bar{q}_{25}^{(k)} & 0 \\ \bar{q}_{31}^{(k)} & \bar{q}_{32}^{(k)} & 0 & 0 & 0 & \bar{q}_{33}^{(k)} \end{bmatrix}, \\
\bar{\mathbf{\Gamma}}_{\mathrm{D}}^{(k)} &= \begin{bmatrix} \bar{d}_{11}^{(k)} & 0 & 0 \\ 0 & \bar{d}_{22}^{(k)} & 0 \\ 0 & 0 & \bar{d}_{33}^{(k)} \end{bmatrix}.
\end{aligned} \tag{7.33}$$

By substituting the harmonic expansions introduced above, the fundamental system of equations for the magneto-electro-mechanical problem is transformed into a linear algebraic system, which is reported below in compact form

$$\sum_{n=1}^{\infty}\sum_{m=1}^{\infty}\left(\sum_{\eta=0}^{N+1}\tilde{\mathbf{L}}_{nm}^{(\tau\eta)}\boldsymbol{\delta}_{nm}^{(\eta)}+\mathbf{Q}_{nm}^{(\tau)}\right)=0. \tag{7.34}$$

Here,

$$\mathbf{Q}_{nm}^{(\tau)}=\left[Q_{1nm}^{(\tau)}\ Q_{2nm}^{(\tau)}\ Q_{3nm}^{(\tau)}\right]^T$$

is the semi-analytical external load vector associated with any n, m, while

$$\boldsymbol{\delta}_{nm}^{(\eta)}=\left[U_{1nm}^{(\eta)}\ U_{2nm}^{(\eta)}\ U_{3nm}^{(\eta)}\ \Phi_{nm}^{(\eta)}\ \Psi_{nm}^{(\eta)}\right]^T$$

contains the semi-analytical amplitudes of each unknown multifield variable. Finally, the semi-analytical fundamental matrix in Eq. (7.34) takes the following expanded form:

$$\tilde{\mathbf{L}}_{nm}^{(\tau\eta)}=\begin{bmatrix}\tilde{L}_{11nm}^{(\tau\eta)\alpha_1\alpha_1} & \tilde{L}_{12nm}^{(\tau\eta)\alpha_1\alpha_2} & \tilde{L}_{13nm}^{(\tau\eta)\alpha_1\alpha_3} & \tilde{L}_{14nm}^{(\tau\eta)\alpha_1\alpha_4} & \tilde{L}_{15nm}^{(\tau\eta)\alpha_1\alpha_5}\\ \tilde{L}_{21nm}^{(\tau\eta)\alpha_2\alpha_1} & \tilde{L}_{22nm}^{(\tau\eta)\alpha_2\alpha_2} & \tilde{L}_{23nm}^{(\tau\eta)\alpha_2\alpha_3} & \tilde{L}_{24nm}^{(\tau\eta)\alpha_2\alpha_4} & \tilde{L}_{25nm}^{(\tau\eta)\alpha_2\alpha_5}\\ \tilde{L}_{31nm}^{(\tau\eta)\alpha_3\alpha_1} & \tilde{L}_{32nm}^{(\tau\eta)\alpha_3\alpha_2} & \tilde{L}_{33nm}^{(\tau\eta)\alpha_3\alpha_3} & \tilde{L}_{34nm}^{(\tau\eta)\alpha_3\alpha_4} & \tilde{L}_{35nm}^{(\tau\eta)\alpha_3\alpha_5}\\ \tilde{L}_{41nm}^{(\tau\eta)\alpha_4\alpha_1} & \tilde{L}_{42nm}^{(\tau\eta)\alpha_4\alpha_2} & \tilde{L}_{43nm}^{(\tau\eta)\alpha_4\alpha_3} & \tilde{L}_{44nm}^{(\tau\eta)\alpha_4\alpha_4} & \tilde{L}_{45nm}^{(\tau\eta)\alpha_4\alpha_5}\\ \tilde{L}_{51nm}^{(\tau\eta)\alpha_5\alpha_1} & \tilde{L}_{52nm}^{(\tau\eta)\alpha_5\alpha_2} & \tilde{L}_{53nm}^{(\tau\eta)\alpha_5\alpha_3} & \tilde{L}_{54nm}^{(\tau\eta)\alpha_5\alpha_4} & \tilde{L}_{55nm}^{(\tau\eta)\alpha_5\alpha_5}\end{bmatrix}. \tag{7.35}$$

The detailed derivation of the expressions for the coefficients in Eq. (7.35) is reported in Ref. [35].

7.3 Recovery Procedure

In this section, an efficient post-processing recovery procedure is presented to reconstruct the actual through-the-thickness distributions of multifield configuration, primary, and secondary variables. The refined 2D solution of Eq. (7.34), based on the higher-order kinematic model of Eq. (7.5), does not automatically satisfy the multifield balance equations along the thickness direction. Therefore, the solution must be updated by directly solving the 3D balance equations. To this end, at each selected point within the rectangular 2D physical domain, a discrete non-uniform grid of I_T sample points is introduced for each k-th layer, following the Chebyshev-Gauss-Lobatto (CGL) harmonic distribution, in line with the relation reported below [35]:

$$\zeta_{\tilde{m}}^{(k)}=\frac{\zeta_{k+1}-\zeta_k}{2}\xi_{\tilde{m}}+\frac{\zeta_{k+1}+\zeta_k}{2}, \tag{7.36}$$

where $\zeta_{\tilde{m}}^{(k)}$ denotes an arbitrary point in this grid, with $k=1,\dots,l$, while $\xi_{\tilde{m}}\in[-1,1]$ represents an arbitrary element of the CGL grid, evaluated as follows

$$\xi_{\tilde{m}}=-\cos\left(\frac{\tilde{m}-1}{I_T-1}\pi\right). \tag{7.37}$$

These points are, then, assembled into the vector $\boldsymbol{\zeta}^{(k)}$ of size $I_T \times 1$:

$$\boldsymbol{\zeta}^{(k)} = \left[\zeta_1^{(k)} \; \dots \; \zeta_{\tilde{m}}^{(k)} \; \dots \; \zeta_{I_T}^{(k)} \right]^T . \tag{7.38}$$

At this point, a global thickness vector of size $lI_T \times 1$, denoted by $\boldsymbol{\zeta}$, is introduced by assembling the vectors from Eq. (7.38). An arbitrary element of this new vector is denoted by ζ_m, with $m = (k-1)I_T + \tilde{m}$. The following definition is, thus, introduced:

$$\boldsymbol{\zeta} = \left[\zeta_1 \; \dots \; \zeta_m \; \dots \; \zeta_{lI_T} \right]^T = \left[\boldsymbol{\zeta}^{(1)T} \; \dots \; \boldsymbol{\zeta}^{(k)T} \; \dots \; \boldsymbol{\zeta}^{(l)T} \right]^T . \tag{7.39}$$

It should be noted that the last element of the vector $\zeta^{(k)}$ coincides with the first element of the vector $\zeta^{(k+1)}$. Once the discretization along the thickness direction is established, the multifield configuration variables are computed at each discrete point ζ_m through Eq. (7.5), where $\boldsymbol{\delta}_{(ij)}^{(\tau)}$ is the vector of generalized unknown variables at the discrete point $(\alpha_{1f}, \alpha_{2g})$ with $f = 1, \dots, I_M$ and $g = 1, \dots, I_N$, on the 2D physical domain. In the same way, each element of the primary variables vector is computed at every point $(\alpha_{1f}, \alpha_{2g}, \zeta_m)$ of the 3D solid. The constitutive relations of Eq. (7.15) through Eq. (7.33) are then used to compute the in-plane 3D stresses $\sigma_1^{(k)}, \sigma_2^{(k)}, \tau_{12}^{(k)}$, the in-plane electric fluxes $\mathcal{D}_1^{(k)}, \mathcal{D}_2^{(k)}$, and the in-plane magnetic fluxes $\mathcal{B}_1^{(k)}, \mathcal{B}_2^{(k)}$. Then, the out-of-plane stresses $\tau_{13}^{(k)}, \tau_{23}^{(k)}, \sigma_3^{(k)}$ are calculated from the 3D equilibrium equations reported below:

$$\begin{aligned} &\frac{\partial \tau_{13}^{(k)}}{\partial \zeta} + \tau_{13}^{(k)} \left(\frac{2}{R_1 + \zeta} + \frac{1}{R_2 + \zeta} \right) = \\ &- \frac{1}{A_1 (1 + \zeta/R_1)} \frac{\partial \sigma_1^{(k)}}{\partial \alpha_1} + \frac{\sigma_2^{(k)} - \sigma_1^{(k)}}{A_1 A_2 (1 + \zeta/R_2)} \frac{\partial A_2}{\partial \alpha_1} - \frac{1}{A_2 (1 + \zeta/R_2)} \frac{\partial \tau_{12}^{(k)}}{\partial \alpha_2} \\ &- \frac{2\tau_{12}^{(k)}}{A_1 A_2 (1 + \zeta/R_1)} \frac{\partial A_1}{\partial \alpha_2}, \end{aligned} \tag{7.40}$$

$$\begin{aligned} &\frac{\partial \tau_{23}^{(k)}}{\partial \zeta} + \tau_{23}^{(k)} \left(\frac{1}{R_1 + \zeta} + \frac{2}{R_2 + \zeta} \right) = \\ &- \frac{1}{A_2 (1 + \zeta/R_2)} \frac{\partial \sigma_2^{(k)}}{\partial \alpha_2} + \frac{\sigma_1^{(k)} - \sigma_2^{(k)}}{A_1 A_2 (1 + \zeta/R_1)} \frac{\partial A_1}{\partial \alpha_2} - \frac{1}{A_1 (1 + \zeta/R_1)} \frac{\partial \tau_{12}^{(k)}}{\partial \alpha_1} \\ &- \frac{2\tau_{12}^{(k)}}{A_1 A_2 (1 + \zeta/R_2)} \frac{\partial A_2}{\partial \alpha_1}, \end{aligned} \tag{7.41}$$

$$\frac{\partial\sigma_3^{(k)}}{\partial\zeta}+\sigma_3^{(k)}\left(\frac{1}{R_1+\zeta}+\frac{1}{R_2+\zeta}\right)=$$
$$-\frac{1}{A_1\left(1+\zeta/R_1\right)}\frac{\partial\tau_{13}^{(k)}}{\partial\alpha_1}-\frac{\tau_{13}^{(k)}}{A_1A_2\left(1+\zeta/R_2\right)}\frac{\partial A_2}{\partial\alpha_1}-\frac{1}{A_2\left(1+\zeta/R_2\right)}\frac{\partial\tau_{23}^{(k)}}{\partial\alpha_2}$$
$$-\frac{\tau_{23}^{(k)}}{A_1A_2\left(1+\zeta/R_1\right)}\frac{\partial A_1}{\partial\alpha_2}+\frac{\sigma_1^{(k)}}{R_1+\zeta}+\frac{\sigma_2^{(k)}}{R_2+\zeta}. \tag{7.42}$$

In particular, Eq.(7.40) is solved for each k-th layer of the structure using the boundary conditions reported below:

$$\begin{cases}\bar{\tau}_{13(ij1)}^{(1)}=q_{1(ij)}^{(-)} & k=1,\\ \bar{\tau}_{13(ij((k)I_T+1))}^{(k)}=\bar{\tau}_{13(ij((k-1)I_T))}^{(k)} & k\neq 1.\end{cases} \tag{7.43}$$

Similarly, the boundary conditions set applied to Eq.(7.41) are defined by the following relations:

$$\begin{cases}\bar{\tau}_{23(ij1)}^{(1)}=q_{2(ij)}^{(-)} & k=1,\\ \bar{\tau}_{23(ij((k)I_T+1))}^{(k)}=\bar{\tau}_{23(ij((k-1)I_T))}^{(k)} & k\neq 1.\end{cases} \tag{7.44}$$

The boundary conditions for the first-order differential equation in Eq. (7.42) are given below:

$$\begin{cases}\bar{\sigma}_{3(ij1)}^{(1)}=q_{3(ij)}^{(-)} & k=1,\\ \bar{\sigma}_{3(ij((k)I_T+1))}^{(k)}=\bar{\sigma}_{3(ij((k-1)I_T))}^{(k)} & k\neq 1.\end{cases} \tag{7.45}$$

Once (7.40) and (7.41) are solved, the external in-plane mechanical loads at the top surface are applied using the following relation [35]:

$$\tau_{13(ijm)}^{(k)}=\bar{\tau}_{13(ijm)}^{(k)}+\frac{q_{1(ij)}^{(+)}-\bar{\tau}_{13(ij(lI_T))}^{(l)}}{h}\left(\zeta_m+\frac{h}{2}\right), \tag{7.46}$$

$$\tau_{23(ijm)}^{(k)}=\bar{\tau}_{23(ijm)}^{(k)}+\frac{q_{2(ij)}^{(+)}-\bar{\tau}_{23(ij(lI_T))}^{(l)}}{h}\left(\zeta_m+\frac{h}{2}\right), \tag{7.47}$$

where $m=1,\ldots lI_T$. It should be noted that the solution of Eqs. (7.40)-(7.42) is not derived monolithically. Indeed, in a first step, $\tau_{13}^{(k)}$ and $\tau_{23}^{(k)}$ are solved. Then, the correction from Eqs. (7.46)-(7.47) is carried out. After that, the numerical solution for $\sigma_3^{(k)}$ is obtained, followed by the linear correction of the out-of-plane normal stress as follows:

$$\sigma_{3(ijm)}^{(k)}=\bar{\sigma}_{3(ijm)}^{(k)}+\frac{q_{3(ij)}^{(+)}-\bar{\sigma}_{3(ij(lI_T))}^{(l)}}{h}\left(\zeta_m+\frac{h}{2}\right). \tag{7.48}$$

Similarly to Eq. (7.48), the recovery procedure is applied to obtain the actual through-the-thickness distribution of the quantities $\mathcal{D}_3^{(k)}$ and $\mathcal{B}_3^{(k)}$. In this case, the following 3D balance equations are used [35]

$$\frac{\partial \mathcal{D}_3^{(k)}}{\partial \zeta} + \mathcal{D}_3^{(k)} \left(\frac{1}{R_1 + \zeta} + \frac{1}{R_2 + \zeta} \right) = \\ - \frac{1}{A_1 (1 + \zeta/R_1)} \frac{\partial \mathcal{D}_1^{(k)}}{\partial \alpha_1} - \frac{\mathcal{D}_1^{(k)}}{A_1 A_2 (1 + \zeta/R_2)} \frac{\partial A_2}{\partial \alpha_1} - \frac{1}{A_2 (1 + \zeta/R_2)} \frac{\partial \mathcal{D}_2^{(k)}}{\partial \alpha_2} \\ - \frac{\mathcal{D}_2^{(k)}}{A_1 A_2 (1 + \zeta/R_1)} \frac{\partial A_1}{\partial \alpha_2}, \tag{7.49}$$

$$\frac{\partial \mathcal{B}_3^{(k)}}{\partial \zeta} + \mathcal{B}_3^{(k)} \left(\frac{1}{R_1 + \zeta} + \frac{1}{R_2 + \zeta} \right) = \\ - \frac{1}{A_1 (1 + \zeta/R_1)} \frac{\partial \mathcal{B}_1^{(k)}}{\partial \alpha_1} - \frac{\mathcal{B}_1^{(k)}}{A_1 A_2 (1 + \zeta/R_2)} \frac{\partial A_2}{\partial \alpha_1} - \frac{1}{A_2 (1 + \zeta/R_2)} \frac{\partial \mathcal{B}_2^{(k)}}{\partial \alpha_2} \\ - \frac{\mathcal{B}_2^{(k)}}{A_1 A_2 (1 + \zeta/R_1)} \frac{\partial A_1}{\partial \alpha_2}. \tag{7.50}$$

The boundary conditions for Eq. (7.49) are given as follows

$$\begin{cases} \bar{\mathcal{D}}_{3(ij1)}^{(1)} = q_{\mathrm{D}(ij)}^{(-)} & k = 1, \\ \bar{\mathcal{D}}_{3(ij((k)I_T+1))}^{(k)} = \bar{\mathcal{D}}_{3(ij((k-1)I_T))}^{(k)} & k \neq 1. \end{cases} \tag{7.51}$$

In the same way, the boundary conditions for (7.50) are reported below

$$\begin{cases} \bar{\mathcal{B}}_{3(ij1)}^{(1)} = q_{\mathrm{B}(ij)}^{(-)} & k = 1, \\ \bar{\mathcal{B}}_{3(ij((k)I_T+1))}^{(k)} = \bar{\mathcal{B}}_{3(ij((k-1)I_T))}^{(k)} & k \neq 1. \end{cases} \tag{7.52}$$

Finally, the distributions of the electric and magnetic fluxes are corrected using the following relation [35]:

$$\mathcal{D}_{3(ijm)}^{(k)} = \bar{\mathcal{D}}_{3(ijm)}^{(k)} + \frac{q_{\mathrm{D}(ij)}^{(+)} - \bar{\sigma}_{3(ij(lI_T))}^{(l)}}{h} \left(\zeta_m + \frac{h}{2} \right), \tag{7.53}$$

$$\mathcal{B}_{3(ijm)}^{(k)} = \bar{\mathcal{B}}_{3(ijm)}^{(k)} + \frac{q_{\mathrm{B}(ij)}^{(+)} - \bar{\sigma}_{3(ij(lI_T))}^{(l)}}{h} \left(\zeta_m + \frac{h}{2} \right). \tag{7.54}$$

Once the out-of-plane stress components and the multifield secondary variables are reconstructed using the procedure outlined above, the out-of-plane 3D strain components and the out-of-plane multifield primary variables are updated by inverting the constitutive relation in Eq. (7.15). In this way, the following linear algebraic system is obtained for each point $(\alpha_{1f}, \alpha_{2g}, \zeta_m)$ of the doubly-curved 3D solid [35]:

$$\mathbf{A}^{(k)}_{(ijm)}\mathbf{x}^{(k)}_{(ijm)} = \mathbf{B}^{(k)}_{(ijm)}. \tag{7.55}$$

Here, $\mathbf{x}^{(k)}_{(ijm)}$ denotes the unknown vector, while $\mathbf{A}^{(k)}_{(ijm)}$ is the coefficient matrix of the linear system. Finally, $\mathbf{B}^{(k)}_{(ijm)}$ is a vector of size 5×1. These quantities are defined as follows:

$$\mathbf{A}^{(k)}_{(ijm)} = \begin{bmatrix} \bar{C}^{(k)}_{44(ijm)} & 0 & 0 & 0 & 0 \\ 0 & \bar{C}^{(k)}_{55(ijm)} & 0 & 0 & 0 \\ 0 & 0 & \bar{C}^{(k)}_{33(ijm)} & -\bar{p}^{(k)}_{33(ijm)} & -\bar{q}^{(k)}_{33(ijm)} \\ 0 & 0 & \bar{p}^{(k)}_{33(ijm)} & \bar{l}^{(k)}_{33(ijm)} & \bar{d}^{(k)}_{33(ijm)} \\ 0 & 0 & \bar{q}^{(k)}_{33(ijm)} & \bar{d}^{(k)}_{33(ijm)} & \bar{m}^{(k)}_{33(ijm)} \end{bmatrix},$$

$$\mathbf{x}^{(k)}_{(ijm)} = \left[\gamma^{(k)}_{13(ijm)}\ \gamma^{(k)}_{23(ijm)}\ \varepsilon^{(k)}_{3(ijm)}\ \mathcal{E}^{(k)}_{3(ijm)}\ \mathcal{H}^{(k)}_{3(ijm)}\right]^T,$$

$$\mathbf{B}^{(k)}_{(ijm)} = \left[b^{(k)}_{1(ijm)}\ b^{(k)}_{2(ijm)}\ b^{(k)}_{3(ijm)}\ b^{(k)}_{4(ijm)}\ b^{(k)}_{5(ijm)}\right]^T. \tag{7.56}$$

The complete expressions for the components of the vector $\mathbf{B}^{(k)}_{(ijm)}$ can be found in Ref. [35]. The solution of the first-order differential equations in Eqs. (7.40)-(7.42) and Eqs. (7.49)-(7.50) is obtained using the GDQ method. As extensively discussed in Ref. [34], this methodology allows the derivatives of a given function to be approximated by a weighted sum of the function values at discrete points within the domain. In particular, if $f = f(x)$ denotes an arbitrary smooth function defined over the closed domain $[a, b]$, the n-th order derivative at a point $x_i \in [a, b]$ is evaluated using the following relation:

$$f^{(n)}(x_i) = \sum_{j=1}^{I_N} \varsigma^{(n)}_{ij} f(x_j). \tag{7.57}$$

Here, $f(x_j)$ denotes the value of the function at the points $x_j \in [a, b]$ with $j = 1, \ldots I_N$, while $\varsigma^{(n)}_{ij}$ represents the weighting coefficients for the n-th order derivative at the point x_i, associated with the quadrature node x_j. These coefficients are evaluated using the following recursive relation

$$\begin{cases} \varsigma^{(n)}_{ij} = n\left(\varsigma^{(1)}_{ij}\varsigma^{(n-1)}_{ii} - \dfrac{\varsigma^{(n-1)}_{ij}}{x_i - x_j}\right), & i \neq j, \\ \varsigma^{(n)}_{ij} = -\displaystyle\sum_{j=1, j\neq i}^{I_N} \varsigma^{(n)}_{ij}, & i = j. \end{cases} \tag{7.58}$$

The GDQ coefficients $\varsigma_{ij}^{(n)}$ with $i, j = 1, ..., I_N$ are conveniently assembled into the matrix $\boldsymbol{\varsigma}^{(n)}$. For the computation of integrals in Eq. (7.19), the GIQ numerical technique is employed. Considering the previously introduced function $f = f(x)$, its integral over the domain $[x_i, x_j]$ is evaluated using the following quadrature formula

$$\int_{x_i}^{x_j} f(x)dx = \sum_{k=1}^{I_N} w_k^{ij} f(x_k) = \sum_{k=1}^{I_N} \left(w_{jk} - w_{ik}\right) f(x_k). \tag{7.59}$$

The coefficients w_{jk} and w_{ik} are the elements of the GIQ matrix $\mathbf{W}$, which is the inverse of the matrix $\bar{\boldsymbol{\varsigma}}^{(1)}$ corresponding to the GDQ shifted coefficients of the first-order derivative over the interval $[a, b]$, introduced above

$$\mathbf{W} = \left(\boldsymbol{\varsigma}^{(1)}\right)^{-1}. \tag{7.60}$$

The GDQ shifted coefficients $\bar{\varsigma}_{ij}^{(1)}$, associated with the interval $[a, b]$, are computed from those, denoted by $\varsigma_{ij}^{(1)}$, referred to the interval $[-1, 1]$

$$\begin{cases} \bar{\varsigma}_{ij}^{(1)} = \dfrac{x_i - \varepsilon}{x_j - \varepsilon} \varsigma_{ij}^{(1)} & i \neq j, \\ \bar{\varsigma}_{ij}^{(1)} = \varsigma_{ii}^{(1)} + \dfrac{1}{x_i - \varepsilon} & i = j \end{cases} \tag{7.61}$$

with $\varepsilon = 1 \times 10^{-10}$. Additional details on the GDQ and GIQ methods can be found in [34].

7.4 Numerical Examples

In this section, the theoretical formulation presented above is applied to predict the stationary response of laminated structures in which each layer exhibits both magnetic and electric properties. The main objective of this section is to demonstrate the accuracy of the formulation compared to 3D simulations developed with FEM-based commercial software. In particular, it is shown that the analytical solution, based on higher-order theories and significantly more computational efficient than the numerical solution, can accurately reproduce the numerical results for the selected case studies. Furthermore, the formulation can be used to perform parametric investigations to evaluate the effects of magneto-electric coupling on the global response of the structure. The first example focuses on a simply-supported (SSSS) rectangular plate with dimensions $L_x = 1.5$ m and $L_y = 1.2$ m. The reference surface equation for the geometric description of the panel is given below [35]

$$\mathbf{r}(\alpha_1, \alpha_2) = \mathbf{r}(x, y) = x\mathbf{e}_1 + y\mathbf{e}_2. \tag{7.62}$$

The structure consists of five layers of different thicknesses ($h_1 = h_3 = h_5 = 0.02$ m and $h_2 = h_4 = 0.03$ m). The lamination scheme is (MAT-I/MAT-IV/MAT-I/MAT-III/MAT-II). The constituent coefficients for each material are obtained via an analytical homogenization procedure based on the Mori-Tanaka scheme, as presented in [35], and are summarized in Table 7.1.

Table 7.1: Homogenized magneto-electro-elastic properties of multiphase materials from Mori-Tanaka scheme.

Quantity	Units	MAT-I	MAT-II	MAT-III	MAT-IV
V_f	-	0.60	0.20	0.10	0.25
m_s	-	1	1	1/15	1/20
ρ	kg/m^3	5600	5400	356.70	271.25
C_{11}	$\times 10^9 N/m^2$	202.08	252.38	17.89	12.25
C_{22}	$\times 10^9 N/m^2$	202.08	252.38	17.89	12.25
C_{66}	$\times 10^9 N/m^2$	48.91	53.83	3.68	2.66
C_{44}	$\times 10^9 N/m^2$	50.81	47.15	3.08	2.38
C_{55}	$\times 10^9 N/m^2$	50.81	47.15	3.08	2.38
C_{33}	$\times 10^9 N/m^2$	196.54	241.14	16.98	11.74
C_{12}	$\times 10^9 N/m^2$	104.26	144.72	10.53	6.93
C_{13}	$\times 10^9 N/m^2$	105.13	143.99	10.43	6.91
C_{23}	$\times 10^9 N/m^2$	105.13	143.99	10.43	6.91
l_{11}	$\times 10^{-10} F/m$	3.12	1.19	0.07	0.07
l_{22}	$\times 10^{-10} F/m$	3.12	1.19	0.07	0.07
l_{33}	$\times 10^{-10} F/m$	76.16	26.10	0.90	1.62
m_{11}	$\times 10^{-4} H/m$	1.52	3.96	0.32	0.18
m_{22}	$\times 10^{-4} H/m$	1.52	3.96	0.32	0.18
m_{33}	$\times 10^{-4} H/m$	0.69	1.28	0.09	0.06
d_{11}	$\times 10^{-12} s/m$	-5.91	-2.56	-0.09	-0.15
d_{22}	$\times 10^{-12} s/m$	-5.91	-2.56	-0.09	-0.15
d_{33}	$\times 10^{-9} s/m$	2.52	2.05	0.08	0.12
p_{14}	C/m^2	0.233	0.038	0.001	0.003
p_{25}	C/m^2	0.233	0.038	0.001	0.003
p_{31}	C/m^2	-3.110	-1.261	-0.044	-0.077
p_{32}	C/m^2	-3.110	-1.261	-0.044	-0.077
p_{33}	C/m^2	10.758	3.394	0.111	0.214
q_{14}	Wb/m^2	138.50	367.81	30.05	16.56
q_{25}	Wb/m^2	138.50	367.81	30.05	16.56
q_{31}	Wb/m^2	0	0	0	0
q_{32}	Wb/m^2	0	0	0	0
q_{33}	Wb/m^2	-332.94	-602.79	-43.69	-28.69

The unit cell of these materials consists of a cylindrical fiber of Barium Titanate ($BaTiO_3$), oriented along the ζ direction, embedded in an infinite isotropic matrix of Cobalt Ferrite ($CoFe_2O_4$). For this reason, the homogenized materials are transversely isotropic, with the plane of symmetry coinciding with the $\alpha_1 - \alpha_2$ plane. The volume fractions of the fiber and the matrix are denoted by V_f and V_m, respectively. These quantities satisfy the following relation

$$V_f + V_m = 1. \tag{7.63}$$

The volume fraction of cylindrical fibers for MAT-I is $V_f = 0.60$, while for MAT-II is $V_f = 0.20$. MAT-III is characterized by $V_f = 0.10$, whereas the values $V_f = 0.25$ is assigned to MAT-IV. In the case of MAT-III and MAT-IV, the homogenized properties are further scaled by a factor equal to $m_s = 1/15$ and $m_s = 1/20$, respectively. The mechanical external load is sinusoidally distributed on the top and bottom surfaces, with magnitudes $\bar{q}_3^{(+)} = -7 \times 10^4 \text{N/m}^2$ and $\bar{q}_3^{(-)} = 3 \times 10^4 \text{N/m}^2$

$$q_3^{(\pm)} = \bar{q}_3^{(\pm)} \sin\left(\frac{\pi}{L_1} s_1\right) \sin\left(\frac{\pi}{L_2} s_2\right), \tag{7.64}$$

where $s_1 = x$ and $s_2 = y$. In addition to mechanical loads, electric and magnetic fluxes are applied at $\zeta = h/2$ and $\zeta = -h/2$, with the same sinusoidal distribution as in Eq. (7.64). The magnitudes of the external electric fluxes are $\bar{q}_D^{(+)} = 4 \times 10^{-8} \text{C/m}^2$ and $\bar{q}_D^{(-)} = 4 \times 10^{-8} \text{C/m}^2$, while the magnetic fluxes are $\bar{q}_B^{(+)} = 2 \times 10^{-2} \text{Wb/m}^2$ and $\bar{q}_B^{(-)} = 3 \times 10^{-2} \text{Wb/m}^2$. A 3D representation of the rectangular plate geometry is shown in Fig. 7.1. The results are presented in terms of through-the-thickness distributions of 3D displacement, strain, and stress components, as well as the configuration, primary, and secondary variables for the electric and magnetic fields, at the point located at $(0.75L_1, 0.25L_2)$ within the 2D physical domain, reported in Figs. 7.2-7.8.

Three different analyses are performed, namely electro-mechanical (E-D), magneto-mechanical (M-D), and magneto-electro-mechanical (M-E-D). For each case, a reference 3D solution is provided from a numerical model developed in FEM-based commercial software to validate the analytical results. This numerical model employs 20-node brick elements with parabolic shape functions. The analytical solution is presented using various ELW kinematic models that adopt the zigzag function described in Eq. (7.8). From Fig. 7.2 it can be observed that the in-plane displacements U_1 and U_2 exhibit slope variations at each interface, reflecting the heterogeneity of the materials in the lamination scheme. Furthermore, increasing the order of the kinematic expansion, does not significantly affect the accuracy of the ELW solution, which shows excellent agreement with the 3D FEM predictions. A similar behavior is observed across the different simulations, indicating that the in-plane components are primarily governed by the mechanical loads. In contrast, the out-of-plane component $U_3^{(k)}$ is influenced by the applied electric and magnetic loads. The through-the-thickness distributions of the 3D strain components are shown in Fig. 7.3. The in-plane strain components ε_1, ε_2 and γ_{12} are largely unaffected

RECTANGULAR PLATE

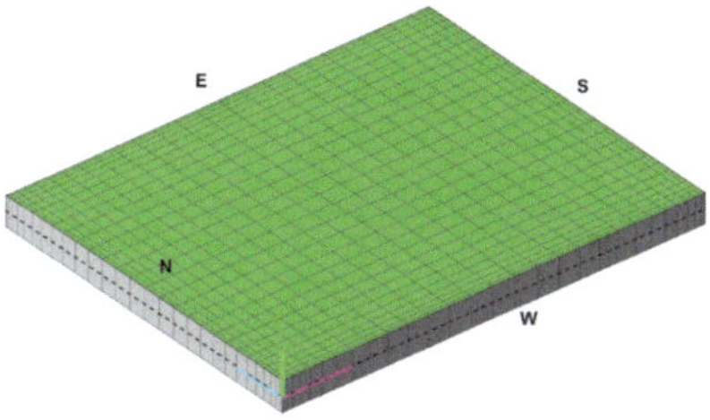

$$L_1 = L_x = 1.50\,\text{m},\; L_2 = L_y = 1.20\,\text{m}$$
$$[\alpha_1^0, \alpha_1^1] \times [\alpha_2^0, \alpha_2^1] = [0, L_x] \times [0, L_y]$$
$$h_1 = h_3 = h_5 = 0.02\,\text{m},\; h_2 = h_4 = 0.03\,\text{m}$$
(MAT-I / MAT-IV / MAT-I / MAT-III / MAT-II)
$$\overline{q}_3^{(+)} = -7\times10^4\ \text{N/m}^2$$
$$\overline{q}_3^{(-)} = 3\times10^4\ \text{N/m}^2$$
$$\overline{q}_D^{(+)} = 4\times10^{-8}\ \text{C/m}^2$$
$$\overline{q}_D^{(-)} = 4\times10^{-8}\ \text{C/m}^2$$
$$\overline{q}_B^{(+)} = 2\times10^{-2}\ \text{Wb/m}^2$$
$$\overline{q}_B^{(-)} = 3\times10^{-2}\ \text{Wb/m}^2$$

CIRCULAR CYLINDER

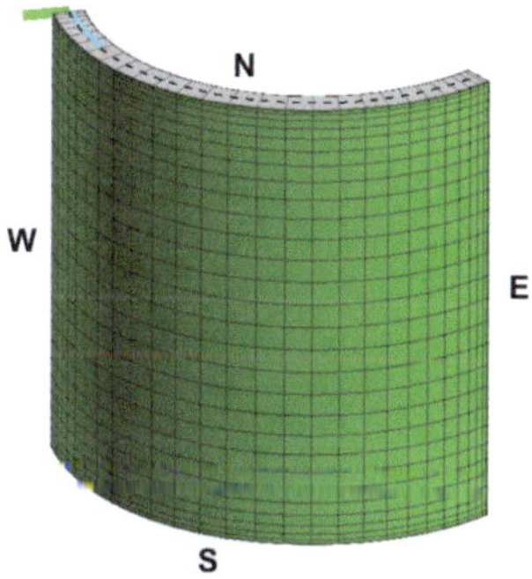

$$[\alpha_1^0, \alpha_1^1] \times [\alpha_2^0, \alpha_2^1] = [0, L_x] \times [\vartheta_0, \vartheta_1] = [0, 1.5] \times [0, 2\pi/3]$$
$$R_2 = 0.8\,\text{m},\; h_1 = h_3 = 0.03\,\text{m},\; h_2 = 0.04\,\text{m}$$
(MAT-I / MAT-III / MAT-II)
$$\overline{q}_3^{(+)} = -7\times10^4\ \text{N/m}^2$$
$$\overline{q}_3^{(-)} = 3\times10^4\ \text{N/m}^2$$
$$\overline{q}_D^{(+)} = -4\times10^{-8}\ \text{C/m}^2$$
$$\overline{q}_D^{(-)} = -4\times10^{-8}\ \text{C/m}^2$$
$$\overline{q}_B^{(+)} = 2\times10^{-5}\ \text{Wb/m}^2$$
$$\overline{q}_B^{(-)} = 3\times10^{-5}\ \text{Wb/m}^2$$

Fig. 7.1: 3D representation of a rectangular plate and a circular cylinder for the numerical examples in the manuscript. Input data for multifield analysis.

by the electric and magnetic fluxes, as evidenced by the similar curves obtained for all cases. On the other hand, the out-of-plane components γ_{13}, γ_{23}, ε_3 exhibit different through-the-thickness distributions in the various simulations. In all cases, excellent agreement is observed between the 3D FEM results and the ELW analytical predictions for any kinematic expansion order. As far as the 3D stress components is concerned, Fig. 7.4 shows that similar profiles are obtained for both in-plane and out-of-plane stresses. In particular, the central lamina, being stiffer than the second and forth laminae, exhibits peculiar stress profiles. In this example, although all higher-order kinematic models yield very accurate results, the highest level of accuracy is achieved with the ELDZL7 kinematic model, especially within the central layer of the plate. The through-the-thickness distributions of the electric and magnetic scalar potentials are presented in Fig. 7.5. It is observed that the presence of a magnetic field induces variations in the electric potential, whereas the electric potential has a negligible effect on the scalar magnetic potential distribution. For M-E-D analyses, these results highlight the importance of using higher-order polynomials in the kinematic expansion to achieve agreement with the 3D reference solution. In the same way, refined kinematic models are necessary to accurately

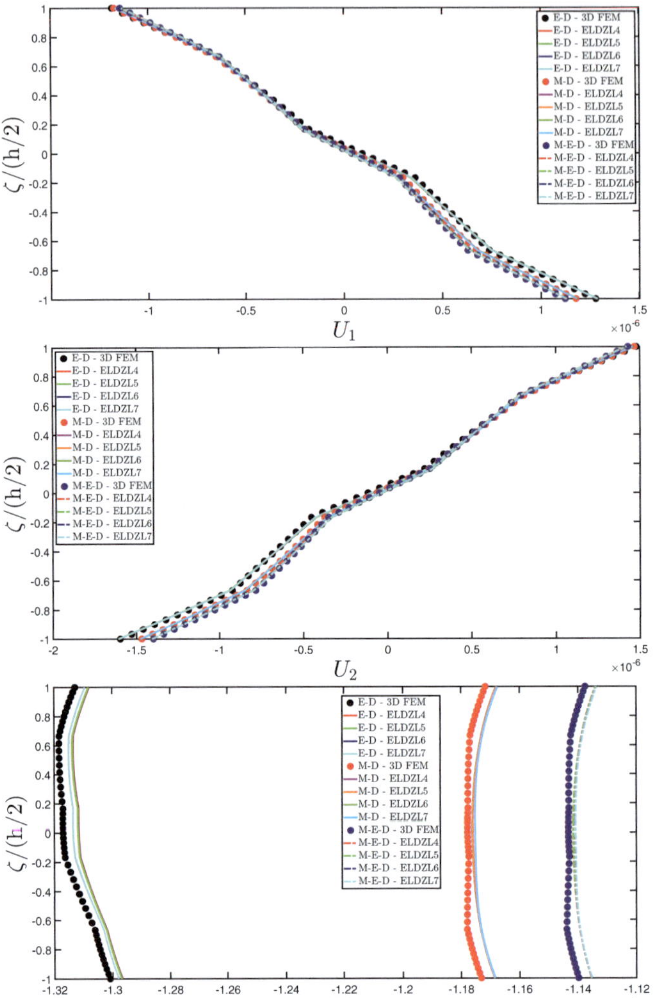

Fig. 7.2: Through-the-thickness distributions of the 3D displacement field components, expressed in [m], of a laminated rectangular plate for the point located at $(0.75L_1, 0.25L_2)$ within the 2D physical domain.

capture the distributions of the electric and magnetic field components, as shown in Fig. 7.6. Here, the in-plane electric field components $\mathcal{E}_1$, $\mathcal{E}_2$, and $\mathcal{E}_3$ predicted by the ELW simulations are in excellent agreement with the 3D FEM results, regardless of the order of the kinematic expansion. In particular, the in-plane components are continuous with slope variations at the interfaces, while the out-of-plane component is discontinuous. On the other hand, the in-plane fluxes $\mathcal{D}_1$ and $\mathcal{D}_2$ are uniformly distributed within the second and third laminae, whereas parabolic profiles are observed in the remaining layers. The out-of-plane component $\mathcal{D}_3$, obtained via

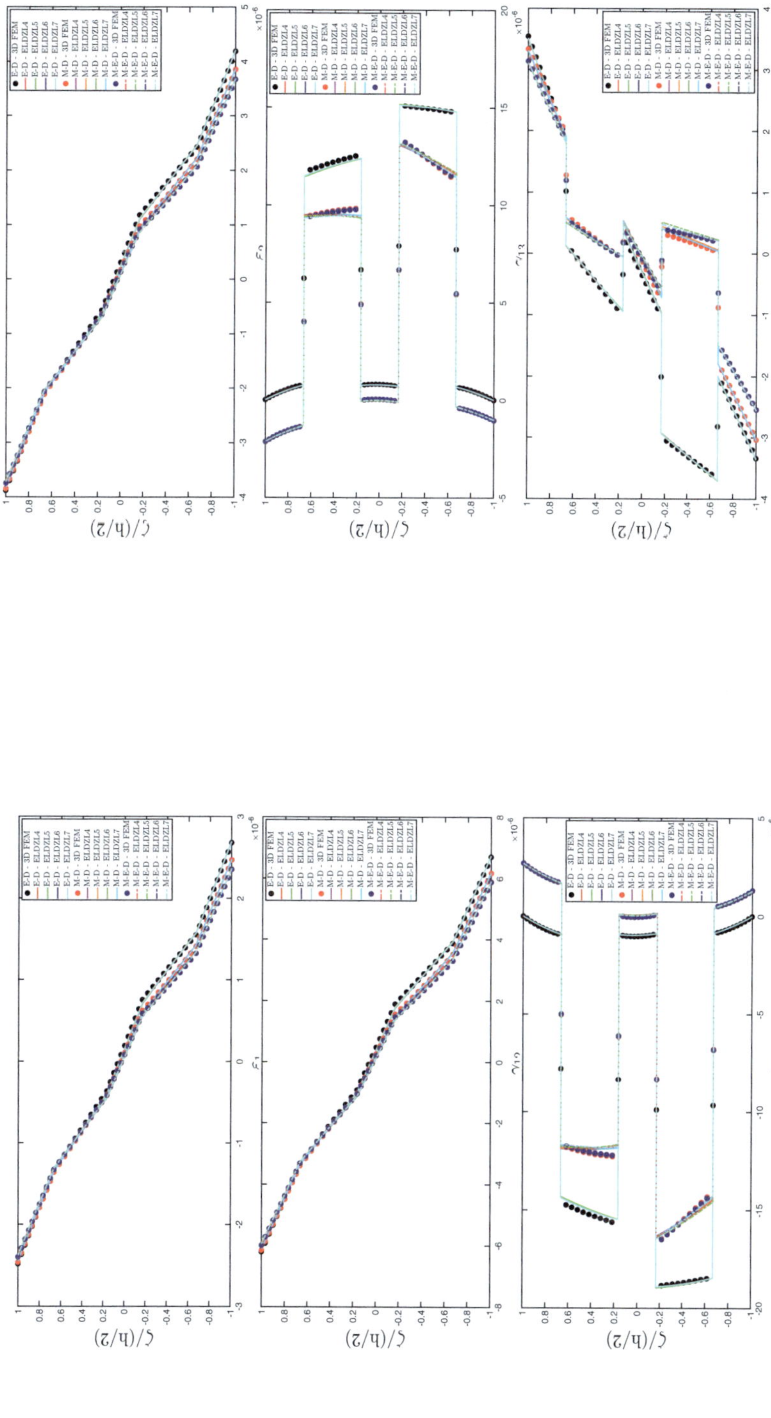

Fig. 7.3: Through-the-thickness distributions of the 3D strain components, expressed in [m/m], of a laminated rectangular plate for the point located at $(0.75L_1, 0.25L_2)$ within the 2D physical domain.

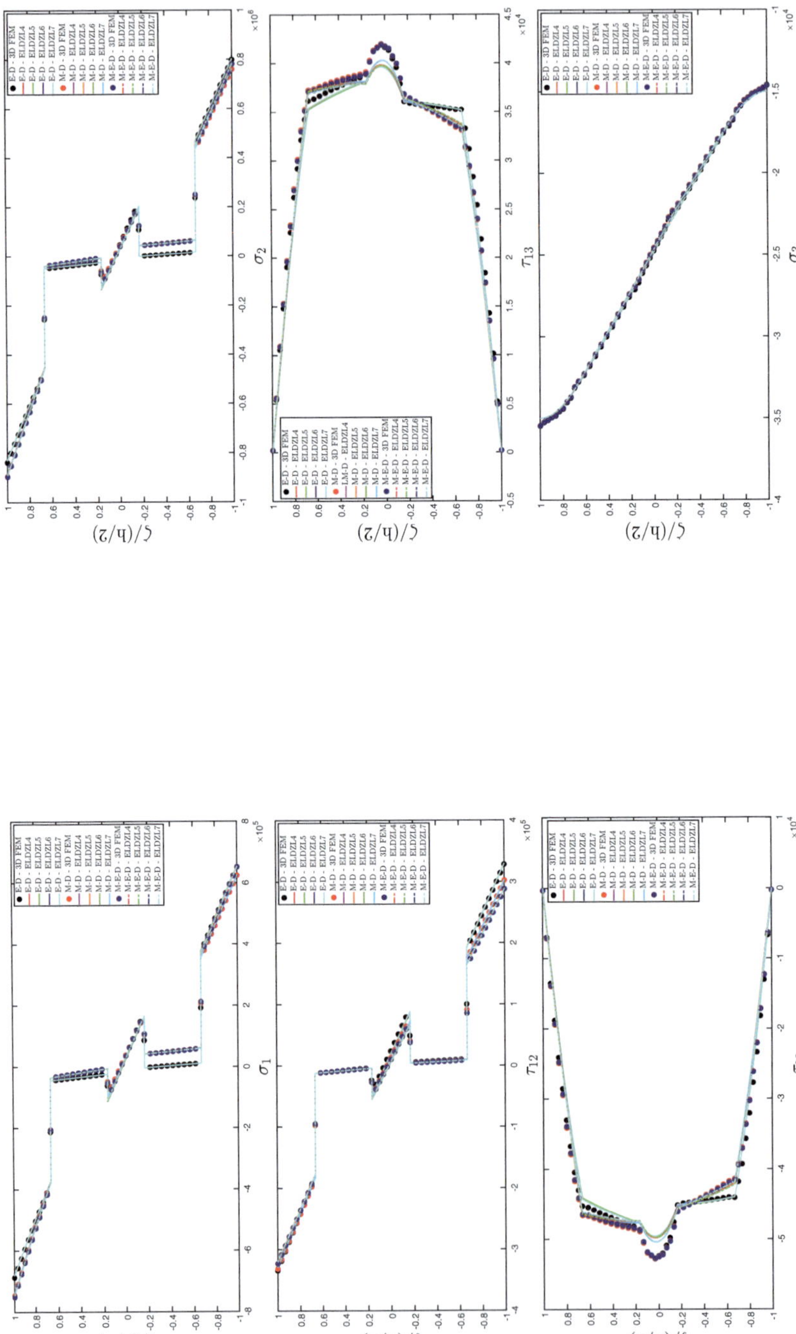

Fig. 7.4: Through-the-thickness distributions of the 3D stress components, expressed in [N/m^2], of a laminated rectangular plate for the point located at $(0.75L_1, 0.25L_2)$ within the 2D physical domain.

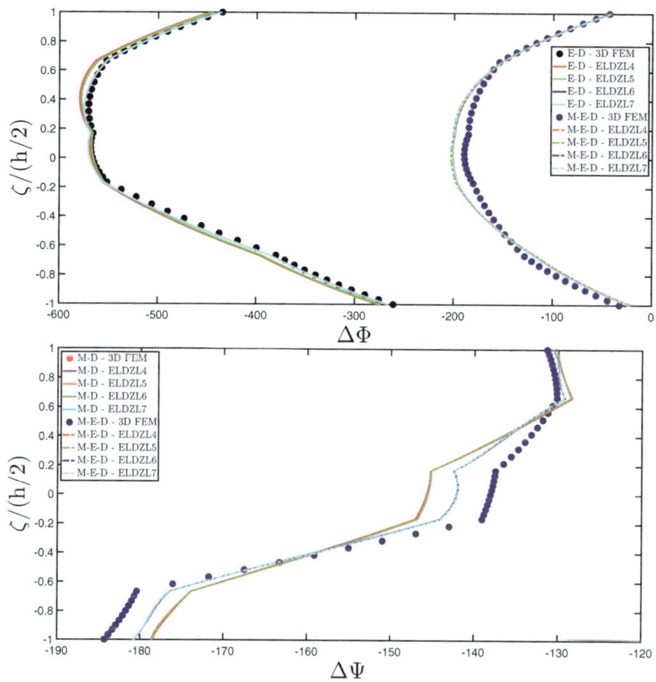

Fig. 7.5: Through-the-thickness distributions of the electric potential, expressed in [V], and of the scalar magnetic potential, expressed in [A], of a laminated rectangular plate made of five layers for the point located at $(0.75L_1, 0.25L_2)$ within the 2D physical domain.

the recovery procedure, matches the applied loads on the top and bottom surfaces and aligns with the 3D FEM predictions. For the magnetic primary and secondary variables shown in Fig. 7.7, the in-plane magnetic fluxes $\mathcal{B}_1$ and $\mathcal{B}_2$ are uniformly distributed within each lamina and exhbit discontinuities at the interfaces. The out-of-plane component $\mathcal{B}_3$, obtained from the recovery procedure, shows a stepwise linear distribution. As far as the magnetic field components is concerned, perfect agreement is observed for $\mathcal{H}_3$, while slightly different curves are obtained for the in-plane components $\mathcal{H}_1$ and $\mathcal{H}_2$. Unlike the electric primary and secondary variables, the results for the magnetic quantities are essentially identical for both M-D and M-E-D analyses, including not only the magnetic potential, but also the magnetic field and flux components. Next example focuses on a laminated circular cylinder composed of three layers with thicknesses $h_1 = h_3 = 0.03$ m and $h_2 = 0.04$ m. The lamination scheme is (MAT-I/MAT-III/MAT-II). The reference surface of the structure is derived from that of a truncated cone [35], reported below, considering a curvature radius $R_2 = 0.8$ m, and $\varphi = \pi/2$

$$\mathbf{r}(\alpha_1, \alpha_2) = \mathbf{r}(x, \vartheta) = R_0(x)\cos\vartheta\mathbf{e}_1 - R_0(x)\sin\vartheta\mathbf{e}_2 + x\sin\varphi\mathbf{e}_3 \tag{7.65}$$

with

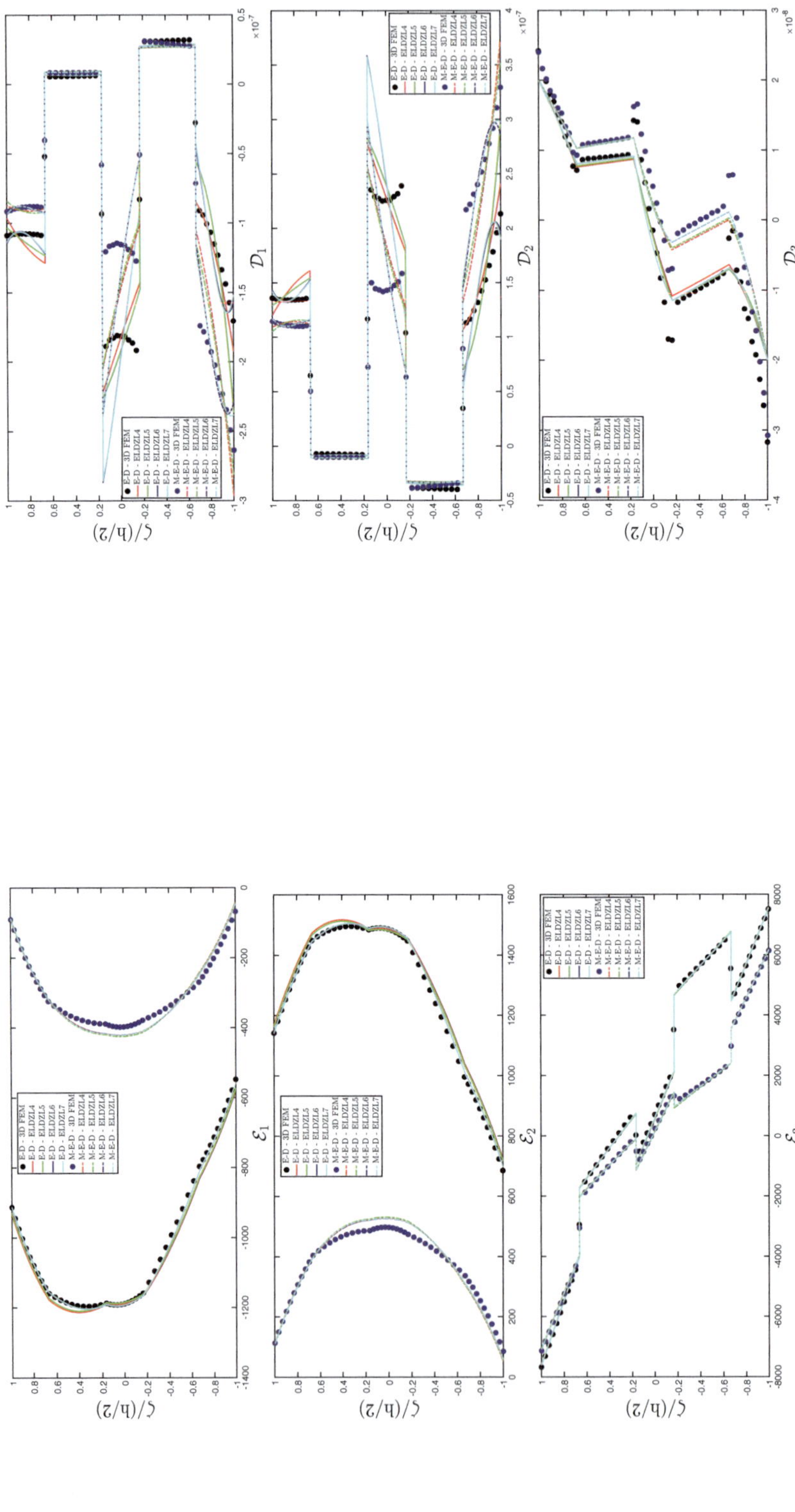

Fig. 7.6: Through-the-thickness distributions of the 3D electric field and flux components, expressed in [V/m] and [C/m^2], of a laminated rectangular plate for the point located at $(0.75L_1, 0.25L_2)$ within the 2D physical domain.

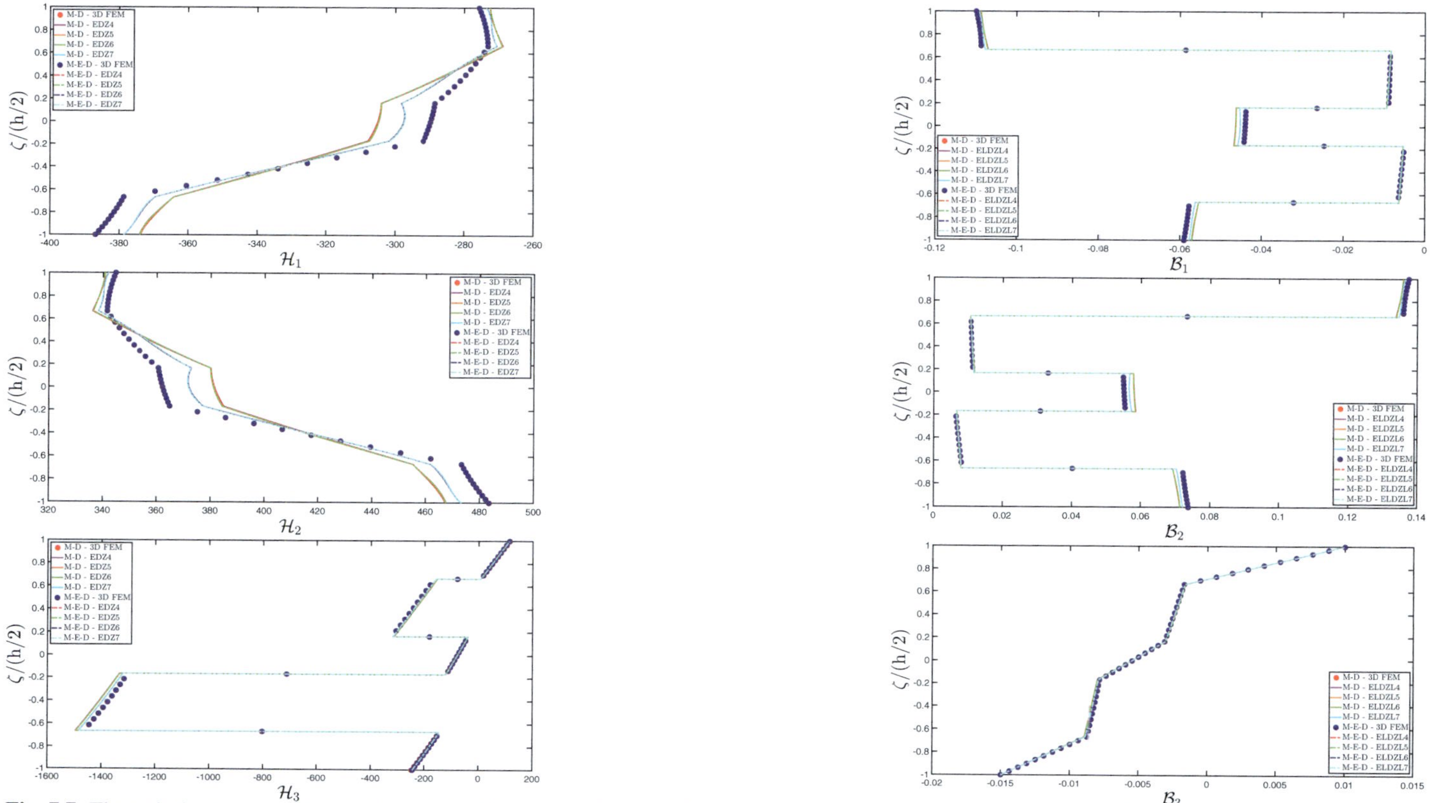

Fig. 7.7: Through-the-thickness distributions of the 3D magnetic field and flux components, expressed in [A/m] and [Wb/m^2], of a laminated rectangular plate for the point located at $(0.75L_1, 0.25L_2)$ within the 2D physical domain.

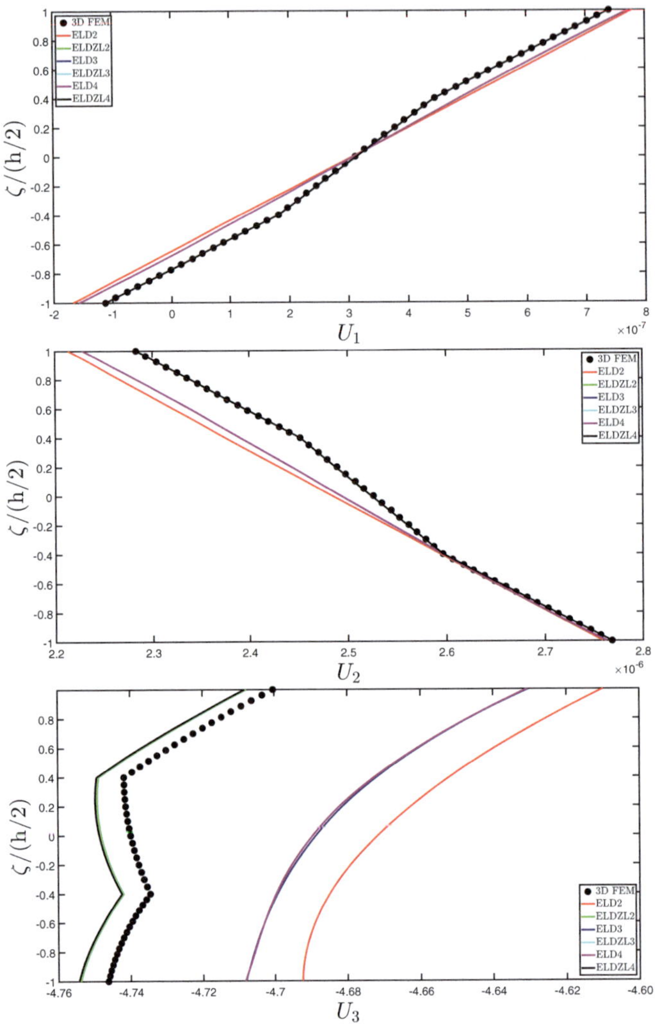

Fig. 7.8: Through-the-thickness distributions of the 3D displacement field components, expressed in [m], of a laminated circular cylinder for the point located at $(0.25L_1, 0.75L_2)$ within the 2D physical domain.

$$R_0(x) = R_2 + x\cos\varphi.$$

A geometric representation of the cylindrical panel is shown in Fig. 7.1. The panel is simply-supported (SSSS) and subjected to sinusoidally-distributed mechanical and multifield loads, in accordance with Eq. (7.64). More specifically, the reference magnitudes of the mechanical loads are $\bar{q}_3^{(+)} = -7 \times 10^4 \mathrm{N/m^2}$ and $\bar{q}_3^{(-)} = 3 \times 10^4 \mathrm{N/m^2}$. The magnitudes of the electric and magnetic fluxes are $\bar{q}_\mathrm{D}^{(+)} = -4 \times 10^{-8} \mathrm{C/m^2}$, $\bar{q}_\mathrm{D}^{(-)} = -4 \times 10^{-8} \mathrm{C/m^2}$ and $\bar{q}_\mathrm{B}^{(+)} = 2 \times 10^{-5} \mathrm{Wb/m^2}$ and $\bar{q}_\mathrm{B}^{(-)} = 3 \times 10^{-5} \mathrm{Wb/m^2}$,

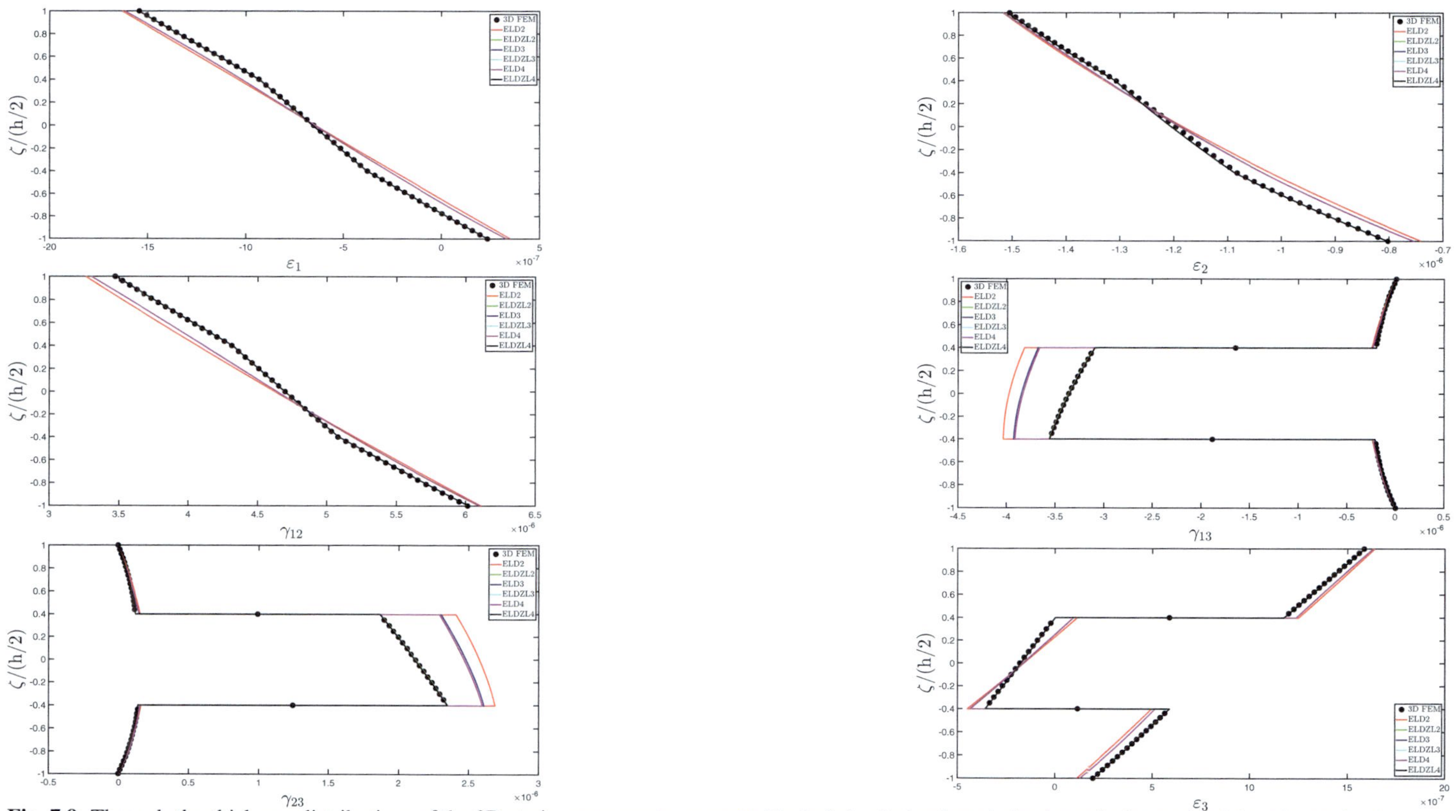

Fig. 7.9: Through-the-thickness distributions of the 3D strain components, expressed in [m/m], of a laminated circular cylinder made of three layers for the point located at $(0.25L_1, 0.75L_2)$ within the 2D physical domain.

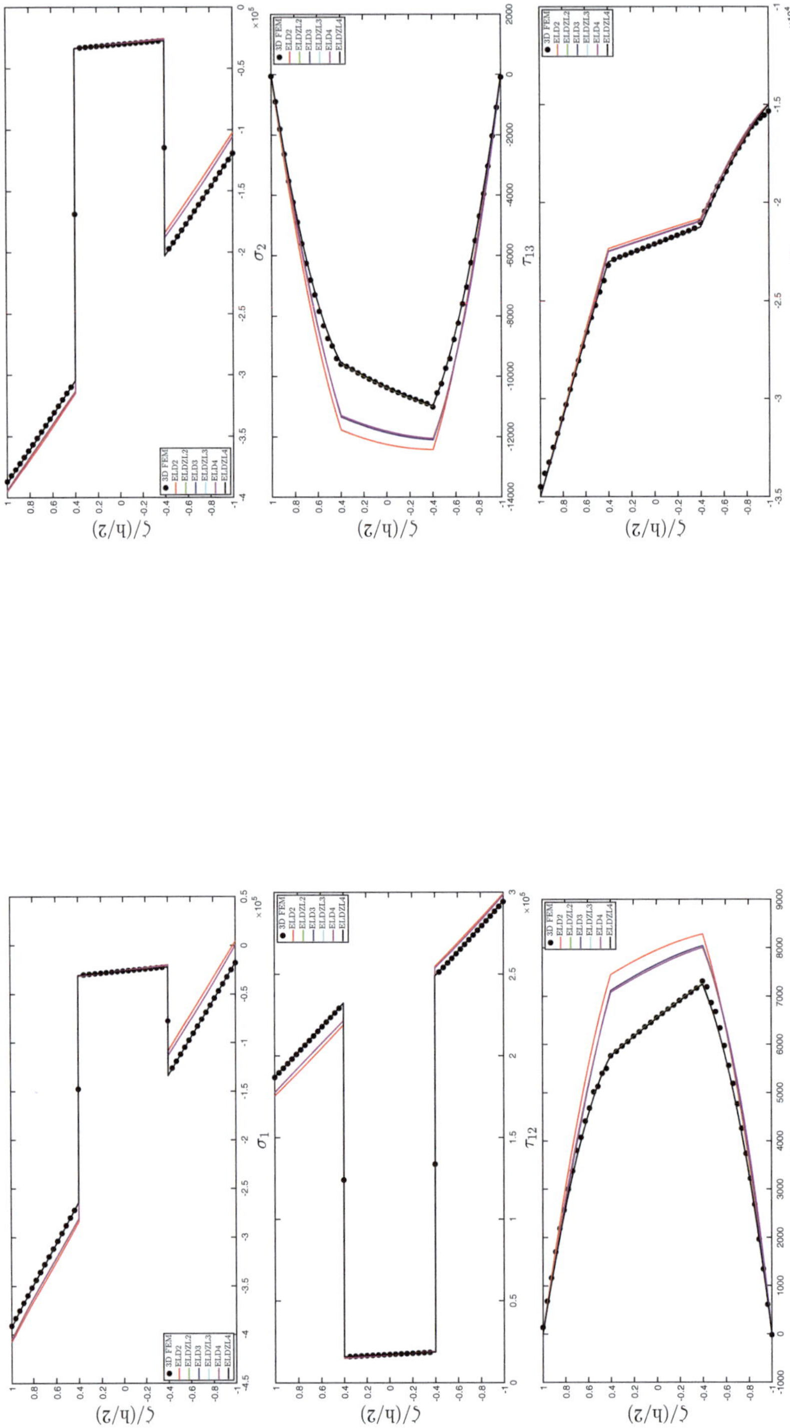

Fig. 7.10: Through-the-thickness distributions of the 3D stress components, expressed in [N/m^2], of a laminated circular cylinder made of three layers for the point located at $(0.25L_1, 0.75L_2)$ within the 2D physical domain.

respectively. The results are obtained using different kinematic models and are validated against 3D FEM simulations developed with commercial software employing 20-node parabolic brick elements. The through-the-thickness distributions of the multifield configuration, primary, and secondary variables, are presented for the point located at $(0.25L_1, 0.75L_2)$ within the 2D physical domain in Figs. 7.8-7.13. In

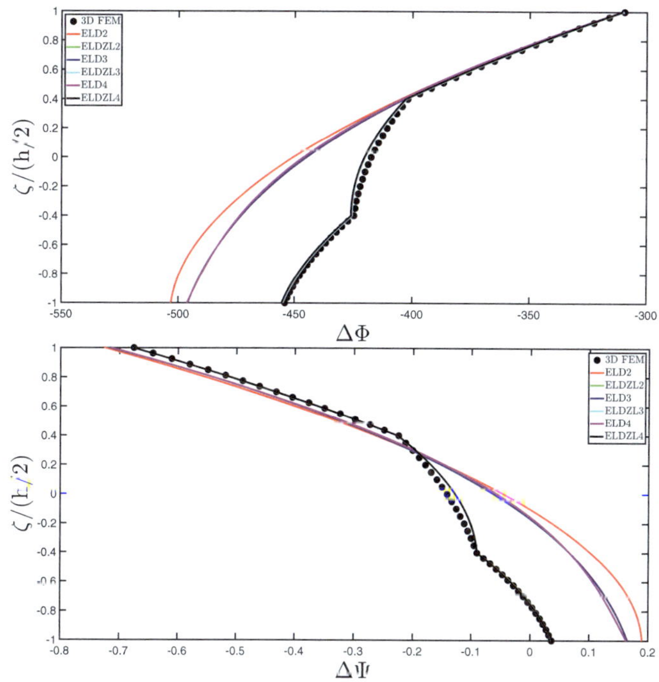

Fig. 7.11: Through-the-thickness distributions of the electric potential, expressed in [V], and of the scalar magnetic potential, expressed in [A], of a laminated circular cylinder made of three layers for the point located at $(0.25L_1, 0.75L_2)$ within the 2D physical domain.

particular, Fig. 7.9 shows the through-the-thickness distributions of the displacement components U_1, U_2, U_3. The reference 3D solution indicates that both in-plane and out-of-plane displacements follow stepwise linear profiles. As a consequence, ELW simulations, in absence of a zigzag function, fail to capture the correct deflection of the structure. In contrast, when the zigzag function is included, very accurate results are obtained even for lower-order kinematic expansion theories. Similar observations apply to the 3D strain components reported in Fig. 7.9. Indeed, the in-plane strains exhibit stepwise linear variations along the thickness, while the out-of-plane strains are discontinuous. When the zigzag function is used in the simulation, the ELW predictions perfectly match the 3D FEM results. Similarly, the 3D stress components show excellent agreement with the 3D FEM reference solution only when the zigzag function is employes, as shown in Fig. 7.10. In contrast, ELW theories without the term associated with $\tau = N + 1$ fail to capture the correct results, particularly in the

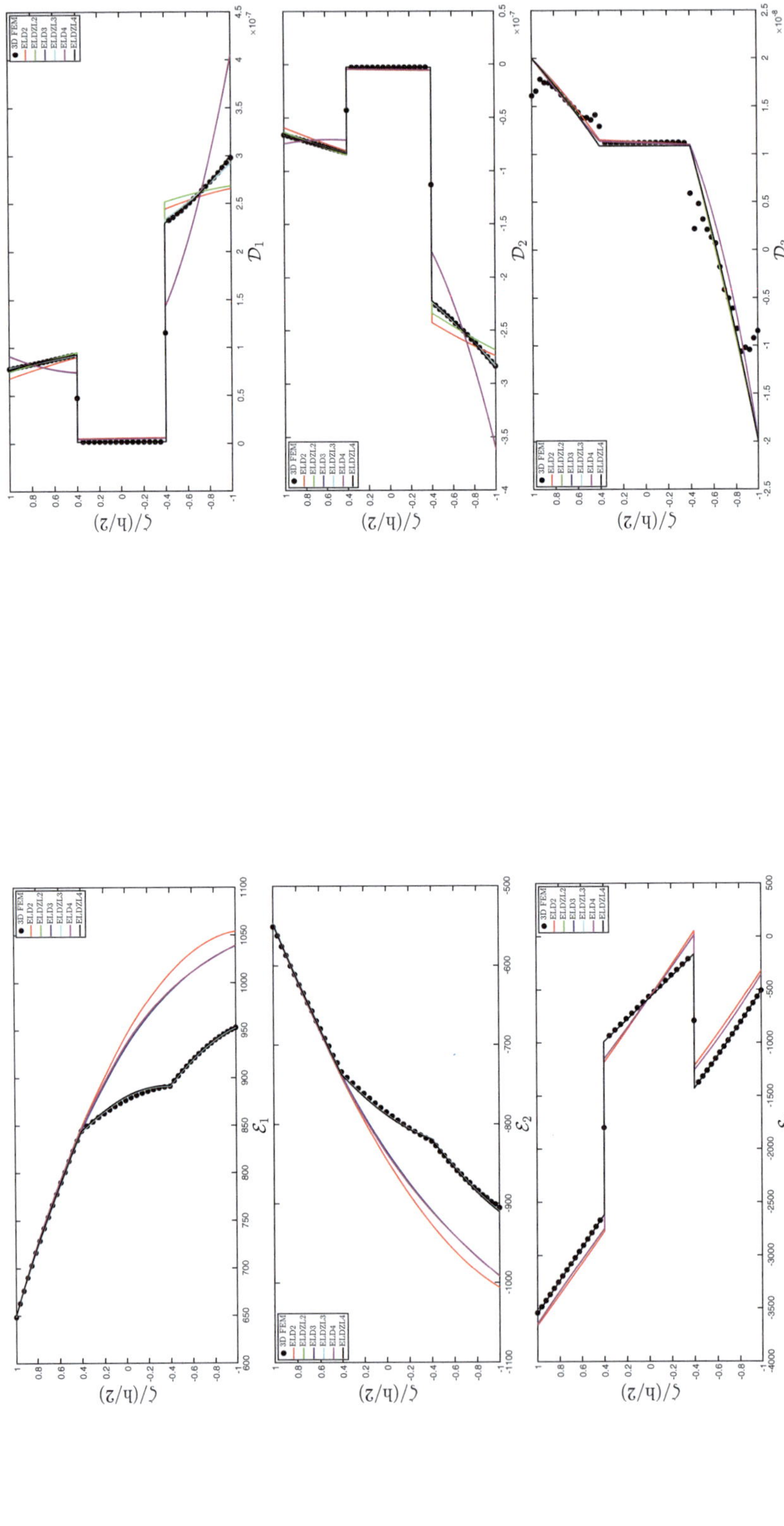

Fig. 7.12: Through-the-thickness distributions of the 3D electric field and flux components, expressed in [V/m] and [C/m^2], of a laminated circular cylinder made of three layers for the point located at $(0.25L_1, 0.75L_2)$ within the 2D physical domain.

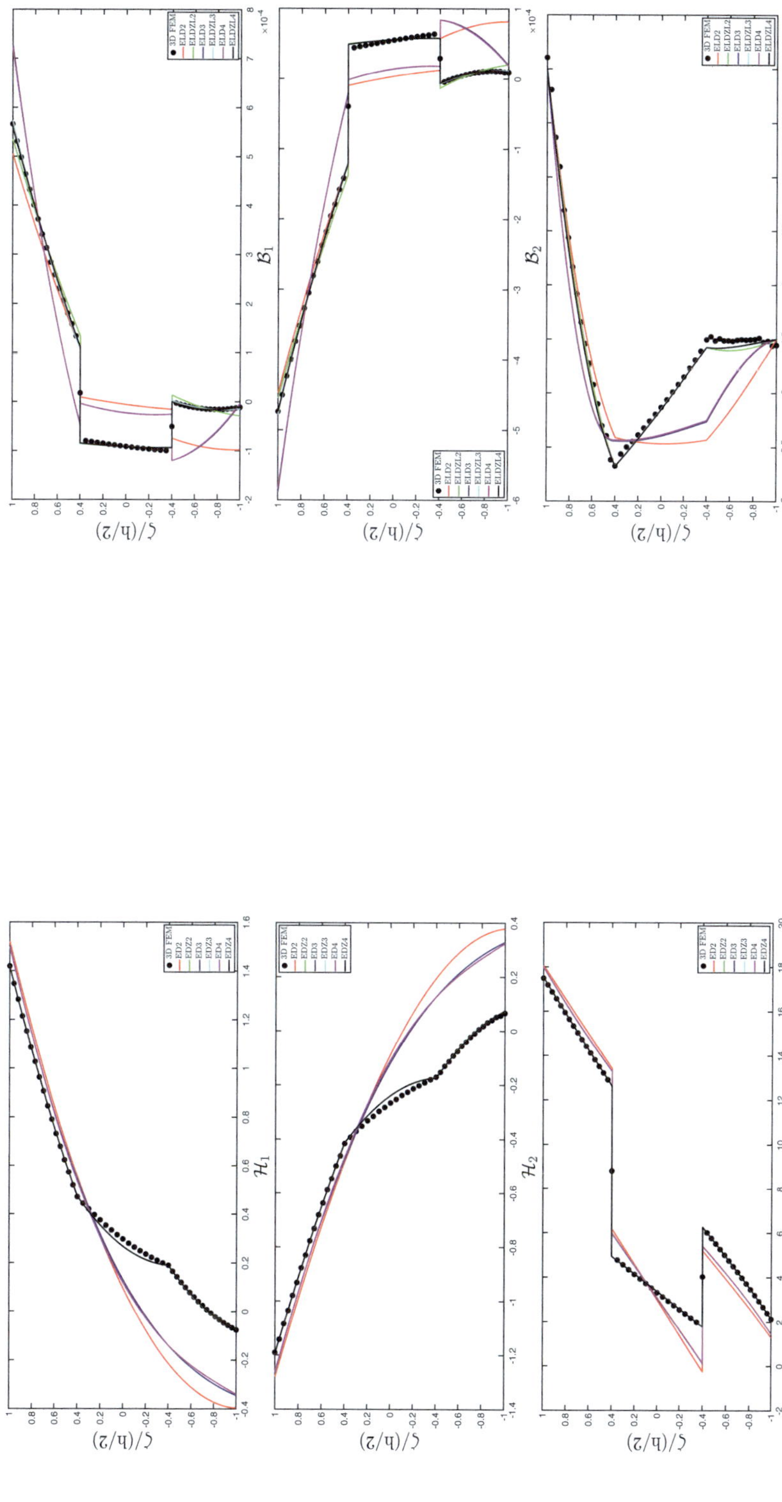

Fig. 7.13: Through-the-thickness distributions of the 3D magnetic field and flux components, expressed in [A/m] and [Wb/m^2], of a laminated circular cylinder made of three layers for the point located at $(0.25L_1, 0.75L_2)$ within the 2D physical domain.

central lamina. In all simulations, the recovery procedure ensures that the through-the-thickness distributions satisfy equilibrium with the applied mechanical loads. The importance of the zigzag function is particularly evident in Fig. 7.11, which shows the distribution along the thickness direction of the electric potential and the scalar magnetic potential. Indeed, the 3D FEM results exhibit parabolic profiles with abrupt slope variations, therefore ELW model reproduces these results accurately only if the zigzag function is used. Otherwise, even higher-order kinematic expansions lead to inaccurate results. For the 3D electric field components of Fig. 7.12, M-E-D simulations reveal that the in-plane components $\mathcal{E}_1$ and $\mathcal{E}_2$ require the zigzag function to match the 3D FEM predictions. In contrast, the out-of-plane component $\mathcal{E}_3$ is well predicted by all kinematic models, although ELDZL2, ELDZL3, and ELDZL4 provide perfect agreement with 3D FEM. The electric flux component $\mathcal{D}_3$ shows similar results for all kinematic models. Finally, Fig. 7.13 demonstrates that the zigzag function is essential for accurately predicting both magnetic primary and secondary variables. Indeed, ELD2, ELD3, and ELD4 yield inaccurate results, whereas ELDZL2, ELDZL3, and ELDZL4 perfectly reproduce the 3D FEM curve. This aspect highlights the crucial role of the zigzag function in multifield analysis. Regarding the maximum kinematic expansion order, lower-order theories are generally sufficient to accurately capture mechanical configuration, primary, and secondary variables. However, higher-order kinematic expansions are necessary for an accurate evaluation of the electric and magnetic quantities.

7.5 Conclusions

In this chapter, a novel 2D formulation has been presented for the analysis of laminated curved shell structures exhibiting magneto-electro-elastic properties under thermodynamic equilibrium conditions. The governing equations have been formulated in principal coordinates, considering a higher-order expansion of the unknown variables along the thickness directions using higher-order polynomials and the unified formulation. Following the ELW framework, this approach allows the imposition of prescribed electric and magnetic potentials on the top and bottom surfaces of the laminate. The governing equations have been derived from the Master Balance principle and are solved in the strong form. A semi-analytical Navier-type solution has been derived for simply-supported shells with uniform curvatures and cross-ply lamination schemes. In the post-processing, a recovery procedure has been introduced that reconstructs the primary and secondary multifield variables with a high level of accuracy relative to 3D solutions. Numerical examples have been presented in which the coupled magneto-electro-elastic response of laminated structures has been computed. The solution has been successfully validated against the predictions from commercial 3D FEM software, demonstrating both its accuracy and computational efficiency in terms of unknown variables. In particular, the results highlight the importance of the adopted kinematic model, which expand not only the mechanical unknown variables, but also the electric and magnetic variables. Furthermore, the analyses demonstrate the effect of the interaction between the electric and magnetic fields on the global response of the structure.

Acknowledgements The authors are grateful to the Department of Innovation Engineering of University of Salento for its continuous support in conducting this research.

References

[1] Aceti P, Calervo L, Bettini P, Sala G (2025) Measurement and Decoupling of Hygrothermal-Mechanical Effects with Optical Fibers: Development of a New Fiber Bragg Grating Sensor. Sensors 25(4):1037, DOI 10.3390/s25041037

[2] Bert CW, Malik M (1996) Differential quadrature method in computational mechanics: a review. Applied Mechanics Reviews 49(1):1-28, DOI 10.1115/1.3101882

[3] Bognet B, Bordeu F, Chinesta F, Leygue A, Poitou A (2012) Advanced simulation of models defined in plate geometries: 3D solutions with 2D computational complexity. Computer Methods in Applied Mechanics and Engineering 201204:1-12, DOI 10.1016/j.cma.2011.08.025

[4] Brischetto S (2009) Effect of the through-the-thickness temperature distribution on the response of layered and composite shells. International Journal of Applied Mechanics 1(4):581-605, DOI 10.1142/S1758825109000393

[5] Brischetto S, Torre R, Cesare D (2023) Three dimensional coupling between elastic and thermal fields in the static analysis of multilayered composite shells. CMES - Computer Modeling in Engineering and Sciences 136(3):2551-2594, DOI 10.32604/cmes.2023.026312

[6] Eastman ED (1928) Theory of the Soret effect, Journal of the American Chemical Society 50(2):283–291, DOI 10.1021/ja01389a007

[7] Ghugal YM, Sayyad AS (2013) Stress analysis of thick laminated plates using trigonometric shear deformation theory. International Journal of Applied Mechanics 5(1):1350003, DOI 10.1142/S1758825113500038

[8] Icardi U, Sola F (2016) Assessment of recent zig-zag theories for laminated and sandwich structures. Composites Part B: Engineering 97:26-52, DOI 10.1016/j.compositesb.2016.04.058

[9] Ingle SE, Horne FH (1973) The Dufour effect, The Journal of Chemical Physics 59(11):5882–5894, DOI 10.1063/1.1679957

[10] Kapuria S, Kumari P, Nath JK (2010) Efficient modeling of smart piezoelectric composite laminates: a review. Acta Mechanica 214(1):31-48, DOI 10.1007/s00707-010-0310-0

[11] Li C, Tian X, He T (2022) New insights on piezoelectric thermoelastic coupling and transient thermo-electromechanical responses of multi-layered piezoelectric laminated composite structure. European Journal of Mechanics - A/Solids 91:104416, DOI 10.1016/j.euromechsol.2021.104416

[12] Li D (2021) Layerwise theories of laminated composite structures and their applications: A Rreview. Archives of Computational Methods in Engineering 28(2):577-600, DOI 10.1007/s11831-019-09392-2
[13] Li JY, Dunn ML (1998) Anisotropic coupled-field inclusion and inhomogeneity problems. Philosophical Magazine A 77(5):1341-1350, DOI 10.1080/01418619808214256
[14] Li JY, Dunn ML (1998) Micromechanics of magnetoelectroelastic composite materials: average fields and effective behavior. Journal of Intelligent Materials Systems and Structures 9(4):404-416, DOI 10.1177/1045389X980090060
[15] Liew KM, Pan ZZ, Zhang LW (2019) An overview of layerwise theories for composite laminates and structures: Development, numerical implementation and application. Composite Structures 216:240-259, DOI 10.1016/j.compstruct.2019.02.074
[16] Liu Z, Lenarda P, Reinoso J, Paggi M (2025) Phase field modeling of anisotropic silicon crystalline cracking in 3D thin-walled photovoltaic laminates. International Journal of Fracture 249(1):19, DOI 10.1007/s10704-024-00821-3
[17] Mongioì F, Selleri G, Brugo TM, Maccaferri E, Fabiani D, Zucchelli A (2024) Multifunctional composite material based on piezoelectric nanofibers and Cu-CFRP electrodes for sensing applications. Composite Structures 337:118076, DOI 10.1016/j.compstruct.2024.118076
[18] Murakami H (1986) Laminated composite plate theory with improved in-plane responses. Journal of Applied Mechanics 53(3):661-666, DOI 10.1115/1.3171828
[19] Onsager L (1931) Reciprocal relations in irreversible processes. I. Physical Review 37(4):405-426, DOI 10.1103/physrev.37.405
[20] Onsager L (1931) Reciprocal relations in irreversible processes. II. Physical Review 38(12):2265–2279, DOI 10.1103/PhysRev.38.2265
[21] Petera J, Pittman JFT (1994) Isoparametric Hermite elements. International Journal for Numerical Methods in Engineering 37(20):3489-3519, DOI 10.1002/nme.1620372006
[22] Pipes RB, Vinson JR, Chou TW (1976) On the hygrothermal response of laminated composite systems. Journal of Composite Materials 10(2):129-148, DOI 10.1177/002199837601000203
[23] Reddy JN (1993) An evaluation of equivalent-single-layer and layerwise theories of composite laminates. Composite Structures 25(1-4):21-35, DOI 10.1016/0263-8223(93)90147-I
[24] Reddy JN (1984) A simple higher-order theory for laminated composite plates. Journal of Applied Mechanics 51(4):745-752, DOI 10.1115/1.3167719
[25] Reddy JN (2003) Mechanics of Laminated Composite Plates and Shells: Theory and Analysis, 2nd ed. CRC Press, Boca Raton, DOI doi.org/10.1201/b12409
[26] Reddy JN (1990) On refined theories of composite laminates. Meccanica 25(4):230-238, DOI 10.1007/BF01559685
[27] Reddy JN, Wang CM (2000) An overview of the relationships between solutions of the classical and shear deformation plate theories. Composites Science and Technology 60(12-13):2327-2335, DOI 10.1016/S0266-3538(00)00028-2

[28] Saravanos DA, Heyliger PR (1999) Mechanics and computational models for laminated piezoelectric beams, plates, and shells. Applied Mechanics Review 52(10):305-320, DOI 10.1115/1.3098918

[29] Shu C, Chew YT, Richards BE (1995) Generalized differential and integral quadrature and their application to solve boundary layer equations. International Journal for Numerical Methods in Fluids 21(9):723–733, DOI 10.1002/fld.1650210903

[30] Sih GC, Michopoulos J, Chou SC (1986) Hygrothermoelasticity. Martinus Nijhoff Publishers, Dordrecht

[31] Sih GC, Shih MT, Chou SC (1980) Transient hygrothermal stresses in composites: Coupling of moisture and heat with temperature varying diffusivity, International Journal of Engineering Science 18(1):19-42, DOI 10.1016/0020-7225(80)90004-X

[32] Soh AK, Liu AJ (2005) On the constitutive equations of magnetoelectroelastic solids. Journal of Intelligent Materials Systems and Structures 16(7-8):597-602, DOI 10.1177/1045389X050516

[33] Tichý J, Erhart J, Kittinger E, Přívratska J (2010) Fundamentals of piezoelectric sensorics: mechanical, dielectric, and thermodynamical properties of piezoelectric materials. Springer Science and Business Media, Berlin, Heidelberg, DOI 10.1007/978-3-540-68427-5

[34] Tornabene F (2023) Generalized Differential and Integral Quadrature. Società Editrice Esculapio, Bologna

[35] Tornabene F (2023) Hygro-Thermo-Magneto-Electro-Elastic Theory of Anisotropic Doubly-Curved Shells: Higher-Order Strong and Weak Formulations for Arbitrarily Shaped Shell Structures. Società Editrice Esculapio, Bologna

[36] Tornabene F, Fantuzzi N, Bacciocchi M, Reddy JN (2016) An equivalent layer-wise approach for the free vibration analysis of thick and thin laminated and sandwich shells. Applied Sciences 7(1):17, DOI 10.3390/app7010017

[37] Tornabene F, Fantuzzi N, Ubertini F, Viola E (2015) Strong formulation finite element method based on differential quadrature: A survey. Applied Mechanics Reviews 67(2):020801, DOI 10.1115/1.4028859

[38] Tornabene F, Viscoti M, Dimitri R (2025) Effect of thermal and electric coupling on the multifield response of laminated shell structures employing higher-order theories. Composite Structures 354:118801, DOI 10.1016/j.compstruct.2024.118801

[39] Tornabene F, Viscoti M, Dimitri R (2024) Equivalent layer-wise theory for the hygro-thermo-magneto-electro-elastic analysis of laminated curved shells. Thin-Walled Structures 198:111751, DOI 10.1016/j.tws.2024.111751

[40] Tornabene F, Viscoti M, Dimitri R (2024) Free vibration analysis of laminated anisotropic doubly-curved shell structures reinforced with three-phase polymer/CNT/fiber material. Engineering Analysis with Boundary Elements 164:105762, DOI 10.1016/j.enganabound.2024.105762

[41] Tornabene F, Viscoti M, Dimitri R (2023) General boundary conditions implementation for the static analysis of anisotropic doubly-curved shells

resting on a Winkler foundation. Composite Structures 322:117198, DOI 10.1016/j.compstruct.2023.117198

[42] Tornabene F, Viscoti M, Dimitri R (2025) Higher-order weak formulation for vibration analysis of anisotropic doubly-curved shells with holes and discontinuities. Engineering Analysis with Boundary Elements 180:106480, DOI 10.1016/j.enganabound.2025.106480

[43] Tornabene F, Viscoti M, Dimitri R (2025) Hygro-Thermo-Mechanical Equivalent Layer-Wise Theory of Laminated Shell Structures. CMES - Computer Modeling in Engineering & Sciences 142(2):1697-1765, DOI 10.32604/cmes.2025.058841

[44] Tornabene F, Viscoti M, Dimitri R (2025) Innovative recovery procedure applied to the static solution of anisotropic doubly-curved shells with holes and irregular shape. Thin-Walled Structures 218:114126, DOI 10.1016/j.tws.2025.114126

[45] Tornabene F, Viscoti M, Dimitri R (2024) On the importance of the recovery procedure in the semi-analytical solution for the static analysis of curved laminated panels: Comparison with 3D finite elements. Materials 17(3):588, DOI 10.3390/ma17030588

[46] Tornabene F, Viscoti M, Dimitri R (2024) Thermo-mechanical analysis of laminated doubly-curved shells: Higher order equivalent layer-wise formulation. Composite Structures 335:117995, DOI 10.1016/j.compstruct.2024.117995

[47] Tornabene F, Viscoti M, Dimitri R, Rabczuk T (2025) Hygro-thermo-electro-mechanical coupled modeling of laminated curved panels. Thin-Walled Structures, 215:113423, DOI 10.1016/j.tws.2025.113423

[48] Tornabene F, Viscoti M, Dimitri R, Rabczuk T (2025) Thermo-magneto-mechanical analysis of curved laminated structures with arbitrary variation of the material properties and novel recovery procedure. Engineering Analysis with Boundary Elements 176:106232, DOI 10.1016/j.enganabound.2025.106232

[49] Tornabene F, Viscoti M, Dimitri R, Reddy JN (2021) Higher order theories for the vibration study of doubly-curved anisotropic shells with a variable thickness and isogeometric mapped geometry. Composite Structures 267:113829, DOI 10.1016/j.compstruct.2021.113829

[50] Vinyas M (2021) Computational analysis of smart magneto-electro-elastic materials and structures: Review and classification. Archives of Computational Methods in Engineering 28(3):1205, DOI 10.1007/s11831-020-09406-4

[51] Wang CM, Reddy JN, Lee KH (2003) Shear Deformable Beams and Plates: Relationships with Classical Solutions, Elsevier, Boca Raton, DOI 10.1016/B978-0-08-043784-2.X5000-X

[52] Washizu K (1975) Variational Methods in Elasticity and Plasticity. Pergamon Press, Oxford

[53] Zienkiewicz OC, Taylor RL (2005) The Finite Element Method for Solid and Structural Mechanics. Elsevier, Amsterdam

Chapter 8
Physics-Informed Extreme Learning Machines for Strain Gradient Models: A Critical Comparison with Finite Element Solutions

Cheng A. Yan, Riccardo Vescovini, Michele Bacciocchi, Nicholas Fantuzzi, and António J.M. Ferreira

Abstract The numerical solution of strain gradient elasticity problems is challenging due to higher-order derivatives and strong nonlocal effects, which can reduce the convergence of conventional methods such as the Finite Element Method (FEM). This work investigates the Physics-Informed Extreme Learning Machine (PIELM), a neural-network-based, mesh-free approach, as an alternative for accurately and efficiently solving these problems.

Three benchmark cases with analytical solutions – static bending, free vibration, and linearized buckling of simply supported isotropic Kirchhoff nanoplates – are used to assess convergence and sensitivity to nonlocality. While FEM exhibits an algebraic convergence with errors increasing with the extent of nonlocality, PIELM achieves exponential convergence independently from the nonlocal scale. Training instabilities due to ill-conditioning can, however, limit the effective range of exponential error decay for large networks. Here, this issue is mitigated by enlarging the initialization of internal parameters.

Overall, the results demonstrate that PIELM can efficiently approximate higher-order elastic fields and represents a promising, mesh-free alternative for the solution of nonlocal continuum models.

Cheng A. Yan · Riccardo Vescovini
DAER Department, Politecnico di Milano, Via La Masa 34, Milan, 20156, Italy, e-mail: chengangelo.yan@polimi.it, riccardo.vescovini@polimi.it

Michele Bacciocchi
DESID Department, Università di San Marino, Via Consiglio dei Sessanta 99, Dogana, 47891, Republic of San Marino, e-mail: michele.bacciocchi@unirsm.sm

Nicholas Fantuzzi (corresponding author)
DICAM Department, Università di Bologna, Viale del Risorgimento 2, Bologna, 40136, Italy, e-mail: nicholas.fantuzzi@unibo.it

António J.M. Ferreira
FEUP Department, Universidade do Porto, Rua Dr. Roberto Frias, Porto, 4200-465, Portugal, e-mail: ferreira@gcloud.fe.up.pt

H. Altenbach et al. (eds.), *Composite Structures and Technology*,
Advanced Structured Materials 255,
https://doi.org/10.1007/978-3-032-24656-1_8

Key words: Kirchhoff nanoplates, Finite element method, Physics-informed neural networks, Extreme learning machine

8.1 Introduction

Strain gradient theories provide an enhanced framework for modeling the mechanical behavior of micro- and nano-scale structures, enabling the capture of size-dependent effects that Classical Continuum Theories (CCTs) fail to predict. These effects are particularly significant in the design and analysis of microelectromechanical systems and advanced composite materials [10, 25, 26], where the characteristic dimensions are comparable to the material's intrinsic length scales. However, the governing equations of strain gradient models involve higher-order Partial Differential Equations (PDEs) with additional boundary conditions compared with their CCT counterparts. This introduces substantial challenges for numerical methods, especially in terms of accuracy and convergence.

The Finite Element Method (FEM) has traditionally served as the primary numerical tool for solving PDEs, offering well-established convergence properties and theoretical foundation [33]. Yet, its application to strain-gradient models is complicated by the need for higher-order continuity in the shape functions. Meeting these requirements is nontrivial and necessitates specialized discretization strategies.

One common approach relies on Hermite-type shape functions to construct polynomial spaces with the required smoothness. Notable examples include the bending formulation for Kirchhoff strain gradient nanoplates developed by Babu and Patel [4], and subsequent extensions by Bacciocchi et al. [8, 9], who coupled these bending elements with high-continuity membrane formulations for laminated nanoplate analysis. Further developments generalized these elements to arbitrarily distorted geometries [6, 7] and incorporated von Kármán nonlinearities [5]. An alternative strategy uses mixed finite element formulations that introduce auxiliary field variables to relax inter-element continuity constraints. Classical examples include formulations that treat displacement gradients as independent unknowns [2, 29, 32], while more recent works further expand the mixed framework by independently discretizing displacements, strains, stresses, and their gradients [11, 12].

Although effective, these approaches inevitably increase implementation and computational complexity by adding either additional nodal degrees of freedom or auxiliary field variables that must be discretized. These limitations motivate the exploration of alternative numerical approaches capable of efficiently solving strain gradient problems while maintaining both accuracy and simplicity.

In recent years, mesh-free approaches have emerged as promising alternatives to conventional finite element schemes [17, 28, 34, 35]. Among these, Physics-Informed Neural Networks (PINNs) [27] have gained significant attention for their ability to solve PDEs without explicit meshing or high-order shape functions. PINNs embed the governing physical laws directly into the network's loss function, enabling the approximation of solution fields while enforcing the underlying PDEs. Within this family, the Physics-Informed Extreme Learning Machine (PIELM) [15] offers a particularly attractive framework, combining the training efficiency of Extreme Learning Machines (ELMs) [20] with the physics-based training setup of PINNs. Recent applications to composite plates and shells [37], linear elasticity [36], strain gradient elasticity [38], and fracture mechanics [40] demonstrate that PIELM can achieve accuracy comparable to or better than FEM at similar computational cost.

Despite this potential, a systematic comparison between FEM and PIELM remains limited in the literature. This paper aims to fill this gap by analyzing the convergence behavior of both methods in the context of strain gradient problems. Specifically, we examine the accuracy and convergence performance of the Hermite-based finite element formulation and PIELM for three representative problems, i.e. a static bending, free vibration and linearized buckling of Kirchhoff strain gradient nanoplates. The results show that PIELM achieves much faster convergence than FEM, highlighting its effectiveness and efficiency as an alternative for solving strain gradient elasticity problems.

Before outlining the structure of the paper, the authors wish to dedicate this contribution to the memory of Professor Cristóvão Manuel Mota Soares, whose pioneering work and inspiring mentorship profoundly influenced research in computational mechanics and advanced continuum theories. His intellectual curiosity and commitment to rigorous analysis continue to guide and inspire new generations of scientists. He was a leading figure in the field, renowned for his innovative approaches to shell and plate analysis, particularly his expertise in higher-order shear deformation theories, finite element methods, and meshless techniques [14, 22, 24, 30, 31]. A part from that, he was also known for his exceptional dedication: always present in the front row at conferences, actively engaging with questions, friendly and keen to discuss with young researchers and colleagues.

The paper is organized as follows: Section 8.2 summarizes the governing equations of the strain gradient theory considered. Sections 8.3 and 8.4 present the numerical formulations for FEM and PIELM, respectively. Section 8.5 discusses the convergence analysis and benchmark results. Finally, Section 8.6 concludes with remarks on comparative performance and perspectives for future developments.

8.2 Strain Gradient Kirchhoff Plate

The geometry and dimensions of the strain gradient plate are shown in Figure 8.1. A cartesian reference system is considered with center O and axes x, y, z. The plate has length a, width b, and total thickness of h.

Based on the classical Kirchhoff's theory, the displacement field

$$\mathbf{U} = \{u_x, u_y, u_z\}^{\mathrm{T}}$$

is approximated as:

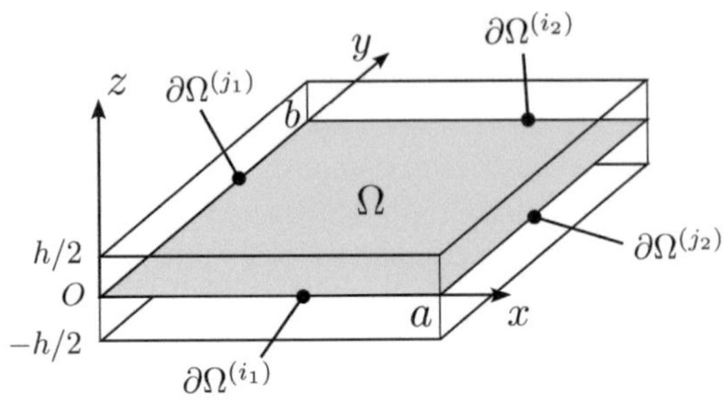

Fig. 8.1 Laminated plate: reference system and geometry.

$$\mathbf{U} = (\mathbf{I} + z\mathbb{D}_s)\,\mathbf{u}, \tag{8.1}$$

where $\mathbf{I}$ is a 3×3 identity matrix, while

$$\mathbf{u} = \{u, v, w\}^{\mathrm{T}}$$

is the vector collecting the displacement components of the plate on the mid-surface Ω. The operator matrix $\mathbb{D}_s$ is defined as:

$$\mathbb{D}_s = \begin{bmatrix} 0 & 0 & (\cdot)_{,x} \\ 0 & 0 & (\cdot)_{,y} \\ 0 & 0 & 0 \end{bmatrix}. \tag{8.2}$$

where $(\cdot)_{,\alpha} = \partial(\cdot)/\partial\alpha$ denotes partial differentiation. The in-plane components of the Green-Lagrange strains tensor

$$\boldsymbol{\epsilon} = \{\epsilon_{xx}, \epsilon_{yy}, \gamma_{xy}\}^{\mathrm{T}}$$

according to von Kármán theory are given by:

$$\boldsymbol{\epsilon} = \boldsymbol{\epsilon}_{\mathrm{m}} + z\boldsymbol{\epsilon}_{\mathrm{b}} + \boldsymbol{\epsilon}_{\mathrm{vk}} = \mathbb{D}_{\mathrm{m}}\mathbf{u} + z\mathbb{D}_{\mathrm{b}}\mathbf{u} + (\mathbb{D}_{\mathrm{v}}\mathbf{u}) \circ (\mathbb{D}_{\mathrm{k}}\mathbf{u}), \tag{8.3}$$

where the symbol $\circ$ denotes the Hadamard product, while

$$\begin{aligned} \boldsymbol{\epsilon}_{\mathrm{m}} &= \mathbb{D}_{\mathrm{m}}\mathbf{u}, \\ \boldsymbol{\epsilon}_{\mathrm{b}} &= \mathbb{D}_{\mathrm{b}}\mathbf{u} \end{aligned}$$

and

$$\boldsymbol{\epsilon}_{\mathrm{vk}} = (\mathbb{D}_{\mathrm{v}}\mathbf{u}) \circ (\mathbb{D}_{\mathrm{k}}\mathbf{u})$$

are the vectors of membrane, bending and von Kármán strains, respectively. These are related to the generalized displacements $\mathbf{u}$ through the following matrix of differential operators:

$$\mathbb{D}_{\mathrm{m}} = \begin{bmatrix} (\cdot)_{,x} & 0 & 0 \\ 0 & (\cdot)_{,y} & 0 \\ (\cdot)_{,y} & (\cdot)_{,x} & 0 \end{bmatrix}, \qquad \mathbb{D}_{\mathrm{b}} = \begin{bmatrix} 0 & 0 & -(\cdot)_{,xx} \\ 0 & 0 & -(\cdot)_{,yy} \\ 0 & 0 & -2(\cdot)_{,xy} \end{bmatrix},$$

$$\mathbb{D}_{\mathrm{v}} = \begin{bmatrix} 0 & 0 & \frac{1}{2}(\cdot)_{,x} \\ 0 & 0 & \frac{1}{2}(\cdot)_{,y} \\ 0 & 0 & (\cdot)_{,x} \end{bmatrix}, \qquad \mathbb{D}_{\mathrm{k}} = \begin{bmatrix} 0 & 0 & (\cdot)_{,x} \\ 0 & 0 & (\cdot)_{,y} \\ 0 & 0 & (\cdot)_{,y} \end{bmatrix}. \tag{8.4}$$

Despite the existence of several nonlocal formulations [1, 3, 13, 16, 23, 39], the present study focuses on the strain gradient one. In this context, under plane-stress conditions, the stress components

$$\boldsymbol{\sigma} = \{\sigma_{xx}, \sigma_{yy}, \sigma_{xy}\}^{\mathrm{T}}$$

are defined through the following constitutive relation:

$$\boldsymbol{\sigma} = \left(1 - l^2 \nabla^2\right) \overline{\mathbf{Q}} \boldsymbol{\epsilon}, \tag{8.5}$$

where l is the nonlocal parameter, $\nabla^2 = (\cdot)_{,xx} + (\cdot)_{,yy}$ is the Laplace operator, whereas $\overline{\mathbf{Q}}$ is the classical matrix of stiffness coefficients which depends on the elastic properties of the nanoplate.

The governing equations are derived starting from the Hamilton's variational principle applied in the time interval $t \in [0, T]$:

$$\int_0^T (\delta U + \delta V - \delta K)\, \mathrm{d}t = 0, \tag{8.6}$$

where δU, δV and δK are the first variations of the potential energy, potential of external forces and kinetic energy, respectively. For a Kirchhoff plate model, these energy contributions read:

$$\delta U = \int_\Omega \left(\delta \boldsymbol{\epsilon}_{\mathrm{m}}^{\mathrm{T}} \mathbf{N} + \delta \boldsymbol{\epsilon}_{\mathrm{b}}^{\mathrm{T}} \mathbf{M} + \delta \boldsymbol{\epsilon}_{\mathrm{vk}}^{\mathrm{T}} \mathbf{N}\right) \mathrm{d}\Omega, \tag{8.7}$$

$$\delta V = \int_\Omega \left(-\delta \mathbf{u}^{\mathrm{T}} \mathbf{q}\right) \mathrm{d}\Omega, \tag{8.8}$$

$$\delta K = \int_\Omega \left(\delta \dot{\mathbf{u}}^{\mathrm{T}} \mathbf{I}_0 \dot{\mathbf{u}} - \delta \dot{\mathbf{u}}^{\mathrm{T}} \mathbf{I}_1 \left(\mathbb{D}_{\mathrm{s}} \dot{\mathbf{u}}\right) - \left(\mathbb{D}_{\mathrm{s}} \delta \dot{\mathbf{u}}\right)^{\mathrm{T}} \mathbf{I}_1 \dot{\mathbf{u}} + \left(\mathbb{D}_{\mathrm{s}} \delta \dot{\mathbf{u}}\right)^{\mathrm{T}} \mathbf{I}_2 \left(\mathbb{D}_{\mathrm{s}} \dot{\mathbf{u}}\right)\right) \mathrm{d}\Omega, \tag{8.9}$$

where $\dot{(\cdot)} = \partial(\cdot)/\partial t$ denotes the time derivative, $\mathbf{I}_\alpha$ is the 3×3 diagonal matrix collecting the mass moment of inertia of the plate (I_0, I_1, I_2),

$$\mathbf{q} = \{q_x, q_y, q_z\}^{\mathrm{T}}$$

is the vector of external surface forces, while

$$\mathbf{N} = \{N_{xx}, N_{yy}, N_{xy}\}^{\mathrm{T}},$$
$$\mathbf{M} = \{M_{xx}, M_{yy}, M_{xy}\}^{\mathrm{T}}$$

are the vectors of force and moment resultants:

$$\mathbf{N} = \int_{-h/2}^{h/2} \sigma \mathrm{d}z \quad \text{and} \quad \mathbf{M} = \int_{-h/2}^{h/2} \sigma z \mathrm{d}z. \tag{8.10}$$

Considering the following expression of the virtual strains:

$$\delta\boldsymbol{\epsilon}_{\mathrm{m}} = \mathbb{D}_{\mathrm{m}}\delta\mathbf{u}, \quad \delta\boldsymbol{\epsilon}_{\mathrm{b}} = \mathbb{D}_{\mathrm{b}}\delta\mathbf{u}, \quad \delta\boldsymbol{\epsilon}_{\mathrm{vk}} = (\mathbb{D}_{\mathrm{v}}\delta\mathbf{u}) \circ (\mathbb{D}_{\mathrm{k}}\mathbf{u}) + (\mathbb{D}_{\mathrm{k}}\delta\mathbf{u}) \circ (\mathbb{D}_{\mathrm{v}}\mathbf{u}), \tag{8.11}$$

and exploiting the property of the Hadamard product

$$(\mathbf{a} \circ \mathbf{b})^{\mathrm{T}} \mathbf{c} = \mathbf{a}^{\mathrm{T}} (\mathbf{b} \circ \mathbf{c}),$$

the governing equations can be obtained by substituting Eq. (8.11) into Eqs. (8.7)-(8.9) and applying integration by parts to remove all derivatives (spatial and temporal) acting on the virtual displacement $\delta\mathbf{u}$. This yields:

$$\int_0^T \int_\Omega \delta\mathbf{u}^{\mathrm{T}}\{-\mathbb{D}_{\mathrm{m}}^{\mathrm{T}}\mathbf{N} + \mathbb{D}_{\mathrm{b}}^{\mathrm{T}}\mathbf{M} - \mathbb{D}_{\mathrm{v}}^{\mathrm{T}}[\mathbf{N}_{\mathrm{d}}(\mathbb{D}_{\mathrm{k}}\mathbf{u})] - \mathbb{D}_{\mathrm{k}}^{\mathrm{T}}[\mathbf{N}_{\mathrm{d}}(\mathbb{D}_{\mathrm{v}}\mathbf{u})] - \mathbf{q} + \mathbf{I}_0\ddot{\mathbf{u}} +$$
$$- \mathbf{I}_1(\mathbb{D}_{\mathrm{s}}\ddot{\mathbf{u}}) + \mathbb{D}_{\mathrm{s}}^{\mathrm{T}}(\mathbf{I}_1\ddot{\mathbf{u}}) - \mathbb{D}_{\mathrm{s}}^{\mathrm{T}}[\mathbf{I}_2(\mathbb{D}_{\mathrm{s}}\ddot{\mathbf{u}})]\}\mathrm{d}\Omega\mathrm{d}t + \boldsymbol{\Gamma}_{\mathrm{bcs}} = 0, \tag{8.12}$$

where

$$\mathbf{N}_{\mathrm{d}} = \mathrm{diag}(\mathbf{N})$$

is a diagonal matrix collecting the in-plane force resultants $\mathbf{N}$, while $\boldsymbol{\Gamma}_{\mathrm{bcs}}$ collects the contribution of boundary terms obtained from the integration by parts.

The strong form of the strain gradient Kirchhoff elasticity problem can be directly extracted from Eq. (8.12) and is expressed as:

$$\mathcal{R} := \mathcal{R}_{\mathrm{K}} + \mathcal{R}_{\mathrm{F}} + \mathcal{R}_{\mathrm{M}} + \mathcal{R}_{\mathrm{G}} = \mathbf{0}, \tag{8.13}$$

where $\mathcal{R}$ is the residual of the governing equations, which is decomposed into four contributions. Specifically, the term $\mathcal{R}_{\mathrm{K}}$ accounts for the linear elastic part, $\mathcal{R}_{\mathrm{F}}$ and $\mathcal{R}_{\mathrm{M}}$ include the effects of external loading and inertia forces, respectively, while $\mathcal{R}_{\mathrm{G}}$ represents the contribution due to geometric nonlinearities. These terms are given by:

$$
\begin{aligned}
\mathcal{R}_{\mathrm{K}} &= \mathbb{D}_{\mathrm{m}}^{\mathrm{T}}\left[\mathbf{A}\mathbb{D}_{\mathrm{m}} + \mathbf{B}\mathbb{D}_{\mathrm{b}} - l^2\left(\mathbf{A}\mathbb{D}_{\mathrm{m},xx} + \mathbf{A}\mathbb{D}_{\mathrm{m},yy} + \mathbf{B}\mathbb{D}_{\mathrm{b},xx} + \mathbf{B}\mathbb{D}_{\mathrm{b},yy}\right)\right]\mathbf{u}+ \\
&\quad - \mathbb{D}_{\mathrm{b}}^{\mathrm{T}}\left[\mathbf{B}\mathbb{D}_{\mathrm{m}} + \mathbf{D}\mathbb{D}_{\mathrm{b}} - l^2\left(\mathbf{B}\mathbb{D}_{\mathrm{m},xx} + \mathbf{B}\mathbb{D}_{\mathrm{m},yy} + \mathbf{D}\mathbb{D}_{\mathrm{b},xx} + \mathbf{D}\mathbb{D}_{\mathrm{b},yy}\right)\right]\mathbf{u}, \\
\mathcal{R}_{\mathrm{F}} &= \mathbf{q}, \\
\mathcal{R}_{\mathrm{G}} &= \mathbb{D}_{\mathrm{v}}^{\mathrm{T}}\left[\mathbf{N}_{\mathrm{d}}\left(\mathbb{D}_{\mathrm{k}}\mathbf{u}\right)\right] + \mathbb{D}_{\mathrm{k}}^{\mathrm{T}}\left[\mathbf{N}_{\mathrm{d}}\left(\mathbb{D}_{\mathrm{v}}\mathbf{u}\right)\right], \\
\mathcal{R}_{\mathrm{M}} &= -\mathbf{I}_0\ddot{\mathbf{u}} + \mathbf{I}_1\left(\mathbb{D}_{\mathrm{s}}\ddot{\mathbf{u}}\right) - \mathbb{D}_{\mathrm{s}}^{\mathrm{T}}\left(\mathbf{I}_1\ddot{\mathbf{u}}\right) - \mathbb{D}_{\mathrm{s}}^{\mathrm{T}}\left[\mathbf{I}_2\left(\mathbb{D}_{\mathrm{s}}\ddot{\mathbf{u}}\right)\right],
\end{aligned}
\tag{8.14}
$$

where $\mathbf{A}$, $\mathbf{B}$, and $\mathbf{D}$ are the classical membrane, membrane-bending coupling, and bending stiffness matrices of the plate, respectively. The extended form of Eq. (8.13) can be found in Ref. [13].

The set of Eqs. (8.13) and (8.14) constitutes a system of three PDEs. The first two correspond to the in-plane equilibrium equations and are of fourth order, whereas the third one represents the bending equilibrium equation and is of order six. These equations are supplemented by the following boundary conditions acting on the edges (i_1, i_2, j_1, j_2):

$$
\begin{aligned}
\mathcal{B}^{(j_k)} &= \begin{cases}
u = \overline{u} \quad \text{or} \quad N_{xx} - N_{xy,y}^{\mathrm{sg},x} = \overline{N}_{xx} - \overline{N}_{xy,y}^{\mathrm{sg},x} \\
v = \overline{v} \quad \text{or} \quad N_{xy} - N_{yy,y}^{\mathrm{sg},x} = \overline{N}_{xy} - \overline{N}_{yy,y}^{\mathrm{sg},x} \\
w = \overline{w} \quad \text{or} \quad V_x = \overline{V}_x \\
u_{,x} = \overline{u}_{,x} \quad \text{or} \quad N_{xx}^{\mathrm{sg},x} = \overline{N}_{xx}^{\mathrm{sg},x} \\
v_{,x} = \overline{v}_{,x} \quad \text{or} \quad N_{xy}^{\mathrm{sg},x} = \overline{N}_{xy}^{\mathrm{sg},x} \\
w_{,x} = \overline{w}_{,x} \quad \text{or} \quad M_{xx} - 2M_{xy,x}^{\mathrm{sg},y} = \overline{M}_{xx} - 2\overline{M}_{xy,x}^{\mathrm{sg},y} \\
w_{,xx} = \overline{w}_{,xx} \quad \text{or} \quad M_{xx}^{\mathrm{sg},x} = \overline{M}_{xx}^{\mathrm{sg},x}
\end{cases} \quad \text{in} \quad \partial\Omega^{(j_k)} \\
\mathcal{B}^{(i_k)} &= \begin{cases}
u = \overline{u} \quad \text{or} \quad N_{xy} - N_{xx,x}^{\mathrm{sg},y} = \overline{N}_{xy} - \overline{N}_{xx,x}^{\mathrm{sg},y} \\
v = \overline{v} \quad \text{or} \quad N_{yy} - N_{xy,x}^{\mathrm{sg},y} = \overline{N}_{yy} - \overline{N}_{xy,x}^{\mathrm{sg},y} \\
w = \overline{w} \quad \text{or} \quad V_y = \overline{V}_y \\
u_{,y} = \overline{u}_{,y} \quad \text{or} \quad N_{xy}^{\mathrm{sg},y} = \overline{N}_{xy}^{\mathrm{sg},y} \\
v_{,y} = \overline{v}_{,y} \quad \text{or} \quad N_{yy}^{\mathrm{sg},y} = \overline{N}_{yy}^{\mathrm{sg},y} \\
w_{,y} = \overline{w}_{,y} \quad \text{or} \quad M_{yy} - 2M_{xy,y}^{\mathrm{sg},x} = \overline{M}_{yy} - 2\overline{M}_{xy,y}^{\mathrm{sg},x} \\
w_{,yy} = \overline{w}_{,yy} \quad \text{or} \quad M_{yy}^{\mathrm{sg},y} = \overline{M}_{yy}^{\mathrm{sg},y}
\end{cases} \quad \text{in} \quad \partial\Omega^{(i_k)}
\end{aligned}
\tag{8.15}
$$

where quantities with an overbar denote prescribed values, while:

$$
\begin{aligned}
V_x &= M_{xx,x} + 2M_{xy,y} - I_1\ddot{u} + I_2\ddot{w}_{,x} + \overline{N}_{xx}w_{,x} + \overline{N}_{xy}w_{,y} - M_{yy,yy}^{\mathrm{sg},x}, \\
V_y &= M_{yy,y} + 2M_{xy,x} - I_1\ddot{v} + I_2\ddot{w}_{,y} + \overline{N}_{xy}w_{,x} + \overline{N}_{yy}w_{,y} - M_{xx,xx}^{\mathrm{sg},y}.
\end{aligned}
\tag{8.16}
$$

and:

$$
\begin{aligned}
\mathbf{N}^{\mathrm{sg,x}} &= \{N_{xx}^{\mathrm{sg,x}}, N_{yy}^{\mathrm{sg,x}}, N_{xy}^{\mathrm{sg,x}}\}^{\mathrm{T}} = l^2\nabla^2\left(\mathbf{A}\mathbb{D}_{\mathrm{m},x} + \mathbf{B}\mathbb{D}_{\mathrm{b},x}\right)\mathbf{u},\\
\mathbf{N}^{\mathrm{sg,y}} &= \{N_{xx}^{\mathrm{sg,y}}, N_{yy}^{\mathrm{sg,y}}, N_{xy}^{\mathrm{sg,y}}\}^{\mathrm{T}} = l^2\nabla^2\left(\mathbf{A}\mathbb{D}_{\mathrm{m},y} + \mathbf{B}\mathbb{D}_{\mathrm{b},y}\right)\mathbf{u},\\
\mathbf{M}^{\mathrm{sg,x}} &= \{M_{xx}^{\mathrm{sg,x}}, M_{yy}^{\mathrm{sg,x}}, M_{xy}^{\mathrm{sg,x}}\}^{\mathrm{T}} = l^2\nabla^2\left(\mathbf{B}\mathbb{D}_{\mathrm{m},x} + \mathbf{D}\mathbb{D}_{\mathrm{b},x}\right)\mathbf{u},\\
\mathbf{M}^{\mathrm{sg,y}} &= \{M_{xx}^{\mathrm{sg,y}}, M_{yy}^{\mathrm{sg,y}}, M_{xy}^{\mathrm{sg,y}}\}^{\mathrm{T}} = l^2\nabla^2\left(\mathbf{B}\mathbb{D}_{\mathrm{m},y} + \mathbf{D}\mathbb{D}_{\mathrm{b},y}\right)\mathbf{u}.
\end{aligned}
\tag{8.17}
$$

The weak form of the governing equations can be obtained by first multiplying the PDEs in Eq. (8.13) by the virtual displacement $\delta\mathbf{u}$ and integrating over the domain Ω, and then applying integration by parts:

$$
\delta\Pi := \int_{\Omega} \delta\mathbf{u}^{\mathrm{T}}\mathcal{R}\,\mathrm{d}\Omega = \delta\Pi_{\mathrm{K}} + \delta\Pi_{\mathrm{F}} + \delta\Pi_{\mathrm{M}} + \delta\Pi_{\mathrm{G}} = 0,
\tag{8.18}
$$

where the different contributions of the variation of the functional are:

$$
\begin{aligned}
\delta\Pi_{\mathrm{K}} = \int_{\Omega} &\Big\{(\mathbb{D}_{\mathrm{m}}\delta\mathbf{u})^{\mathrm{T}}(\mathbf{A}\mathbb{D}_{\mathrm{m}} + \mathbf{B}\mathbb{D}_{\mathrm{b}})\,\mathbf{u} + (\mathbb{D}_{\mathrm{b}}\delta\mathbf{u})^{\mathrm{T}}(\mathbf{B}\mathbb{D}_{\mathrm{m}} + \mathbf{D}\mathbb{D}_{\mathrm{b}})\,\mathbf{u}+\\
&+l^2\Big[\left(\mathbb{D}_{\mathrm{m},x}\delta\mathbf{u}\right)^{\mathrm{T}}\left(\mathbf{A}\mathbb{D}_{\mathrm{m},x} + \mathbf{B}\mathbb{D}_{\mathrm{b},x}\right)\mathbf{u} + \left(\mathbb{D}_{\mathrm{b},x}\delta\mathbf{u}\right)^{\mathrm{T}}\left(\mathbf{B}\mathbb{D}_{\mathrm{m},x} + \mathbf{D}\mathbb{D}_{\mathrm{b},x}\right)\mathbf{u}+\\
&+\left(\mathbb{D}_{\mathrm{m},y}\delta\mathbf{u}\right)^{\mathrm{T}}\left(\mathbf{A}\mathbb{D}_{\mathrm{m},y} + \mathbf{B}\mathbb{D}_{\mathrm{b},y}\right)\mathbf{u} + \left(\mathbb{D}_{\mathrm{b},y}\delta\mathbf{u}\right)^{\mathrm{T}}\left(\mathbf{B}\mathbb{D}_{\mathrm{m},y} + \mathbf{D}\mathbb{D}_{\mathrm{b},y}\right)\mathbf{u}\Big]\Big\}\,\mathrm{d}\Omega,\\
\delta\Pi_{\mathrm{F}} &= \int_{\Omega}\left(-\delta\mathbf{u}^{\mathrm{T}}\mathbf{q}\right)\mathrm{d}\Omega,\\
\delta\Pi_{\mathrm{M}} &= \int_{\Omega}\left[\delta\mathbf{u}^{\mathrm{T}}\mathbf{I}_0\ddot{\mathbf{u}} - \delta\mathbf{u}^{\mathrm{T}}\mathbf{I}_1(\mathbb{D}_{\mathrm{s}}\ddot{\mathbf{u}}) - (\mathbb{D}_{\mathrm{s}}\delta\mathbf{u})^{\mathrm{T}}\mathbf{I}_1\ddot{\mathbf{u}} + (\mathbb{D}_{\mathrm{s}}\delta\mathbf{u})^{\mathrm{T}}\mathbf{I}_2(\mathbb{D}_{\mathrm{s}}\ddot{\mathbf{u}})\right]\mathrm{d}\Omega,\\
\delta\Pi_{\mathrm{G}} &= \int_{\Omega}\left\{(\mathbb{D}_{\mathrm{v}}\delta\mathbf{u})^{\mathrm{T}}[\mathbf{N}_{\mathrm{d}}(\mathbb{D}_{\mathrm{k}}\mathbf{u})] + (\mathbb{D}_{\mathrm{k}}\delta\mathbf{u})^{\mathrm{T}}[\mathbf{N}_{\mathrm{d}}(\mathbb{D}_{\mathrm{v}}\mathbf{u})]\right\}\mathrm{d}\Omega.
\end{aligned}
\tag{8.19}
$$

Next, the weak and strong forms of the governing equations described by Eqs. (8.18) and (8.13) are solved using the FEM in Section 8.3, and the PIELM in Section 8.4, respectively.

8.3 Finite Element Method

The Finite Element Method (FEM) is used to solve the weak form of the strain gradient, i.e. Eq. (8.18). The formulation involves second-order derivatives of the in-plane displacements (u, v) and third-order derivatives of the transverse displacement (w). Accordingly, C^1-continuity is required for the in-plane components and C^2-continuity for the bending one. The Hermite finite elements proposed by Ref. [8] represent a suitable formulation to satisfy these requirements. This formulation is briefly discussed next.

The nanoplate is modeled with quadrilateral finite elements, as shown in Fig. 8.2, where $\boldsymbol{\xi} = \{\xi, \eta\}^{\mathrm{T}}$ represent the nondimensional coordinates which are related to the physical coordinates $\mathbf{x} = \{x, y\}^{\mathrm{T}}$ through the mapping functions $\boldsymbol{\chi} = \{\chi_x, \chi_y\}^{\mathrm{T}}$:

$$x = \chi_x(\xi, \eta) \quad \text{and} \quad y = \chi_y(\xi, \eta). \tag{8.20}$$

Using a conforming formulation, the finite element approximation $\mathbf{u}_h$ of the displacement field $\mathbf{u}$ is given by:

$$\mathbf{u} \simeq \mathbf{u}_h = \begin{bmatrix} \mathbf{f}^{\mathrm{T}} & \mathbf{0}_{1\times16} & \mathbf{0}_{1\times36} \\ \mathbf{0}_{1\times16} & \mathbf{f}^{\mathrm{T}} & \mathbf{0}_{1\times36} \\ \mathbf{0}_{1\times16} & \mathbf{0}_{1\times16} & \mathbf{g}^{\mathrm{T}} \end{bmatrix} \begin{Bmatrix} \mathbf{d}_{\mathrm{u}}^e \\ \mathbf{d}_{\mathrm{v}}^e \\ \mathbf{d}_{\mathrm{w}}^e \end{Bmatrix} = \mathbf{N}^e \mathbf{d}^e, \tag{8.21}$$

where $\mathbf{0}_{a\times b} \in \mathbb{R}^{a\times b}$ is a matrix of zeros of size $a \times b$, $\mathbf{d}^e \in \mathbb{R}^{(16+16+36)\times1}$ is the vector of elemental degrees of freedom, and $\mathbf{N}^e \in \mathbb{R}^{3\times(16+16+36)}$ is the corresponding matrix of interpolating functions. Note, nonconforming formulations reduce the size of these matrices and vectors and, consequently, the number of degrees of freedom, leading to improved convergence rates [4, 8]. However, they do not enforce the required C^1- and C^2-continuity exactly. The conforming formulation is therefore adopted here to ensure consistency with the regularity requirements of the governing equations.

According to the different continuity requirements required by the weak form, distinct shape functions are adopted for representing the membrane and bending displacement components:

$$\begin{aligned} \mathbf{f} &= \left\{f_1(\xi)f_1(\eta), \ldots, f_i(\xi)f_j(\eta), \ldots, f_4(\xi)f_4(\eta)\right\}^{\mathrm{T}} \quad \text{for } i, j = 1, 2, 3, 4. \\ \mathbf{g} &= \left\{g_1(\xi)g_1(\eta), \ldots, g_p(\xi)g_q(\eta), \ldots, g_6(\xi)g_6(\eta)\right\}^{\mathrm{T}} \quad \text{for } p, q = 1, 2, 3, 4, 5, 6. \end{aligned} \tag{8.22}$$

The one-dimensional shape functions $f_\alpha(\xi)$ used for the membrane displacements ensure C^1-continuity. Their expressions read:

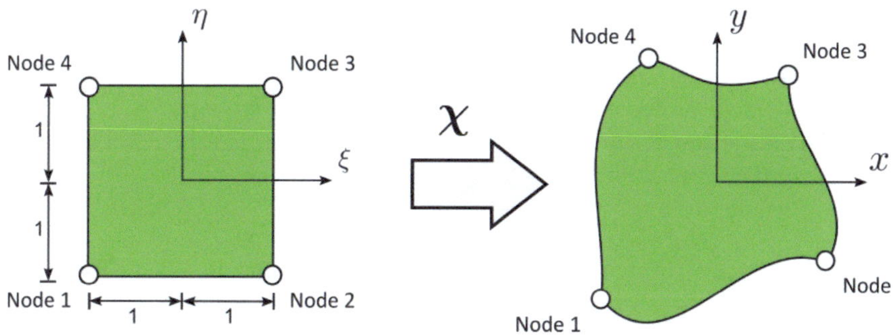

Fig. 8.2: Finite element's physical and computational domains.

$$
\begin{aligned}
f_1(\xi) &= \frac{1}{4}(2 - 3\xi + \xi^3),\\
f_2(\xi) &= \frac{1}{4}(1 - \xi - \xi^2 + \xi^3),\\
f_3(\xi) &= \frac{1}{4}(2 + 3\xi - \xi^3),\\
f_4(\xi) &= \frac{1}{4}(-1 - \xi + \xi^2 + \xi^3),
\end{aligned}
\tag{8.23}
$$

where f_1 and f_3 correspond to nodal values at the element endpoints, while f_2 and f_4 are associated with the first derivatives at the same nodes.

For the bending displacement, the shape functions $g_\alpha(\xi)$ enforce C^2-continuity, ensuring the displacement and its derivatives up to order two are continuous across elements. In this case, the expressions are:

$$
\begin{aligned}
g_1(\xi) &= \frac{1}{2} - \frac{15}{16}\xi + \frac{5}{8}\xi^3 - \frac{3}{16}\xi^5,\\
g_2(\xi) &= \frac{5}{16} - \frac{7}{16}\xi - \frac{3}{8}\xi^2 + \frac{5}{8}\xi^3 + \frac{1}{16}\xi^4 - \frac{3}{16}\xi^5,\\
g_3(\xi) &= \frac{1}{16} - \frac{1}{16}\xi - \frac{1}{8}\xi^2 + \frac{1}{8}\xi^3 + \frac{1}{16}\xi^4 - \frac{1}{16}\xi^5,\\
g_4(\xi) &= \frac{1}{2} + \frac{15}{16}\xi - \frac{5}{8}\xi^3 + \frac{3}{16}\xi^5,\\
g_5(\xi) &= -\frac{5}{16} - \frac{7}{16}\xi + \frac{3}{8}\xi^2 + \frac{5}{8}\xi^3 - \frac{1}{16}\xi^4 - \frac{3}{16}\xi^5,\\
g_6(\xi) &= \frac{1}{16} + \frac{1}{16}\xi - \frac{1}{8}\xi^2 - \frac{1}{8}\xi^3 + \frac{1}{16}\xi^4 + \frac{1}{16}\xi^5,
\end{aligned}
\tag{8.24}
$$

where g_1 and g_4 correspond to the nodal values at the endpoints, g_2 and g_5 to the first derivatives, and g_3 and g_6 to the second derivatives.

Based on the finite element approximation in Eq. (8.21), the weak form in Eq. (8.18) can be discretized as:

$$
\delta\Pi \simeq \bigcup_{e=1}^{N_\mathrm{e}} \left(\delta\mathbf{d}^{e\mathrm{T}}\mathbf{K}^e\mathbf{d}^e - \delta\mathbf{d}^{e\mathrm{T}}\mathbf{f}^e - \delta\mathbf{d}^{e\mathrm{T}}\mathbf{M}^e\ddot{\mathbf{d}}^e + \delta\mathbf{d}^{e\mathrm{T}}\mathbf{G}^e\mathbf{d}^e \right) = 0,
\tag{8.25}
$$

where the symbol $\bigcup$ denotes assembly over the N_e finite elements, $\mathbf{K}^e$, $\mathbf{G}^e$, $\mathbf{M}^e$, and $\mathbf{f}^e$ are the elemental stiffness, geometric, mass matrices, and nodal force vector, respectively. These quantities are defined as:

$$
\begin{aligned}
\mathbf{K}^e &= \int_{\Omega^e} \Big(\mathbf{B}_{\mathrm{m}}^{e\mathrm{T}} \mathbf{A} \mathbf{B}_{\mathrm{m}}^e + \mathbf{B}_{\mathrm{m}}^{e\mathrm{T}} \mathbf{B} \mathbf{B}_{\mathrm{b}}^e + \mathbf{B}_{\mathrm{b}}^{e\mathrm{T}} \mathbf{B} \mathbf{B}_{\mathrm{m}}^e + \mathbf{B}_{\mathrm{b}}^{e\mathrm{T}} \mathbf{D} \mathbf{B}_{\mathrm{b}}^e + \mathbf{B}_{\mathrm{m},x}^{e\mathrm{T}} \mathbf{A} \mathbf{B}_{\mathrm{m},x}^e + \\
&\qquad + \mathbf{B}_{\mathrm{m},x}^{e\mathrm{T}} \mathbf{B} \mathbf{B}_{\mathrm{b},x}^e + \mathbf{B}_{\mathrm{b},x}^{e\mathrm{T}} \mathbf{B} \mathbf{B}_{\mathrm{m},x}^e + \mathbf{B}_{\mathrm{b},x}^{e\mathrm{T}} \mathbf{D} \mathbf{B}_{\mathrm{b},x}^e + \mathbf{B}_{\mathrm{m},y}^{e\mathrm{T}} \mathbf{A} \mathbf{B}_{\mathrm{m},y}^e + \mathbf{B}_{\mathrm{m},y}^{e\mathrm{T}} \mathbf{B} \mathbf{B}_{\mathrm{b},y}^e + \\
&\qquad + \mathbf{B}_{\mathrm{b},y}^{e\mathrm{T}} \mathbf{B} \mathbf{B}_{\mathrm{m},y}^e + \mathbf{B}_{\mathrm{b},y}^{e\mathrm{T}} \mathbf{D} \mathbf{B}_{\mathrm{b},y}^e \Big) \, \mathrm{d}\Omega^e, \\
\mathbf{f}^e &= \int_{\Omega^e} \left(-\mathbf{N}^{e\mathrm{T}} \mathbf{q} \right) \mathrm{d}\Omega^e, \\
\mathbf{G}^e &= \int_{\Omega^e} \left(\mathbf{B}_{\mathrm{v}}^{e\mathrm{T}} \mathbf{N}_{\mathrm{d}} \mathbf{B}_{\mathrm{k}}^e + \mathbf{B}_{\mathrm{k}}^{e\mathrm{T}} \mathbf{N}_{\mathrm{d}} \mathbf{B}_{\mathrm{v}}^e \right) \mathrm{d}\Omega^e, \\
\mathbf{M}^e &= \int_{\Omega^e} \left(\mathbf{N}^{e\mathrm{T}} \mathbf{I}_0 \mathbf{N}^e - \mathbf{N}^{e\mathrm{T}} \mathbf{I}_1 \mathbf{B}_{\mathrm{s}}^e - \mathbf{B}_{\mathrm{s}}^{e\mathrm{T}} \mathbf{I}_1 \mathbf{N}^e + \mathbf{B}_{\mathrm{s}}^{e\mathrm{T}} \mathbf{I}_2 \mathbf{B}_{\mathrm{s}}^e \right) \mathrm{d}\Omega^e,
\end{aligned}
\tag{8.26}
$$

where the following matrices have been introduced:

$$
\begin{aligned}
\mathbf{B}_{\mathrm{m}}^e &= \mathbb{D}_{\mathrm{m}} \mathbf{N}^e, & \mathbf{B}_{\mathrm{m},x}^e &= l \mathbb{D}_{\mathrm{m},x} \mathbf{N}^e, & \mathbf{B}_{\mathrm{m},y}^e &= l \mathbb{D}_{\mathrm{m},y} \mathbf{N}^e, \\
\mathbf{B}_{\mathrm{b}}^e &= \mathbb{D}_{\mathrm{b}} \mathbf{N}^e, & \mathbf{B}_{\mathrm{b},x}^e &= l \mathbb{D}_{\mathrm{b},x} \mathbf{N}^e, & \mathbf{B}_{\mathrm{b},y}^e &= l \mathbb{D}_{\mathrm{b},y} \mathbf{N}^e, \\
\mathbf{B}_{\mathrm{s}}^e &= \mathbb{D}_{\mathrm{s}} \mathbf{N}^e, & \mathbf{B}_{\mathrm{v}}^e &= \mathbb{D}_{\mathrm{v}} \mathbf{N}^e, & \mathbf{B}_{\mathrm{k}}^e &= \mathbb{D}_{\mathrm{k}} \mathbf{N}^e.
\end{aligned}
\tag{8.27}
$$

By imposing the stationarity condition $\delta\Pi = 0$, Eq. (8.25) reduces to:

$$
\mathbf{K}\mathbf{d} - \beta_1 \mathbf{f} - \beta_2 \mathbf{M}\ddot{\mathbf{d}} + \beta_3 \mathbf{G}\mathbf{d} = 0,
\tag{8.28}
$$

where

$$
\mathbf{K} = \bigcup_{e=1}^{N_{\mathrm{e}}} \mathbf{K}^e
$$

is the global stiffness matrix,

$$
\mathbf{G} = \bigcup_{e=1}^{N_{\mathrm{e}}} \mathbf{G}^e
$$

is the global geometric stiffness matrix,

$$
\mathbf{M} = \bigcup_{e=1}^{N_{\mathrm{e}}} \mathbf{M}^e
$$

is the global mass matrix. The vector

$$
\mathbf{d} = \bigcup_{e=1}^{N_{\mathrm{e}}} \mathbf{d}^e
$$

defines the global nodal degrees of freedom, while

$$\mathbf{f} = \bigcup_{e=1}^{N_e} \mathbf{f}^e$$

represents the global external nodal force vector. The coefficients β_i act as Boolean flags that give different physical interpretations of the same general equation. Depending on their value, the equation describes one of the three standard finite element problems:

$$\text{for } \beta_i = \delta_{i1}: \qquad \mathbf{Kd} = \mathbf{f}, \tag{8.29}$$

$$\text{for } \beta_i = \delta_{i2}: \qquad \left(\mathbf{K} - \omega^2\mathbf{M}\right)\hat{\mathbf{d}} = 0, \tag{8.30}$$

$$\text{for } \beta_i = \delta_{i3}: \qquad (\mathbf{K} + \lambda\mathbf{G})\,\tilde{\mathbf{d}} = 0, \tag{8.31}$$

where δ_{ij} is the Kronecker's delta. Specifically, Eq. (8.29) corresponds to the linear static equilibrium problem with solution $\mathbf{d}$, Eq. (8.30) defines the free vibration problem with mode shapes $\hat{\mathbf{d}}$ and squared natural frequencies ω^2, and Eq. (8.31) describes the linearized buckling problem, where $\tilde{\mathbf{d}}$ and λ denote the buckling mode shapes and the corresponding eigenvalue.

8.4 Physics-Informed Extreme Learning Machine

The Physics-Informed Extreme Learning Machine (PIELM) [15] is used to solve the strong form of Eq. (8.13). In this approach, the solution of the PDEs is approximated via Physics-Informed Neural Networks (PINNs) [27], whose training is guided by a loss function constructed from the residuals of the governing equations $\mathcal{R}$ and the boundary conditions $\mathcal{B} = \{\mathcal{B}^{(i_k)}, \mathcal{B}^{(j_k)}\}^{\mathrm{T}}$. The network parameters are then optimized using the Extreme Learning Machine (ELM) training strategy [20].

Without lack of generality, the present study considers PINNs with a single hidden-layer architecture, as illustrated in Fig. 8.3. This has no restriction on the representation capability of the PINN, as the universal approximation theorem ensures that neural networks with one sufficiently large hidden layer can approximate any smooth function to arbitrary accuracy [19].

Within this setting, the normalized spatial coordinates $\boldsymbol{\xi} = \{\xi, \eta\}^{\mathrm{T}}$ are provided as inputs to the network. These are processed by a single hidden layer of n neurons to produce the displacement field $\mathbf{u} = \{u, v, w\}^{\mathrm{T}}$ as output. The mapping between inputs and outputs is expressed as:

$$\mathbf{u} = \mathbf{C}\sigma(\mathbf{W}\boldsymbol{\xi} + \mathbf{b}), \tag{8.32}$$

where $\mathbf{W} \in \mathbb{R}^{n\times 2}$ and $\mathbf{b} \in \mathbb{R}^{n\times 1}$ are the internal weights and biases and $\mathbf{C} \in \mathbb{R}^{3\times n}$ are the output weights. These matrices take the form:

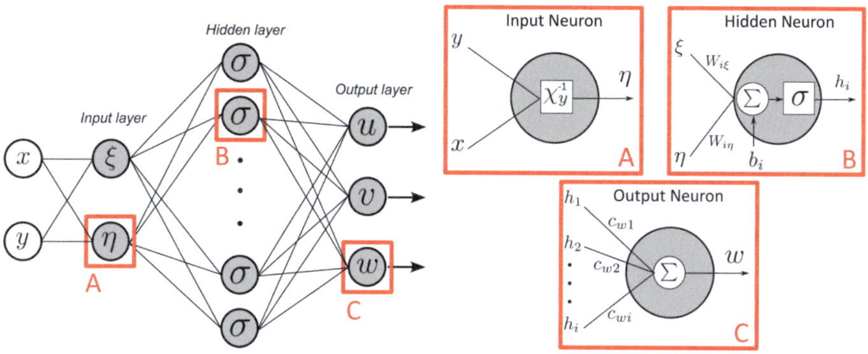

(a) Single hidden-layer network architecture.

(b) Computational neuron units in different layers.

Fig. 8.3: Physics-informed extreme learning machine.

$$\mathbf{W} = \begin{bmatrix} W_{1\xi} & W_{1\eta} \\ W_{2\xi} & W_{2\eta} \\ \vdots & \vdots \\ W_{n\xi} & W_{n\eta} \end{bmatrix}, \quad \mathbf{b} = \begin{bmatrix} b_1 \\ b_2 \\ \vdots \\ b_n \end{bmatrix} \quad \mathbf{C} = \begin{bmatrix} c_{\mathrm{u}1} & c_{\mathrm{u}2} & \dots & c_{\mathrm{u}n} \\ c_{\mathrm{v}1} & c_{\mathrm{v}2} & \dots & c_{\mathrm{v}n} \\ c_{\mathrm{w}1} & c_{\mathrm{w}2} & \dots & c_{\mathrm{w}n} \end{bmatrix} = \begin{bmatrix} \mathbf{c}_{\mathrm{u}}^{\mathrm{T}} \\ \mathbf{c}_{\mathrm{v}}^{\mathrm{T}} \\ \mathbf{c}_{\mathrm{w}}^{\mathrm{T}} \end{bmatrix}, \tag{8.33}$$

where $W_{k\alpha}$ is the internal weight connecting input α with the k-th hidden neuron with bias b_k and activation function σ, while $c_{\beta k}$ specifies the connection between that neuron and output displacement component β.

In the ELM-based training strategy, the parameters $\mathbf{W}$ and $\mathbf{b}$ are randomly initialized and kept fixed, while only the output weights $\mathbf{C}$ are trainable. Under this assumption, the neural network prediction in Eq. (8.32), hereafter denoted by $\mathbf{u}_n$, can be rewritten in a form analogous to the finite element approximation in Eq. (8.21):

$$\mathbf{u} \simeq \mathbf{u}_n = \begin{bmatrix} \mathbf{h}^{\mathrm{T}} & \mathbf{0}_{1\times n} & \mathbf{0}_{1\times n} \\ \mathbf{0}_{1\times n} & \mathbf{h}^{\mathrm{T}} & \mathbf{0}_{1\times n} \\ \mathbf{0}_{1\times n} & \mathbf{0}_{1\times n} & \mathbf{h}^{\mathrm{T}} \end{bmatrix} \begin{Bmatrix} \mathbf{c}_{\mathrm{u}} \\ \mathbf{c}_{\mathrm{v}} \\ \mathbf{c}_{\mathrm{w}} \end{Bmatrix} = \mathbf{H}\mathbf{c}, \tag{8.34}$$

where $\mathbf{c} = \{\mathbf{c}_{\mathrm{u}}, \mathbf{c}_{\mathrm{v}}, \mathbf{c}_{\mathrm{w}}\}^{\mathrm{T}} \in \mathbb{R}^{3n\times 1}$ collects the trainable output weights, and $\mathbf{H} \in \mathbb{R}^{3\times 3n}$ is the neural network counterpart of the finite element shape function matrix $\mathbf{N}^e$. Its blocks are built from the vector:

$$\mathbf{h} = \{h_1(\xi,\eta), h_2(\xi,\eta), \dots, h_k(\xi,\eta) \dots, h_n(\xi,\eta)\}^{\mathrm{T}}, \tag{8.35}$$

where each entry is a neural-network shape function of the form:

$$h_k(\xi,\eta) = \sigma\left(W_{k\xi}\xi + W_{k\eta}\eta + b_k\right). \tag{8.36}$$

These shape functions depend on the activation function σ and on the randomly initialized input weights ($W_{k\xi}$, $W_{k\eta}$) and biases (b_k). In this work, the hyperbolic tangent $\sigma(s) = \tanh(s)$ is used owing to its infinite differentiability – required for the higher-order derivatives in Eq. (8.13) – and its demonstrated superior accuracy and robustness in both classical and strain-gradient plate problems [37, 38]. The parameters **W** and **b** are sampled from a uniform random distribution on $[-1, 1]$, unless otherwise specified. It has been proved that such randomization, combined with a nonlinear activation, produces basis functions that are linearly independent with high probability [21].

The network is trained considering the following physics-based loss function:

$$\mathcal{L} = \sum_{p=1}^{N_\mathrm{f}} \frac{|\mathcal{R}^{(p)}|^2}{2N_\mathrm{f}} + \sum_{q=1}^{N_\mathrm{b}} \frac{|\mathcal{B}^{(q)}|^2}{2N_\mathrm{b}}, \tag{8.37}$$

where $|\cdot|$ is the Euclidean norm, N_f and N_b are the number of training points sampled in the computational domain Ω and along its boundary $\partial\Omega$, respectively, so that the total number is given by $N_\mathrm{c} = N_\mathrm{f}+N_\mathrm{b}$. The quantities $\mathcal{R}^{(p)}$ and $\mathcal{B}^{(q)}$ are the residual of the governing equations (Eq. (8.13)) and the boundary conditions (Eq. (8.15)) evaluated at the p-th and q-th training points using the network approximation of Eq. (8.34).

Within the ELM framework, minimizing Eq. (8.37) reduces to solving a least-squares system:

$$\mathbf{L}_\mathrm{K}\mathbf{c} - \beta_1\mathbf{t} - \beta_2\mathbf{L}_\mathrm{M}\ddot{\mathbf{c}} + \beta_3\mathbf{L}_\mathrm{G}\mathbf{c} = \mathbf{0}, \tag{8.38}$$

where the different building blocks $\mathbf{L}_\mathrm{K}, \mathbf{L}_\mathrm{M}, \mathbf{L}_\mathrm{G} \in \mathbb{R}^{3N_\mathrm{c}\times 3n}$ and $\mathbf{t} \in \mathbb{R}^{3N_\mathrm{c}\times 1}$ are defined by:

$$\mathbf{L}_\mathrm{K} = \begin{bmatrix} \frac{\partial \mathcal{R}_\mathrm{K}^{(1)}}{\partial \mathbf{c}} \\ \vdots \\ \frac{\partial \mathcal{R}_\mathrm{K}^{(N_\mathrm{f})}}{\partial \mathbf{c}} \\ \frac{\partial \mathcal{B}^{(1)}}{\partial \mathbf{c}} \\ \vdots \\ \frac{\partial \mathcal{B}^{(N_\mathrm{b})}}{\partial \mathbf{c}} \end{bmatrix}, \quad \mathbf{L}_\mathrm{M} = \begin{bmatrix} \frac{\partial \mathcal{R}_\mathrm{M}^{(1)}}{\partial \mathbf{c}} \\ \vdots \\ \frac{\partial \mathcal{R}_\mathrm{M}^{(N_\mathrm{f})}}{\partial \mathbf{c}} \\ \mathbf{0} \\ \vdots \\ \mathbf{0} \end{bmatrix}, \quad \mathbf{L}_\mathrm{G} = \begin{bmatrix} \frac{\partial \mathcal{R}_\mathrm{G}^{(1)}}{\partial \mathbf{c}} \\ \vdots \\ \frac{\partial \mathcal{R}_\mathrm{G}^{(N_\mathrm{f})}}{\partial \mathbf{c}} \\ \mathbf{0} \\ \vdots \\ \mathbf{0} \end{bmatrix}, \quad \mathbf{t} = \begin{Bmatrix} \mathbf{q}^{(1)} \\ \vdots \\ \mathbf{q}^{(N_\mathrm{f})} \\ \mathbf{0} \\ \vdots \\ \mathbf{0} \end{Bmatrix}. \tag{8.39}$$

In analogy with the finite element formulation, three types of problems are defined using Boolean selectors β_i:

$$\text{for } \beta_i = \delta_{i1}: \quad \mathbf{L}_\mathrm{K}\mathbf{c} = \mathbf{t} \quad \Rightarrow \quad \mathbf{c} = \mathbf{L}_\mathrm{K}^{\dagger}\mathbf{t}, \tag{8.40}$$

$$\text{for } \beta_i = \delta_{i2}: \quad \left(\mathbf{L}_\mathrm{K} - \omega^2\mathbf{L}_\mathrm{M}\right)\hat{\mathbf{c}} = \mathbf{0} \quad \Rightarrow \quad (\mathbf{P}\mathbf{L}_\mathrm{K})\,\hat{\mathbf{c}} = \omega^2\,(\mathbf{P}\mathbf{L}_\mathrm{M})\,\hat{\mathbf{c}}, \tag{8.41}$$

$$\text{for } \beta_i = \delta_{i3}: \quad (\mathbf{L}_\mathrm{K} + \lambda\mathbf{L}_\mathrm{G})\,\tilde{\mathbf{c}} = \mathbf{0} \quad \Rightarrow \quad (\mathbf{P}\mathbf{L}_\mathrm{G})\,\tilde{\mathbf{c}} = -\lambda\,(\mathbf{P}\mathbf{L}_\mathrm{G})\,\tilde{\mathbf{c}}. \tag{8.42}$$

The static problem described by Eq. (8.40) is solved by computing the pseudoinverse of $\mathbf{L}_\mathrm{K}$, here denoted as $\mathbf{L}_\mathrm{K}^{\dagger} \in \mathbb{R}^{3n\times 3N_\mathrm{c}}$, via Singular Value Decomposition (SVD). The solutions of the vibration Eq. (8.41) and buckling Eq. (8.42) require first the construction of a preconditioning matrix $\mathbf{P} \in \mathbb{R}^{3n \times 3N_c}$, used to transform both equations into numerically well-conditioned generalized eigenvalue problems – here this matrix is taken as the pseudoinverse of the stiffness least square matrix $\mathbf{P} = \mathbf{L}_K^{\dagger}$. After preconditioning, the resulting systems are solved using a Schur decomposition-based eigensolver.

8.5 Results

The convergence behavior of the Finite Element Method (FEM) and the Physics-Informed Extreme Learning Machine (PIELM) is assessed by comparing their numerical predictions with analytical solutions for strain gradient nanoplate problems. Three closed-form solutions are considered: (I) static bending, (II) free vibration, and (III) linearized buckling of an isotropic rectangular nanoplate.

The boundary conditions considered in the analysis are simply supported boundary conditions with movable edges:

$$\mathcal{B}_{ik} = \begin{cases} w = w_{,xx} = 0 \\ N_{xx} - N^{sg,x}_{xy,y} - N_{xy} - N^{sg,x}_{yy,y} = N^{sg,x}_{xx} = N^{sg,x}_{xy} = M_{xx} - 2M^{sg,y}_{xy,x} = 0 \end{cases},$$
$$\mathcal{D}_{ik} = \begin{cases} w = w_{,yy} = 0 \\ N_{xy} - N^{sg,y}_{xx,x} = N_{yy} - N^{sg,y}_{xy,x} = N^{sg,y}_{xy} = N^{sg,y}_{yy} = M_{yy} - 2M^{sg,x}_{xy,y} = 0 \end{cases} \tag{8.43}$$

The nanoplate dimensions are $a = 10$ nm, $b = 10$ nm, $h = 0.34$ nm, while the material parameters are given by: $E = 1100$ GPa, $\nu = 0.3$, and $\rho = 2700$ kg/m^3.

For the static bending problem, the analytical deflection field is:

$$w(x, y) = \frac{16 q_z a^4}{D\pi^6} \sum_{i=1,3,5,\ldots}^{\infty} \sum_{j=1,3,5,\ldots}^{\infty} \frac{\sin \alpha_i \sin \beta_j}{ij\left(k_{ij}^2 + \frac{l^2\pi^2}{a^2} k_{ij}^3\right)}, \tag{8.44}$$

where $D = Eh^2/12(1 - \nu^2)$ is the bending stiffness parameter, while:

$$\alpha_i = \frac{i\pi x}{a}, \quad \beta_j = \frac{j\pi y}{b}, \quad k_{ij} = i^2 + \frac{j^2 a^2}{b^2}. \tag{8.45}$$

The analytical expression for the natural frequencies for the free vibration problem is:

$$\omega_{ij}^2 = \frac{D\pi^4}{a^4}\left(\frac{k_{ij}^2 + \frac{l^2\pi^2}{a^2} k_{ij}^3}{I_0 + \frac{I_2\pi^2}{a^2} k_{ij}}\right), \tag{8.46}$$

where i and j denote the number of half wavelengths in the x- and y-directions, respectively, of the corresponding mode shape.

Regarding the linearized buckling problem, the analytical critical load is given by:

$$\lambda_{ij} = \frac{D\pi^2}{a^2}\left(\frac{k_{ij}^2 + \frac{l^2\pi^2}{a^2}k_{ij}^3}{i^2 + \mu\frac{j^2a^2}{b^2}}\right), \tag{8.47}$$

where $\mu = \overline{N}_{11}/\overline{N}_{22}$ is the membrane resultant ratio. Here, this value is taken as $\mu = 1$, corresponding to a biaxial loading condition where the direct and transverse components share the same intensity.

The results for the three problems are reported in terms of nondimensional maximum transverse displacement ($\check{w}$), fundamental frequency ($\check{\omega}$), and critical buckling load ($\check{\lambda}$):

$$\check{w} = \frac{D}{q_z a^4} w\,(0,0)\,, \quad \check{\omega} = \sqrt{\frac{a^4 I_0}{D}}\omega_{11}, \quad \check{\lambda} = \frac{a^2}{D}\lambda_{11}. \tag{8.48}$$

Errors are expressed as percentage differences:

$$E_{\%} = \frac{\Delta - \Delta_\circ}{\Delta} \times 100, \tag{8.49}$$

where $\Delta \in \{\hat{w}, \hat{\omega}, \hat{\lambda}\}$ is the exact solution, while $\Delta_\circ \in \{\hat{w}_\circ, \hat{\omega}_\circ, \hat{\lambda}_\circ\}$ is the corresponding numerical approximation computed either by FEM or PIELM. Because the internal parameters of PIELM are randomly initialized, this error is evaluated as the mean over 10 independent Monte Carlo realizations, and the reported values represent the mean approximation accuracy.

The convergence of the FEM solution is evaluated using a sequence of uniformly refined meshes of square elements with increasing resolutions (h). This strategy is commonly referred to as h-refinement and is expected to yield an algebraic convergence rate of the numerical solution [33]:

$$\lim_{h\to\infty} \|\mathbf{u} - \mathbf{u}_h\|_{\mathcal{V}} = O\left(\frac{k}{N^\beta}\right), \tag{8.50}$$

where $\|\cdot\|_{\mathcal{V}}$ denotes an appropriate norm in the FE function space $\mathcal{V}$, N is the number of degrees of freedom of the FE model, k is a constant independent of the mesh, while β is the convergence rate.

The convergence for the PIELM is assessed considering different network architectures, where the number of hidden neurons (n) is progressively increased. Training points are sampled on a sufficiently dense grid $N_c = 30 \times 30$ to avoid any possible overfitting problem on the training process. The Chebyshev-Gauss-Lobatto distribution is employed to generate these points:

$$\xi_i = \cos\left(\frac{i-1}{\sqrt{N_c}-1}\pi\right) \quad \text{for} \quad i = 1, ..., \sqrt{N_c}. \tag{8.51}$$

The PIELM approximation is expected to converge at an exponential rate, provided that the internal parameters are selected appropriately. As shown in Ref. [18], under a suitable choice of the random weights and biases, the error satisfies:

$$\lim_{n\to\infty} \|\mathbf{u} - \mathbf{u}_n\|_p = O\left(r^{-N}\right), \tag{8.52}$$

where $||\cdot||_p$ denotes an appropriate norm in the Banach space, while r is a constant which depends on the smoothness of the exact solution $\mathbf{u}$.

However, it is important to note that Eq. (8.52) provides only an upper bound. Exponential decay is not guaranteed for arbitrary initializations, and poorly chosen internal parameters may lead to significantly slower convergence, or even divergence, as it will be shown by the numerical results for the static bending, free vibration, and linearized buckling problems in Subsects. 8.5.1, 8.5.2, and 8.5.3, respectively.

8.5.1 Static Bending Problem

The bending benchmark is obtained by setting $\beta_i = \delta_{i1}$ in Eq. (8.29) and Eq. (8.40) for the FEM and PIELM, respectively. The convergence results are shown in Figure 8.4 for two values of the nonlocal parameter, i.e. $(l/a)^2 = 0.01$ (mild nonlocality) and $(l/a)^2 = 1.00$ (strong nonlocality).

The finite element solution exhibits algebraic convergence in agreement with the theoretical estimate in Eq. (8.50). For both levels of nonlocality, the observed convergence rate is approximately $\beta \approx 0.5$. However, the overall magnitude of the error increases significantly as the nonlocal parameter l grows, as evidenced by

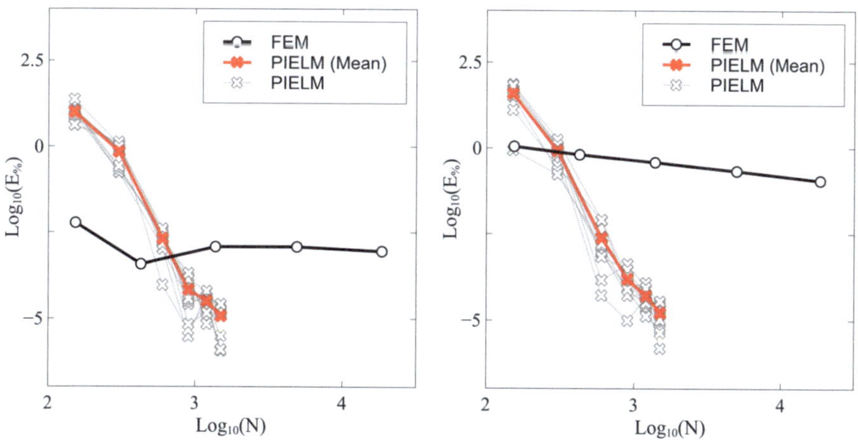

(a) Mild nonlocality ($(l/a)^2 = 0.01$). (b) Strong nonlocality ($(l/a)^2 = 1.00$).

Fig. 8.4: Convergence comparison between FEM and PIELM for the static bending problem. Percentage errors for PIELM are averaged over 10 random initializations of the internal parameters, with the mean values highlighted in red.

the upward shift of the convergence curves from Fig. 8.4(a) to Fig. 8.4(b). This degradation is due to the growing influence of higher-order gradients, which cannot be effectively captured by low-order, piece-wise polynomial approximations.

In contrast, the PIELM solution initially exhibits exponential error decay, which eventually transitions to an algebraic regime with an observed rate of $\beta \approx 4.35$. Note, both the exponential phase and the asymptotic algebraic rate remain almost insensitive to the value of the nonlocal parameter. This insensitivity is attributed to the high smoothness and infinitely differentiability of the neural-network trial space, which can better accommodate the higher-order continuity requirements than the low-degree polynomial FEM bases.

These results highlight a significant computational advantage of PIELM over FEM for problems dominated by gradient effects, where low-order/piece-wise discretizations typically suffer from reduced accuracy and slower convergence.

8.5.2 Free Vibration Problem

The second benchmark deals with the free vibration problem. For this case, the choice $\beta = \delta_{i2}$ in Eq. (8.30) and Eq. (8.41) yields the FEM and PIELM equations for the dynamic equilibrium. The corresponding convergence curves for both mild and strong nonlocality are presented in Figs. 8.5.

As in the bending benchmark, the FEM solution shows algebraic convergence with rates $\beta \approx 0.5$. The same upward shift in the error curves is observed as the nonlocal

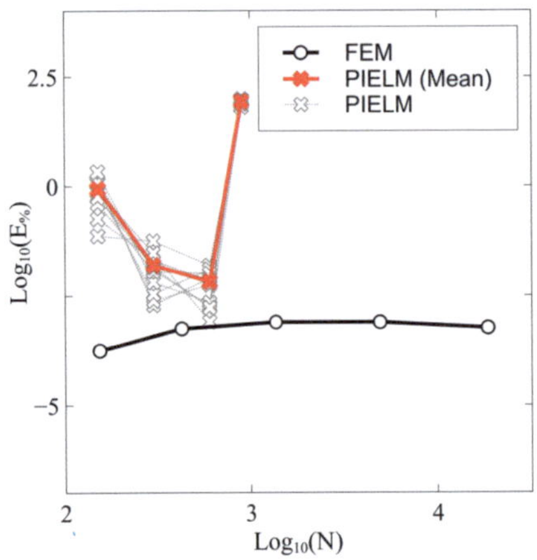

(a) Mild nonlocality ($(l/a)^2 = 0.01$).

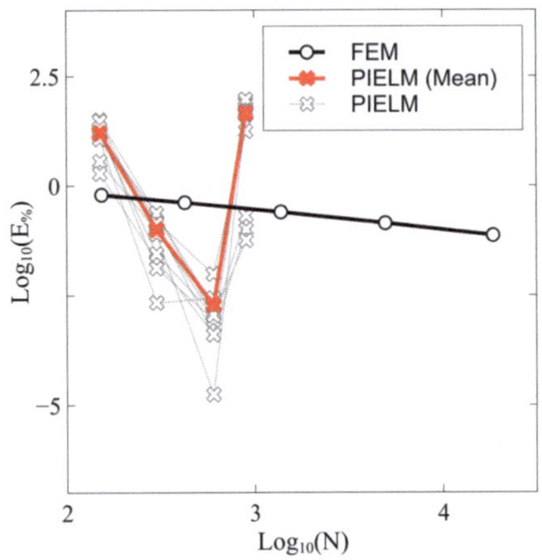

(b) Strong nonlocality ($(l/a)^2 = 1.00$).

Fig. 8.5: Convergence comparison between FEM and PIELM for the free vibration problem. Percentage errors for PIELM are averaged over 10 random initializations of the internal parameters, with the mean values highlighted in red.

parameter increases, reflecting the reduced effectiveness of the FE interpolation in capturing higher-order gradients when strain gradient effects become important.

For the PIELM, the error decreases at first exponentially with the number of neurons, consistent with the results of the static bending problem. However, after a certain network size the errors begin to increase, leading to a global loss of accuracy. This divergence occurs for both values of nonlocal parameters ($(l/a)^2 = 0.01$ and $(l/a)^2 = 1.00$), indicating that it is not related to nonlocality but to numerical instability in the training process. One possible explanation is that, unlike the bending problem, which involves solving a least-squares system, the vibration problem requires solving a generalized eigenvalue problem, which is more sensitive to matrix conditioning, as shown in previous studies [37].

A practical remedy is to adjust the initialization range of the internal parameters of the network. A similar strategy has been adopted in previous studies, see e.g. [18]. Figure 8.6 shows the convergence behavior for the strong nonlocality case ($(l/a)^2 = 1.00$) as the initialization interval is progressively enlarged from $[-1.0, 1.0]$ to $[-1.5, 1.5]$ and $[-2.0, 2.0]$. As seen, enlarging this range delays the onset of ill-conditioning, shifting the divergence point to larger network sizes and recovering exponential convergence over a broader interval.

Although the instability cannot be completely removed without more advanced regularization techniques, this simple modification significantly improves robustness of the approach.

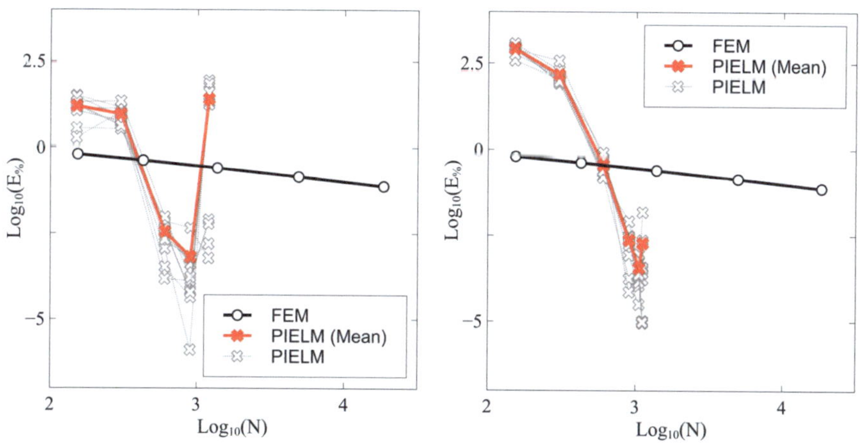

(a) $[-1.5, 1.5]$.

(b) $[-2.0, 2.0]$.

Fig. 8.6: Convergence comparison between FEM and PIELM for the free vibration problem with strong nonlocality, using different initialization ranges. Percentage errors for PIELM are averaged over 10 random initializations of the internal parameters, with the mean values highlighted in red.

8.5.3 Linearized Buckling Problem

The third benchmark considers the linearized buckling problem which is obtained by setting $\beta_i = \delta_{i3}$ in Eqs. (8.31) and (8.42). Convergence results for mild and strong nonlocality are shown in Figure 8.7.

The FEM solution exhibits again algebraic convergence with $\beta \approx 0.5$ and a similar upward shift in the error curves as the nonlocal parameter l increases. This confirms that the deterioration in FEM accuracy with increasing nonlocality is a consequence of the higher-order derivatives inherent in strain gradient elasticity, independent of the type of governing equations.

For PIELM, a behavior similar was observed for the free vibration problem: when the network sizes grow beyond a certain threshold, the solution diverges due to ill-conditioning. This occurs because the training in both the buckling and vibration problems requires the solution of a generalized eigenvalue problem, which is more sensitive to matrix conditioning. To circumvent this issue, the initialization range of the internal parameters was set to $[-2.0, 2.0]$ to ensure stable training, up to an accuracy of approximately 10^{-5}.

As seen from Figure 8.7, the PIELM approach retains exponential convergence, and its performance remains essentially insensitive to the value of the nonlocal parameter. This makes PIELM particularly advantageous for problems with strong nonlocal effects, where FEM would require a substantially higher number of degrees of freedom to achieve comparable accuracy due to the increase of errors with increasing l.

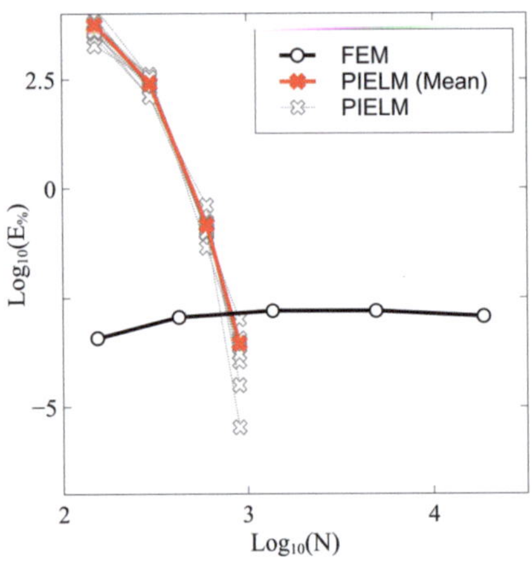

(a) Mild nonlocality ($(l/a)^2 = 0.01$).

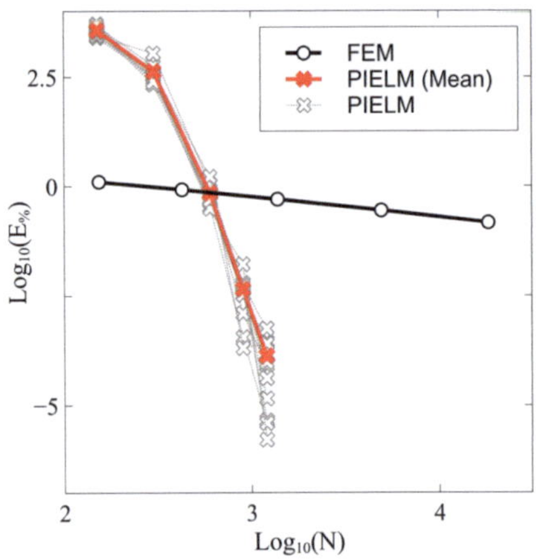

(b) Strong nonlocality ($(l/a)^2 = 1.00$).

Fig. 8.7: Convergence comparison between FEM and PIELM for the linearized buckling problem. Percentage errors for PIELM are averaged over 10 random initializations of the internal parameters, with the mean values highlighted in red.

8.6 Conclusions

A comparative analysis between a conforming Finite Element Method (FEM) based on Hermite functions and Physics-Informed Extreme Learning Machine (PIELM) has been presented for three benchmark problems in strain gradient elasticity: (I) static bending, (II) free vibration, and (III) linearized buckling of simply supported isotropic Kirchhoff nanoplates. The study assessed the convergence properties of both numerical approaches against analytical solutions, with particular emphasis on the influence of the nonlocal length parameter on their performance.

The results confirm that the FEM exhibits algebraic convergence with mesh refinement, consistent with theoretical results. The convergence rate for all problems is approximately $\beta \approx 0.5$. However, as the nonlocal parameter increased, the errors tend to increase globally. This deterioration is attributed to the higher-order derivative effects which are difficult to capture using polynomial interpolation functions.

In contrast, the PIELM approach demonstrates exponential convergence with respect to the number of hidden neurons, in agreement with theoretical predictions for Extreme Learning Machines with smooth activation functions. In addition, the convergence rate of PIELM remains almost unaffected by the degree of nonlocality, highlighting the inherent ability of neural-network-based trial spaces to satisfy the higher-order continuity requirements of strain gradient elasticity.

Although PIELM offers a clear advantage in terms of convergence, it can be affected by ill-conditioning during training. This effect was observed in the free vibration and linearized buckling problems, where the solution experienced a loss of accuracy once the number of neurons exceeded a certain threshold. In this work, the ill-conditioning issues are mitigated by adjusting the initialization range of the internal parameters, which delayed the onset of instability and extended the regime of exponential convergence.

Overall, the comparative study demonstrates that PIELM achieves substantially lower errors than FEM for a comparable number of degrees of freedom. The method combines high accuracy, rapid convergence, and robustness against strong gradient effects, without requiring mesh refinement or specialized finite elements. These findings position PIELM as a promising mesh-free alternative to conventional finite element approaches for strain gradient models. Future work will focus on developing advanced regularization strategies to further improve the robustness of PIELM, particularly for eigenvalue-based problems.

References

[1] Aifantis EC (1992) On the role of gradients in the localization of deformation and fracture. International Journal of Engineering Science 30(10):1279–1299, DOI 10.1016/0020-7225(92)90141-3

[2] Amanatidou E, Aravas N (2002) Mixed finite element formulations of strain-gradient elasticity problems. Computer Methods in Applied Mechanics and

Engineering 191(15–16):1723–1751, DOI 10.1016/S0045-7825(01)00353-X
[3] Apuzzo A, Bartolomeo C, Luciano R, Scorza D (2020) Novel local/nonlocal formulation of the stress-driven model through closed form solution for higher vibrations modes. Composite Structures 252:112688, DOI 10.1016/j.compstruct.2020.112688
[4] Babu B, Patel BP (2019) A new computationally efficient finite element formulation for nanoplates using second-order strain gradient Kirchhoff's plate theory. Composites Part B: Engineering 168:302–311, DOI 10.1016/j.compositesb.2018.12.066
[5] Babu B, Patel BP (2020) An improved quadrilateral finite element for nonlinear second-order strain gradient elastic kirchhoff plates. Meccanica 55(1):139–159, DOI 10.1007/s11012-019-01087-z
[6] Bacciocchi M, Fantuzzi N (2023) On the mapping procedure based on higher-order hermite polynomials for laminated thin plates with arbitrary domains in gradient elasticity. International Journal for Numerical Methods in Engineering 124(11):2552–2570, DOI 10.1002/nme.7220
[7] Bacciocchi M, Fantuzzi N (2025) Free vibrations and bending of laminated thin plates with distorted domains in gradient elasticity: an isoparametric finite element formulation based on Hermite mapping. Acta Mechanica pp 1–21, DOI 10.1007/s00707-025-04254-4
[8] Bacciocchi M, Fantuzzi N, Ferreira AJM (2020) Conforming and nonconforming laminated finite element Kirchhoff nanoplates in bending using strain gradient theory. Computers & Structures 239:106322, DOI 10.1016/j.compstruc.2020.106322
[9] Bacciocchi M, Fantuzzi N, Luciano R, Tarantino AM (2021) Linear eigenvalue analysis of laminated thin plates including the strain gradient effect by means of conforming and nonconforming rectangular finite elements. Computers & Structures 257:106676, DOI 10.1016/j.compstruc.2021.106676
[10] Baughman R, , Cui C, Zakhidov AA, Iqbal Z, Barisci JN, Spinks GM, Wallace GG, Mazzoldi A, Rossi DD, Rinzler AG, Jaschinski O, Roth S, Kertesz M (1999) Carbon nanotube actuators. Science 284(5418):1340–1344, DOI 10.1126/science.284.5418.1340
[11] Chirkov AY, Nazarenko L, Altenbach H (2024) Plane crack problems within strain gradient elasticity and mixed finite element implementation. Computational Mechanics 74(3):703–721, DOI 10.1007/s00466-024-02451-x
[12] Chirkov AY, Nazarenko L, Altenbach H (2025) Mixed fem implementation of three-point bending of the beam with an edge crack within strain gradient elasticity theory. Continuum Mechanics and Thermodynamics 37:1, DOI 10.1007/s00161-024-01333-0
[13] Cornacchia F, Fabbrocino F, Fantuzzi N, Luciano R, Penna R (2021) Analytical solution of cross-and angle-ply nano plates with strain gradient theory for linear vibrations and buckling. Mechanics of Advanced Materials and Structures 28(12):1201–1215, DOI 10.1080/15376494.2019.1655613

[14] Correia IFP, Soares CMM, Soares CAM, Herskovits J (2003) Analysis of laminated conical shell structures using higher order models. Composite Structures 62(3–4):383–390, DOI 10.1016/j.compstruct.2003.09.009

[15] Dwivedi V, Srinivasan B (2020) Physics informed extreme learning machine (PIELM) - a rapid method for the numerical solution of partial differential equations. Neurocomputing 391:96–118, DOI 10.1016/j.neucom.2019.12.099

[16] Eringen AC (1983) On differential equations of nonlocal elasticity and solutions of screw dislocation and surface waves. Journal of Applied Physics 54(9):4703–4710, DOI 10.1063/1.332803

[17] Fabbrocino F, Saitta S, Vescovini R, Fantuzzi N, Luciano R (2022) Meshless computational strategy for higher order strain gradient plate models. Mathematical and Computational Applications 27(2):19, DOI 10.3390/mca27020019

[18] Fabiani G (2025) Random projection neural networks of best approximation: Convergence theory and practical applications. SIAM Journal on Mathematics of Data Science 7(2):385–409

[19] Hornik K, Stinchcombe M, White H (1989) Multilayer feedforward networks are universal approximators. Neural Networks 2(5):359–366

[20] Huang G, Zhu Q, Siew C (2006) Extreme learning machine: theory and applications. Neurocomputing 70(1-3):489–501

[21] Igelnik B, Pao Y (1995) Stochastic choice of basis functions in adaptive function approximation and the functional-link net. IEEE transactions on Neural Networks 6(6):1320–1329

[22] Lage RG, Soares CMM, Soares CAM, Reddy JN (2004) Layerwise partial mixed finite element analysis of magneto-electro-elastic plates. Computers & Structures 82(17–19):1293–1301, DOI 10.1016/j.compstruc.2004.03.026

[23] Mindlin RD, Tiersten H (1962) Effects of couple-stresses in linear elasticity. Archive for Rational Mechanics and Analysis 11(1):415–448, DOI 10.1007/BF00253946

[24] Neves AMA, Ferreira AJM, Carrera E, Cinefra M, Roque CMC, Jorge RMN, Soares CMM (2013) Static, free vibration and buckling analysis of isotropic and sandwich functionally graded plates using a quasi-3d higher-order shear deformation theory and a meshless technique. Composites Part B: Engineering 44(1):657–674, DOI 10.1016/j.compositesb.2012.01.089

[25] Niu C, Sichel E, Hoch R, Moy D, Tennent H (1997) High power electrochemical capacitors based on carbon nanotube electrodes. Applied Physics Letters 70(11):1480–1482

[26] Power A, Gorey B, Chandra S, Chapman J (2018) Carbon nanomaterials and their application to electrochemical sensors: a review. Nanotechnology Reviews 7(1):19–41

[27] Raissi M, Perdikaris P, Karniadakis GE (2019) Physics-informed neural networks: A deep learning framework for solving forward and inverse problems involving nonlinear partial differential equations. Journal of Computational Physics 378:686–707, DOI 10.1016/j.jcp.2018.10.045

[28] Roque CMC, Zur KK (2022) On the static, vibration, and transient responses of micro-plates made of materials with different microstructures. Engineering

Analysis with Boundary Elements 143:579–590, DOI 10.1016/j.enganabound. 2022.07.007

[29] Shu JY, King WE, Fleck NA (1999) Finite elements for materials with strain gradient effects. International Journal for Numerical Methods in Engineering 44(3):373–391, DOI 10.1002/(SICI)1097-0207(19990130)44:3<373::AID-NME508>3.0.CO;2-7

[30] Soares CMM, Correia VMF, Mateus H, Herskovits J (1995) A discrete model for the optimal design of thin composite plate-shell type structures using a two-level approach. Composite Structures 30(2):147–157, DOI 10.1016/0263-8223(94)00042-5

[31] Soares CMM, Soares CAM, Correia VMF (1997) Multiple eigenvalue optimization of composite structures using discrete third order displacement models. Composite Structures 38(1–4):99–110, DOI 10.1016/S0263-8223(97)00046-9

[32] Soh AK, Wanji C (2024) Finite element formulations of strain gradient theory for microstructures and the C^{0-1} patch test. International Journal for Numerical Methods in Engineering 61(3):433–454, DOI 10.1002/nme.1075

[33] Szabó B, Babuška I (2011) Introduction to finite element analysis: formulation, verification and validation. John Wiley & Sons, Southern Gate, UK

[34] Thai CH, Ferreira AJM, Nguyen-Xuan H, Nguyen LB, Phung-Van P (2023) A nonlocal strain gradient analysis of laminated composites and sandwich nanoplates using meshfree approach. Engineering with Computers 39(1):5–21, DOI 10.1007/s00366-021-01501-9

[35] Wang B, Lu C, Fan C, Zhao M (2021) A meshfree method with gradient smoothing for free vibration and buckling analysis of a strain gradient thin plate. Engineering Analysis with Boundary Elements 132:159–167, DOI 10.1016/j.enganabound.2021.07.014

[36] Wang Q, Li C, Zhang S, Zhou C, Zhou Y (2025) Physics-Informed Extreme Learning Machine framework for solving linear elasticity mechanics problemshttps://doi.org/10.1016/j.ijsolstr.2024.113157. International Journal of Solids and Structures 309:113157, DOI https://doi.org/10.1016/j.ijsolstr.2024.113157

[37] Yan C, Vescovini R, Dozio L (2022) A framework based on physics-informed neural networks and extreme learning for the analysis of composite structures. Computers & Structures 265(106761):517–530

[38] Yan C, Vescovini R, Fantuzzi N (2023) A neural network-based approach for bending analysis of strain gradient nanoplates. Engineering Analysis with Boundary Elements 146:517–530

[39] Yang FACM, Chong ACM, Lam DCC, Tong P (2002) Couple stress based strain gradient theory for elasticity. International Journal of Solids and Structures 39(10):2731–2743, DOI 10.1016/S0020-7683(02)00152-X

[40] Zhu B, Li H, Zhang Q (2025) Extended physics-informed extreme learning machine for linear elastic fracture mechanics. Computer Methods in Applied Mechanics and Engineering 435:117655, DOI 10.1016/j.cma.2024.117655

MIX
Papier aus verantwortungsvollen Quellen
Paper from responsible sources
FSC® C105338

If you have any concerns about our products,
you can contact us on
ProductSafety@springernature.com

In case Publisher is established outside the EU,
the EU authorized representative is:
Springer Nature Customer Service Center GmbH
Europaplatz 3, 69115 Heidelberg, Germany

Printed by Libri Plureos GmbH
in Hamburg, Germany